Dianion Chemistry in Organic Synthesis

NEW DIRECTIONS in ORGANIC and BIOLOGICAL CHEMISTRY

Series Editor : C.W. Rees, FRS
Imperial College of Science, Technology and Medicine
London, UK

New and Forthcoming Titles

Chirality and the Biological Activity of Drugs
Roger J. Crossley

Enzyme-Assisted Organic Synthesis
Manfred Schneider and Stefano Servi

C-Glycoside Synthesis
Maarten Postema

Organozinc Reagents in Organic Synthesis
Ender Erdik

Activated Metals in Organic Synthesis
Pedro Cintas

Capillary Electrophoresis: Theory and Practice
Patrick Camilleri

Cyclization Reactions
C. Thebtaranonth and Y. Thebtaranonth

Mannich Bases: Chemistry and Uses
Maurilio Tramontini and Luigi Angiolini

Vicarious Nucleophilic Substitution and Related Processes in Organic Synthesis
Mieczyslaw Makosza

Radical Cations and Anions
M. Chanon, S. Fukuzumi, and F. Chanon

Chlorosulfonic Acid: A Versatile Reagent
R. J. Cremlyn and J. P. Bassin

Aromatic Fluorination
James H. Clark and Tony W. Bastock

Selectivity in Lewis Acid Promoted Reactions
M. Santelli and J.-M. Pons

Dianion Chemistry
Charles M. Thompson

Asymmetric Methodology in Organic Synthesis
David J. Ager and Michael B. East

Synthesis Using Vilsmeier Reagents
C. M. Marson and P. R. Giles

The Anomeric Effect
Eusebio Juaristi

Chiral Sulfur Reagents
M. Mikołajczyk, J. Drabowicz, and P. Kiełbasiński

Dianion Chemistry in Organic Synthesis

Charles M. Thompson
Professor of Chemistry
University of Montana
Missoula, Montana

CRC Press
Boca Raton Ann Arbor London Tokyo

Library of Congress Cataloging-in-Publication Data

Thompson, Charles M. (Charles Mark), 1956-
Dianion chemistry in organic synthesis / author, Charles M. Thompson.
p. cm. -- (New directions in organic and biological chemistry)
Includes bibliographical references and index.
ISBN 0-8493-7868-0
1. Organic compounds--Synthesis. 2. Anions. I. Title. II. Series.
QD262.T52 1994
547′.2--dc20 94-17629
CIP

International Standard Book Number 0-8493-7868-0
Library of Congress Card Number 94-17629
Printed in the United States of America 1 2 3 4 5 6 7 8 9 0
Printed on acid-free paper

PREFACE

Much of the rapid progress in dianion chemistry can necessarily be attributed to those pioneering researchers who studied simple organometallic reagents, charge-stabilizing groups and unique deprotonating bases. Using this combinatorial library of successful "monocarbanion" chemistry experiments, dianion hybrids of these transformations began to appear. For example, investigations emerged that used dianion reagents or dianion-based tactics to achieve precise control of the reaction regioselectivity, stereoselectivity, chemoselectivity or an amalgamation of these effects. Unfortunately, many laboratories were reluctant at first to embrace dianion chemistry as part of the standard reaction repertoire. There were several possible reasons for this aversion, including: a. dianions required expensive, inert atmosphere equipment, b. dianions were relegated to the specialist, c. dianions were highly reactive and therefore, non-selective, d. dianion reagents were costly, e. dianions were ersatz monocarbanion equivalents, and others. Initially, these comments and observations were justified; however, contemporary dianion chemistry no longer fits any of these claims, and researchers can comfortably draw upon the dianion chemistry literature to choose among an abundance of reagents and strategies that are reliable, effective and in many cases, the best answer to a synthetic problem.

This book endeavors to introduce, survey and consolidate carbon-based dianion chemistry for the reader interested in the remarkable versatility of dianions and to overtake some of the misconceptions associated with this field. Using these goals as an outline, I have surveyed a majority of the literature contributions through 1992 covering the realm of dianion structure and properties, methodology, use in total synthesis and select experimental procedures that aspire to familiarize the reader with dianion technology and to augment strategies available to the practitioner of organic chemistry. Chapter 1 has been designed to serve as an introduction and as an index of the various dianions types that will be subsequently covered in this book. Chapters 2 through 5 cover the ensemble of dianion types designated by their first deprotonation site, namely, C,C-, O,C-, N,C-, and S,C-dianions. Each chapter contains a modest experimental section at the end that is intended to inform the reader of relevant protocols. This book should be a practical addition to the library of students, teachers, academic and industrial researchers and all my fellow practitioners of organic synthesis.

I cannot remember the first time I thought of applying dianion chemistry to a synthetic problem, but it most likely came to mind during a capuccino-laden conversation with one of my colleagues while a post-doctoral at U.C. Berkeley. I am indebted to those friends.

ACKNOWLEDGMENTS

There are several important people that have provided professional and personal support over the years for which I am truly thankful. Professor's T. Roy Fukuto, E.J. Corey, and Henry Rapoport challenged my skills and cultivated my enthusiasm for organic chemistry. My lab colleagues and friends at U.C. Riverside, Harvard, and U.C. Berkeley taught me how to have fun conducting experiments, and furnished the needed "outside" distractions. I genuinely appreciate the family spirit imparted by my own group of graduate students, post-doctorals, and undergraduates. They continue to make research a pleasure. I also would like to acknowledge the assistance of Dr. Diana L.C. Green, who coauthored a review article of dianion chemistry with me that ultimately led to the writing of this book.

A very special thanks is owed to John and Kathy who have now survived 15 years of my friendship. I am extremely grateful to my wife Susan and my parents for their love and patience particularly during this undertaking, but for so much before and beyond.

THE AUTHOR

Charles M. Thompson, Ph.D., is a Professor of Chemistry at the University of Montana in Missoula, Montana. Dr. Thompson received his B.A. from Rutgers University (Newark) in 1982, and his M.S. (1980) and Ph.D. (1982) from the University of California, Riverside. The focus of his Ph.D. thesis work was in the area of mechanism of cholinesterase inhibition by organophosphates. From 1982 to 1985, Dr. Thompson conducted post-doctoral studies in organic synthesis under the auspices of Prof. E.J. Corey (Harvard) and Prof. Henry Rapoport (U.C. Berkeley). Following his post-doctoral work, Dr. Thompson was a chemistry professor at Loyola University of Chicago (1985-1994). In 1994, he took a position at the University of Montana.

Dr. Thompson has worked in several areas of research. These have included the synthesis and anti-cholinesterase action of enantioenriched organophosphorus compounds, the photochemistry of arylimines, dianion methodology and its use in natural product synthesis, and the synthesis of transition-state analogs. Dr. Thompson is a member of the American Chemical Society (Divisions of Organic and Agrochemicals), and was granted a FIRST Award from the NIH (1987-1992). He has published over 35 research articles. His outside interests include home-brewing, skiing, running, and listening to the music of the Grateful Dead.

TABLE OF CONTENTS

CHAPTER 1. INTRODUCTION

1.1 Scope of Coverage 1
1.2 Organization and Presentation of the Chapters 2
1.3 Background Needed 3
References 6

CHAPTER 2. DIANIONS FROM C,C-BIS DEPROTONATIONS

2.1 Introduction 9
2.2 1,1-(C,C)-Dianions (Geminal Dianions) 11
2.2.1 α-Halocarbonyl-stabilized 1,1-(C,C)-dianions 13
2.2.2 Nitrile Stabilized 1,1-(C,C)-dianions 16
2.2.3 Nitro-Stabilized 1,1-(C,C)-dianions 16
2.2.4 Sulfonyl-Stabilized 1,1-(C,C)-dianions 18
2.3 1,2-(C,C)-Dianions (Vicinal Dianions) 21
2.3.1 Acyclic vicinal C,C-dianions 22
2.3.2 Cyclic vicinal C,C-dianions 23
2.4 1,3-(C,C)-Dianions (β-Dicarbonyl Dianions) 27
2.4.1 Alkylation of β-dicarbonyl dianions 31
Alkylation of β-dicarbonyl dianions with alkyl halides 31
Alkylation of β-dicarbonyl dianions with epoxides 36
2.4.2 Condensation reactions of β-dicarbonyl dianions 37
2.4.3 Acylation of β-dicarbonyl dianions 40
2.4.4 Addition reactions of β-dicarbonyl dianions 40
2.4.5 Miscellaneous 1,3-(C,C)-dianions derived from ketones other than β-dicarbonyl 42
2.5 Sulfinyl and Sulfonyl-Stabilized 1,3-(C,C)-Dianions 44
2.5.1 β-Ketosulfone 1,3-dianions 45
2.5.2 1,3-Sulfoxide dianions 48
2.5.3 β-Ketosulfoxide 1,3-dianions 48
2.6 Phosphorus-Stabilized 1,3-(C,C)-Dianions 51
2.7 Remote C,C-Dianions 54
2.7.1 Remote 1,3-(C,C)-dianions 55
Alkene-based 1,3-(C,C)-dianions 56
Alkyne-based 1,3-(C,C)-dianions 56
(*Exo*)-distabilized-1,3-(C,C)-dianions 57
2.7.2 Remote 1,4-(C,C)-dianions 58
(*Endo*)-1,4-(C,C)-dianions 58
(*Exo*)-1,4-(C,C)-dianions 60
Ring-metalated 1,4-(C,C)-dianions 62
2.7.3 Remote 1,5-(C,C)-dianions 62
2.7.4 Remote 1,Z-(C,C)-dianions 64
2.8 Delocalized C,C-Dianions 65
2.8.1 Allylic-based delocalized dianions 66
2.8.2 Benzylic-based delocalized dianions 68
2.9 Experimental Procedures 71
References 75

CHAPTER 3. DIANIONS FROM O,C-BIS DEPROTONATIONS

3.1 Introduction 87
3.2 Carboxylic Acid-Based Dianions 88
3.2.1 Generation of carboxylic acid-based dianions 90
3.2.2 Reactions of α-carboxylate O,C-dianions (α,α′-carboxylate dianions) . . . 91
α-Carboxylate dianions with no added stabilizing groups 91
β-Carbonyl-based α-carboxylate dianions 104
α-Carboxylate dianions bearing carbanion stabilizing groups other than carbonyl 105
3.2.3 Reactions of alkenoic and alkynoic acid-based O,C-dianions 111
Alkynoic acid dianions 117
3.2.4 Reactions of benzoic acid-based O,C-dianions 119
3.2.5 Reactions of remote carboxylic acid-based O,C-dianions 124
1,4-Dianions of carboxylic acids 125
1,5-Dianions of carboxylic acids 127
1,Z-Dianions of carboxylic acids 129
3.3 Hydroxyl-Based dianions 130
3.3.1 Generation of hydroxy-based dianions 131
3.3.2 Reactions of hydroxyl-based 1,2-dianions (vicinal O,C-dianions) 131
3.3.3 Reactions of hydroxyl-based 1,3-dianions (β-Hydroxyl-α-stabilized O,C-dianions) 133
β-Hydroxycarbonyl 1,3-dianions 134
β-Hydroxysulfinyl and β-hydroxysulfonyl 1,3-dianions 139
Miscellaneous β-hydroxy-1,3-dianions 145
3.3.4 Reactions of allylic and propargylic alcohol-based O,C-dianions 146
Allylic site deprotonation of allylic alcohols leading to O,C-dianions 147
Vinylic site deprotonation of allylic alcohols leading to O,C-dianions 151
Propargylic alcohol-based dianions 152
3.3.5 Reactions of remote, hydroxyl-based dianions 153
3.3.6 Preparation and reactions of O,C-dianions not derived through initial . . 154
3.4 Phenol- and Cresol-Based Dianions 156
3.4.1 Generation of phenol-based dianions 157
3.4.2 Reactions of phenol-based dianions 158
3.5 Oxime-Based dianions 160
3.5.1 Generation of oxime-based dianions 161
3.5.2 Reactions of oxime 1,4-dianions 161
Syn/Anti isomerism and oxime dianion formation 161
Reactions of oxime 1,4-dianions 162
3.5.3 Reactions of remote oxime-based dianions 166
3.6 Experimental procedures 167
References 172

Chapter 4. DIANIONS FROM N,C-BIS DEPROTONATIONS

4.1 Introduction 185
4.2 Carboxamide-, Thioamide-, Sulfinamide- and Sulfonamide-Based Dianions 186
4.2.1 Generation of carboxamide, thioamide, sulfinamide, and sulfonamide N,C-dianions 186
4.2.2 Reactions of carboxamide-based N,C-dianions 187
α,α′-Carboxamide dianions 187
Vicinal carboxamide Dianions 190
Remote carboxamide Dianions 191
4.2.3 Reactions of thioamide-based N,C-dianions 197

4.2.4 Reactions of sulfinamide and sulfonamide-based N,C-dianions 199
Sulfinamides 199
Sulfonamides 199
4.3 Hydrazone- and Sulfonylhydrazone-Based Dianions 203
4.3.1 Generation of hydrazone- and sulfonylhydrazone-based N,C-dianions .. 204
4.3.2 Reactions of hydrazone-based N, C-dianions 204
4.3.3 Reactions of sulfonylhydrazone-based N,C-dianions 205
Sulfonylhydrazone dianions and the Shapiro reaction 205
α-Methylene lactones via sulfonylhydrazone dianions 207
1,2-Carbonyl transposition using sulfonylhydrazone dianions 207
Remote metalation via sulfonylhydrazone dianions 208
Other 208
4.4 Carbamate-Based Dianions 208
4.4.1 Generation of carbamate-based dianions 209
4.4.2 Reactions of carbamate-based dianions 209
Vicinal carbamate dianions 209
Remote carbamate dianions 209
4.5 Enaminone-Based Dianions 212
4.5.1 Generation of enaminone-based N,C-dianions 212
4.5.2 Reactions of enaminone-based N,C-dianions 213
4.6 Pyrrole- and Benzazole-Based Dianions 216
4.6.1 Generation of indole- and pyrrole-based dianions 216
4.6.2 Reactions of pyrrole-based dianions 216
4.6.3 Reactions of indole- and indazole-based dianions 216
4.7 Imide- and N-acylsulfonamide-Based Dianions 218
4.7.1 Generation of imide- and N-acylsulfonamide-based N,C-dianions 218
4.7.2 Reactions of imide-based dianions 218
4.7.3 Reactions of N-acylsulfonamide dianions 221
4.8 Miscellaneous Amine- and Aniline-Based Dianions 222
4.8.1 Reactions of amine-based dianions 222
4.9 Experimental Procedures 224
References 227

Chapter 5. DIANIONS FROM S,C-BIS DEPROTONATIONS

5.1 Introduction 235
5.2 Thiol-Based dianions 236
5.2.1 Generation of thiol-based S,C-dianions 236
5.2.2 Reactions of "vicinal delocalized" S,C-dianions 236
5.2.3 Reactions of vicinal "fixed" S,C-dianions 238
5.2.4 Reactions of remote S,C-dianions 240
5.3 Thioic-Acid Based Dianions 241
5.3.1 Generation of thioic acid-based S,C-dianions 241
5.3.2 Reactions of thioic acid-based S,C-dianions 242
5.4 Experimental procedures 242
References 243

Index 245

Chapter 1

INTRODUCTION

1.1 SCOPE OF COVERAGE

Many carbon-based reactive intermediates are used routinely in organic synthesis of which carbocations, carbanions, carbenes, and radicals have been the most widely explored. To this select group of useful, reactive species now can be added the "dianions." Dianions, which are molecules that bear two anionic centers, have become increasingly popular in organic synthesis particularly in novel carbon-carbon bond-forming reactions. It is this important aspect of organic dianion chemistry that will be the focus of this book.

The term dianion is somewhat general, and literature searching of the term will reveal a plethora of molecule types, classes, and charged states, many of which do not contain a carbanionic center. Such examples of non-organic dianions include doubly charged inorganic species like phosphate groups [$RP(O)O_2^{-2}$], or related metal derivatives. It is therefore important to first limit the field of dianion entries that will be included in this book. Similarly, searching only the term "dianion" is likely to overlook citations that have used analogous terms for these intermediates including dimetalated (e.g., dilithio-, disodio-, dipotassio-), dicarbanions, dienolate, polymetalated, etc. The goal of this book is to cover organic-based dianion species that have been used exclusively in synthetic maneuvers, specifically as carbon nucleophiles, and have been prepared by bis-deprotonations of certain organic molecules. This restriction on the dianion type will exclude transition metal, electrochemically produced, theoretically interesting dianions (e.g., dianions that result in the formation of aromatic or antiaromatic compounds), and heteroatom-only dianions. Likewise, this restriction adjusts the scope of this coverage to only dianion species that react at a carbanionic site resulting from the second deprotonation of a given molecule. Within this framework, there are four major types of dianions used in organic synthesis that result from two sequential deprotonations, namely, C,C-dianions (Chapter 2), O,C-dianions (Chapter 3), N,C-dianions (Chapter 4) and S,C-dianions (Chapter 5), which vary only in the nature of the first deprotonation. For example, a C,C-dianion results from deprotonations occurring at two C-H positions, whereas a N,C-dianion results from first deprotonation at an N-H functional group followed by deprotonation at a C-H (Figure 1.1).

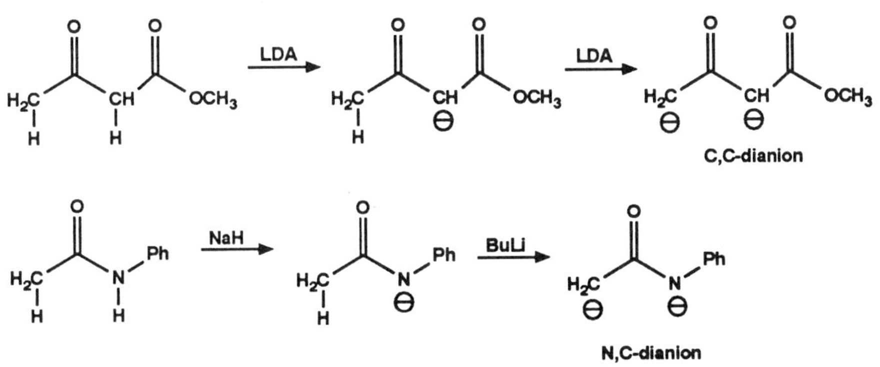

Figure 1.1 Comparison of C,C- and N,C-dianions

Each of the four types represent a unique class of dianions and their use in synthetic operations is covered. Each chapter will identify and review the various subtypes of dianions based upon the functional group that undergoes the deprotonation, the functional groups serving as stabilizing groups, the distance appearing between the anionic centers, the resultant reactivity, and combinations of these features.

To date, several general and specific reviews of dianions have appeared.[1-12] Some reviews focus upon one dianion type (e.g., carboxylic acid-based dianions)[3,7] or stabilizing group (e.g., sulfur-stabilized).[6] Some reviews provide theoretical features of dianions,[8-10] which furnish useful background to better understand the nature and reactivity of these intermediates. Other reviews highlight the synthetic aspects of dianions,[1-7,11,12] and have served to provide essential foundation material for this book. Each chapter will cover the major uses of the specific dianion type in synthesis with most attention given to variety in application. Representative experimental procedures are collected at the end of each chapter to provide an elementary understanding of the experiment that was conducted. The reader should review the primary literature before any experiment is attempted.

The reactions of dianions are not difficult to understand once the nature of the intermediate has been grasped. In all cases, *the site of the second (C-H) deprotonation is the first site to undergo reaction with an electrophile.*

1.2 ORGANIZATION AND PRESENTATION OF THE CHAPTERS

Because dianions have a long and diverse history without rigidly enforced guidelines for nomenclature and classification, a major task for this book was identifying and organizing the various types. Two issues needed to be clarified. First, how would the dianions be pictorially represented so both novice and expert could easily understand their reactivity, and second, how would the major classes of dianions be best divided?

Dianions, like their monoanion counterparts, represent molecules where some or all of the negative charges reside at a carbon atom. The net charge existing on the carbon (or carbons) may be very small when neighboring carbanion stabilizing groups furnish a measure of π-conjugation, and permit the negative charge to relocate to a more electronegative atom (e.g., a nitrogen or oxygen atom). This phenomena is generally appreciated in enolate chemistry, and is fully applicable to chemically equivalent dianion intermediates. Although the depiction of an α-carbonyl deprotonation in the oxyanion, enolate form is more chemically and theoretically correct, the common feature in the use of carbanions or dicarbanions is the formation of a new bond at the reactive carbanion center. It is therefore most useful to "tag" or "symbolize" this reactive site, so that the new bond can be readily identified. To this end, most dianion reactions in this text are depicted in the [carbanionic] form corresponding directly with the sites of each deprotonation, and each site has been indicated by a circled minus charge (⊖) (Figure 1.1). At the beginning of each chapter, and in certain instances throughout, the resonance representation also has been shown.

The first level of classification, which separates the various dianions according to the nature of the first deprotonation (C,C-, O,C-, N,C- and S,C-dianions), was straightforward, and as mentioned, each of these major dianion types have been placed into individual chapters. Each of these types are further divided according to some structural or reactivity difference within the dianion class.

There have been several nomenclature and organizational themes used in the literature to describe the various dianion types. Although no distinct nomenclature preference has emerged, there has been a move toward numerical descriptors that indicate the precise sites of the deprotonations. For example, the term "vicinal dianion" may be applied to any intermediate that has been bis-deprotonated at neighboring atoms that bear active or acidic hydrogens. These intermediates also have been termed α,β-dianions, consistent with positional assignments that are relative to an anchor or stabilizing group. The more accurate numerical descriptor would be a 1,2-dianion, and more precisely, a 1,2-(N,C)-dianion when the sequential deprotonations occur at neighboring (vicinal) N-H and C-H positions, respectively. The numerical system has largely been adopted for this text, but the generic and greek aliases also have been assigned to the structure in the introductory sections of each chapter to aid recognition, and provide a measure of continuity with the primary literature. In a similar manner, geminal dianions become 1,1-(X,C)-dianions (where X = C, O, N, or S). "Remote" dianions has been used to specify bis-deprotonations that occur at sites insulated by two or more atoms, and their numerical equivalent are termed 1,Z-(X,C)-dianions (where Z = 3, 4, etc.). In this manner, the C,C-dianions (Chapter 2) have been separated into sections according to the distances appearing between the carbanion sites.

Further distinctions are needed to identify dianions within a class that result from deprotonations from different functional groups. These distinctions are particularly important for the heteroatom-carbon dianions that may be prepared from the various functional groups. O,C-Dianions, which result from sequential deprotonation of O-H and C-H groups, may be prepared from carboxylic acids, alcohols, phenols or oximes. Each of these O,C-dianion groups are given a distinct subsection within the O,C-dianion chapter. Similarly, N,C-dianions are divided into carboxamide-, hydrazone-, carbamate-, enamine-, azole-, imide-, and amine-based dianions, and S,C-dianions have been classified as dianion derivatives of thiols and thioic acids. C,C-Dianions pose a different classification problem since most C,C-dianions result from deprotonation of an aliphatic or sp^3 carbon. As a result, the various C,C-dianion types have been further categorized according to the type of stabilization group.

Although the index serves as a preliminary guide to the dianions, a graphical summary of select dianions is presented to accelerate searches for specific types (Figure 1.2).

1.3 BACKGROUND NEEDED

Anyone wishing to explore the use of dianions in organic synthesis should have prior experience in air-sensitive experimental techniques and possess ample theoretical background of elementary carbanion chemistry. Although some dianion experiments are straightforward requiring only minimal preparation, most experiments require rigorously purified solvents, precisely titrated alkyl-metal reagents, no-air transfer techniques, and carefully controlled temperatures. Of pivotal importance is a comprehensive understanding of the reactivity and handling of certain alkyllithium reagents, which can react violently with certain compounds and/or be pyrophoric. The novice experimenter should first review the use of alkyllithium reagents before attempting any dianion experiment. Many excellent monographs are available to assist the design of experimental protocols using alkyllithiums including, "*The Manipulation of Air-Sensitive Compounds*," by Shriver and Drezdzon[13], and "*Laboratory Operations with Air-Sensitive Substances: Survey*," by Kramer, Levy and Midland found as the concluding chapter in H.C. Brown's "*Organic Synthesis via Boranes*."[14] More common techniques and purifications can be found in any undergraduate laboratory text, although the advanced experimenter will probably find other sources more applicable to dianions.[15]

Several topics in organic chemistry are worth reviewing prior to studying dianion chemistry including: pKa of carbon acids and carbanions, enolate chemistry, [carb]anion-stabilizing groups, C- vs. O-alkylation, metal hydrides, alkyl-metal reagents, resonance-stabilization, carbanion titrating reagents, kinetic and thermodynamic equilibrium acidities, etc. Most of these topics are covered in some detail in most graduate level organic texts.

Dianion Structure	Name	Section
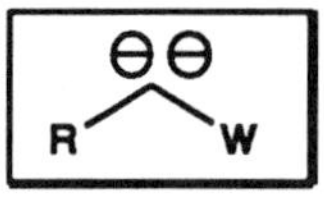	1,1-(C,C)-dianion (geminal or α,α-dianion)	2.2
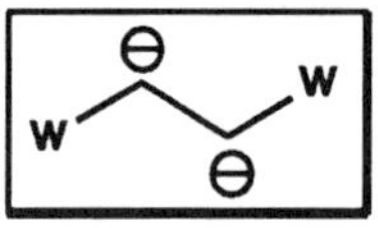	1,2-(C,C)-dianion (vicinal or α,β-dianion)	2.3
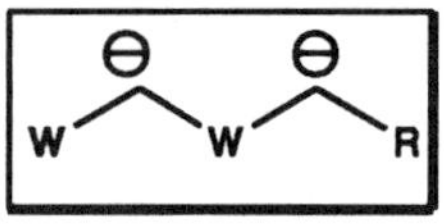	1,3-(C,C)-dianion (α,γ-dianion)	2.4, 2.5, 2.6
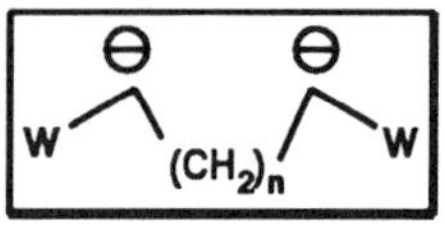	1,Z-(C,C)-dianion (N > 2) (remote dianion)	2.7
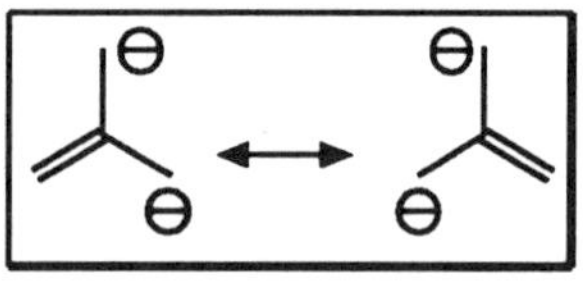	delocalized (C,C)-dianion resonance-stabilized dianion	2.8
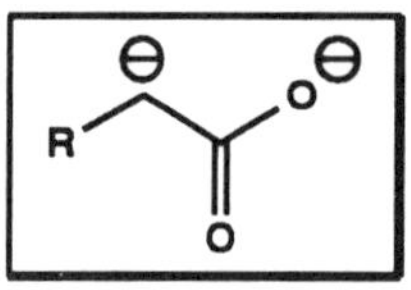	1,3-(O,C)-dianion α-carboxylate dianion α,α′-(O,C)-dianion	3.2
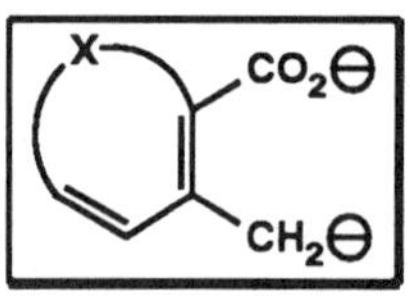	Remote 1,5-(O,C)-dianion "lateral deprotonation"	3.2.4
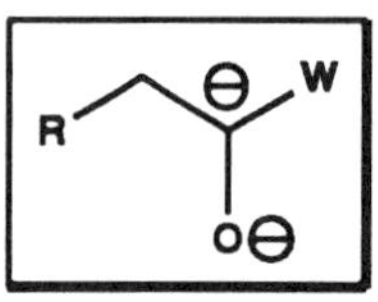	Hydroxy-based 1,2-(O,C)-dianion (vicinal or α,β-dianion)	3.3

W = C(O)R, SO_nR, NO_2, $P(O)OR_2$, CN, etc.

Figure 1.2. Structures and designations of certain dianions

Dianion Structure	Name	Section
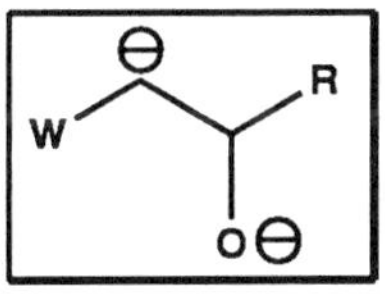	Hydroxy-based 1,3-(O,C)-dianion β-hydroxyl-(O,C)-dianion	3.3.3
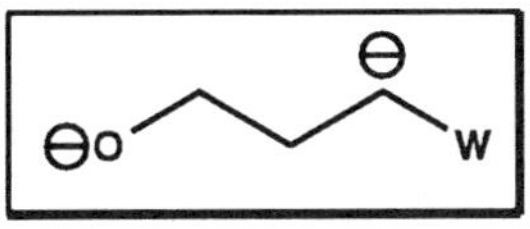	Hydroxy-based 1,4-(O,C)-dianion Remote O,C-dianion	3.3.5
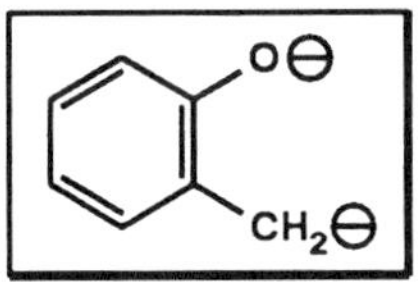	Phenol-based (O,C)-dianion 1,4-(O,C)-dianion	3.4.1
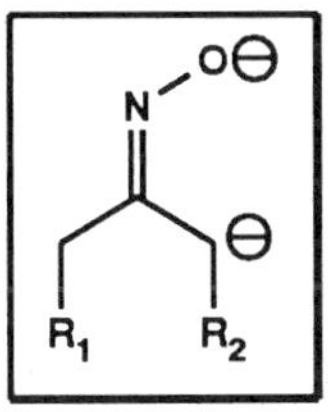	Oxime-based (O,C)-dianion 1,4-(O,C)-dianion	3.5
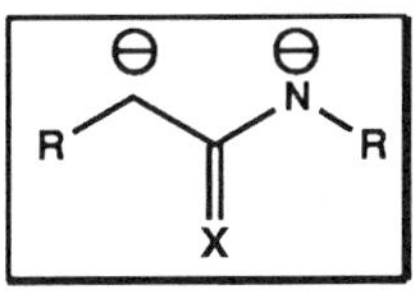	Carboxamide N,C-dianion (X=O) Thioamide N,C-dianion (X=S) 1,3-(N,C)-dianion	4.2
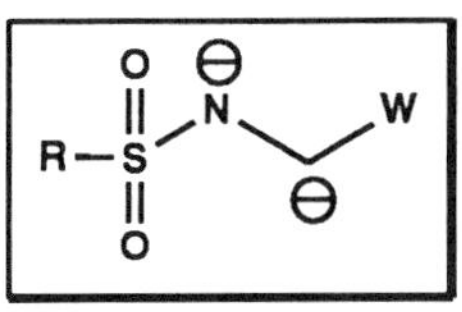	Sulfonamide N,C-dianion Vicinal sulfonamide dianion 1,2-(N,C)-dianion	4.2
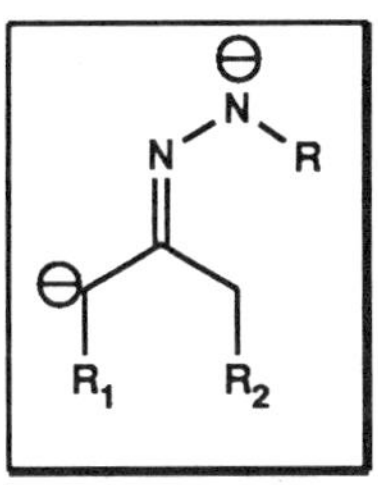	Hydrazone-based (N,C)-dianion 1,4-(N,C)-dianion	4.3

W = C(O)R, SO_nR, NO_2, $P(O)OR_2$, CN, etc.

Figure 1.2 (cont.). Structures and designations of certain dianions

Dianion Structure	Name	Section
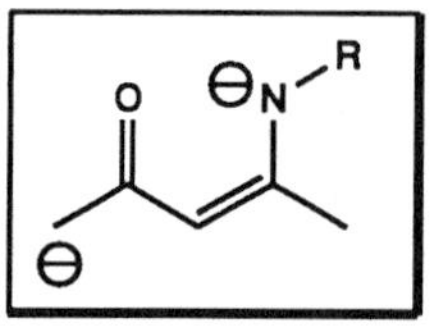	**Enaminone-based N,C-dianion** **Amine-based 1,5-(N,C)-dianion**	**4.5**
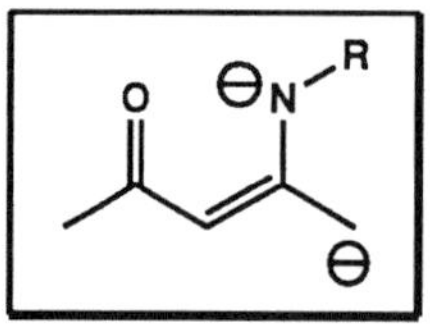	**Enaminone-based N,C-dianion** **Amine-based 1,3-(N,C)-dianion**	**4.5**
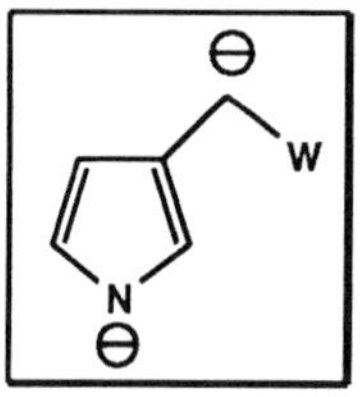	**Pyrrole-based (N,C)-dianion** **1,4-(N,C)-dianion**	**4.6**
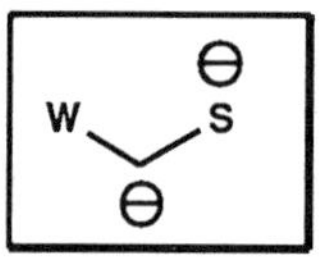	**Thiol-based S,C-dianion** **Vicinal thiol-based 1,2-(S,C)-dianion**	**5.2**
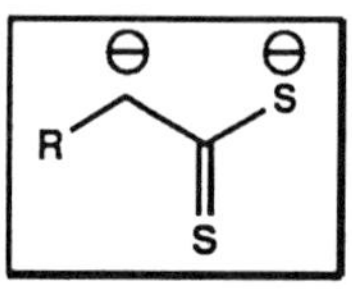	**Dithioic acid-based S,C-dianion** **Dithioic acid 1,3-(S,C)-dianion**	**5.3**

W = C(O)R, SO_nR, NO_2, $P(O)OR_2$, CN, etc.

Figure 1.2 (cont.). Structures and designations of certain dianions

References

1. **Harris, T.M. and Harris, C.M.** The γ-alkylation and γ-arylation of dianions of β-dicarbonyl compounds, *Org. React.* **17**, 155 (1969).
2. **Angelo, B. and Normant, H.** Dianions fonctionnels et leurs applications en synthese organique, *Isv. Khim.* **8**, 5 (1975).
3. **Ivanov, D., Vassilev, G. and Panayotev, I.** Syntheses and reactions of organolithium reagents derived from weakly acidic carbon-hydrogen compounds, *Synthesis* 83 (1975).
4. **Kaiser, E.M., Petty, J.D. and Knutson, P.L.A.** Di- and polyalkali metal derivatives of heterofunctionally substituted molecules, *Synthesis* 509 (1979).
5. **Stowell, J.C.** *Carbanions in Organic Synthesis* J. Wiley & Sons, New York, 1979.
6. **Tanaka, K. and Kaji, A.** The synthetic utility of sulfur-containing dianions, *Sulfur Rep.* **1**, 97 (1980).
7. **Petragnani, N. and Yonashiro, M.** The reactions of dianions of carboxylic acid and ester enolates, *Synthesis* 521 (1982).

8. **von Rague Schleyer, P.** Lithium synthetic reagents: dimerization and intramolecular association. Double bridging in "dianions," *Pure Appl. Chem.* **55**, 355 (1983).
9. **Grovenstein, E. Jr.** Structures of organodialkali metal compounds ("dianions"), *Stud. Org. Chem.* **5**, 175 (1987).
10. **Rabinovitz, M. and Cohen, Y.** Probing the nature of polycyclic conjugated dianions; from carbocyclic to heterocyclic dianions: NMR Studies, π-delocalization and electronic structure, *Tetrahedron* **44**, 6957 (1988).
11. **Thompson, C.M. and Green, D.L.C.** Recent advances in dianion chemistry, *Tetrahedron* **47**, 4223 (1991).
12. **Simpkins, N.S.** *Sulphones in organic synthesis*, Pergamon Press, Oxford, U.K., 1993.
13. **Shriver, D.F. and Drezdzon, M.A.** *The Manipulation of Air-Sensitive Compounds*, Wiley-Interscience, J. Wiley & Sons, NY, 1986.
14. **Brown, H.C.** *Organic Syntheses Via Boranes*, Wiley Interscience, John Wiley & Sons, NY, 1975, p. 191-261.
15. **Perrin, D.D. and Armarego, W.L.F.** *Purification of Laboratory Chemicals*, Pergamon Press, Oxford, 1988.

Chapter 2

DIANIONS FROM C,C-BIS DEPROTONATIONS

2.1 INTRODUCTION

C,C-Dianions are generated by two sequential C-H deprotonations. They have a long and diverse history, and continue to serve as the cornerstone dianion for synthetic use. The development and contemporary acceptance of C,C-dianions was probably initiated by the report of Hauser and Harris in 1958,[1] who showed that the dipotassio salt of benzoyl- or acetoacetone undergoes alkylation at the terminal position rather than between the two carbonyl groups as was the case for the monoanion. However, the spark for broadened synthetic application likely followed the landmark review by Harris and Harris,[2] who summarized a ten-year period of activity in the scope of alkylation and arylation reactions of β-dicarbonyl-based dianions. The review by Harris and Harris served as a conduit for new and inventive variations of the C,C-dianion theme, which have now blossomed into an arsenal of synthetic tools.[2-9]

Among the four major types of dianions derived from bis-deprotonations (C,C-, O,C-, N,C-, and S,C) the C,C-dianion type generally requires a stabilizing group for the first deprotonation, whereas -OH, -NHR, and -SH containing molecules are acidic enough to be deprotonated without stabilizing groups. This requirement of C,C-dianions alters the selection of both base type and strength.

The needed carbanion-stabilizing group, as will be seen in the following sections, guides the formation of the carbanion sites, but it too can be used to advantage in synthesis. The choice of stabilizing group, therefore, is governed largely by the length or type of carbon component to be installed. Alternatively, the stabilizing group may be an essential part of the starting material and subsequent functional group interconversion may follow dianion use.

Prior reviews grouped C,C-dianions by functional group type.[2,3,5,6] Because vastly different C,C-dianion types may be formed using identical stabilizing groups, C,C-dianions are not easily categorized with respect to the nature of the functional group appended (carbonyl, sulfonyl, alkene, etc.) or hybridization state, a chemical identity or priority was sought that would enable the novice and expert dianion user to recognize and contrast C,C-dianion types. During review of the literature for this chapter it was recognized that a system that divided the various C,C-dianion types into categories based upon the locations that the deprotonations took place had already evolved. Specifically, sequential deprotonations occurring at the same carbon atom are named 1,1-dianions (geminal), at adjacent carbons named 1,2-dianions (vicinal), at carbon sites one atom apart named 1,3-dianions, and so forth. This classification was adopted for this chapter. Generic and/or Greek symbol (equivalent) nomenclatures found in the primary literature were reassigned to the numerical system (*vide infra*) where appropriate.

In some instances, the dianion type may be further distinguished by the appended functional groups or the particular synthetic use to better focus individual dianion types. For example, 1,3-(C,C)-dianions have been accessed from β-dicarbonyl, β-ketosulfone, and other bifunctional systems. These dianions are found in the 1,3-dianion major headings (Sections 2.4, 2.5 and 2.6) with the stabilizing group further defining the sub-category. The nature of the reaction and/or stabilizing group becomes an added refinement to be considered in order to accommodate differences in generation, synthetic use, and reactivity. The classifications of C,C-dianions by the position of the deprotonation is in accord with our prior review.[3] However, coverage in this book will differ in that the presentation order will more strictly follow an increasing [numerical] progression of C,C-dianion types, namely, 1,1-dianions will be covered first, followed by 1,2-dianions, 1,3-dianions, etc. Because delocalized C,C-dianions (those existing through resonance-delocalization) can give rise to a mixture of carbanion site assignments, their categorization does not easily lend itself to the numerical system, and are covered last (Section 2.8). When the delocalized-dianion system is part of an already identified system, these examples are found under that general heading.

Representative examples of each of the C,C-dianion types are shown in Figure 2.1 accompanied both by the numerical assignment used to organize them in this chapter, and generic names where appropriate. The dianions are shown in their carbanionic form rather than the resonance-contributor or enolate form so that the reader can identify the various C,C-dianions by their deprotonation site, and hence, their reactive positions. This representation is for instructive purposes only.

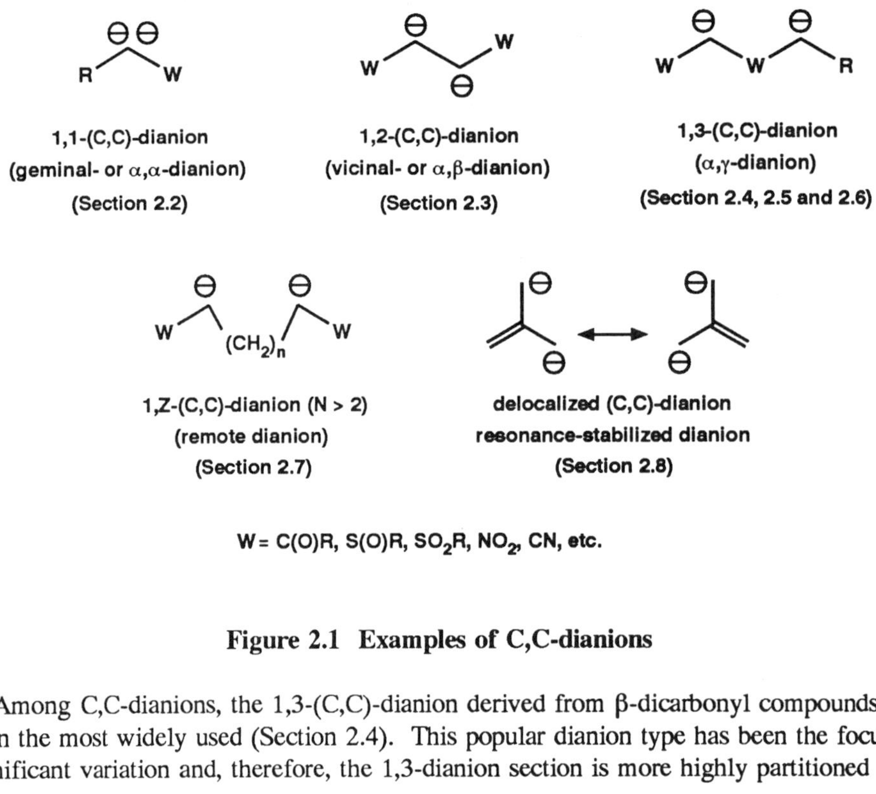

Figure 2.1 Examples of C,C-dianions

Among C,C-dianions, the 1,3-(C,C)-dianion derived from β-dicarbonyl compounds has been the most widely used (Section 2.4). This popular dianion type has been the focus of significant variation and, therefore, the 1,3-dianion section is more highly partitioned than others.

The purpose of this chapter is to acquaint the reader with the various C,C-dianion types with particular emphasis on the identification and use of several new and unique C,C-dianion types that have emerged following the pioneering review by Harris and Harris, which focused solely on β-dicarbonyl compounds. Increased attention in the form of expanded text and equations has been given to the less popular C,C-dianions, and the more popular C,C-dianions have received reduced coverage in this text (in the form of a tabular survey). This preference is intended to bring more diversity to the coverage and not reflect the relative importance of the dianion types.

Many of the early studies of C,C-dianions were conducted with β-dicarbonyl compounds. Because most carbonyl-stabilizing groups are subject to attack by organometallics, metal amides (MNH_2; M = Li, Na, K) were employed to conduct the bis-deprotonation. Today, most investigators use hindered organic amide bases (e.g., LDA, LiHMDS, LiTMP, etc.) because they are convenient to prepare and less nucleophilic. However, metal amides can give superlative product yields and their use should not be dismissed. Representative experimentals using metal amides or hindered amide bases to access the dianions are given in Section 2.9. Since many 1,3-(C,C)-dianions result from initial deprotonation at a highly acidic, doubly stabilized C-H (pKa < 16), simple hydride bases (NaH, KH, etc.) have been found to give excellent conversion to monoanion in route to dianion species (following reaction with a second, stronger base).

In the following sections, a survey of C,C-dianions shall be conducted with emphasis upon methods in carbon-carbon bond formation. In some cases, application to total synthesis will be noted, but this is not a goal of the section. Likely, oversights of dianion applications in specific accounts have occurred. In these instances, it is hoped that the general methods for use have been covered. Select experimental conditions are collected in Section 2.9, which should furnish the reader with a brief understanding of the experimental design and execution. The reader is encouraged to review the primary literature before attempting the desired experiments.

2.2 1,1-(C,C)-DIANIONS (GEMINAL DIANIONS)

Several reviews have covered the generation and reactions of dianions formed from stepwise deprotonations from a sole methylene group.[3,5,6] This process, leading to geminal or 1,1-(C,C)-dianions (also known as α,α-dianions), usually occurs when there is little or no opportunity for second deprotonation to occur elsewhere in the molecule. The interest in geminal dianions may have been prompted by a need for increased reactivity since many simple α-stabilized monoanions are slow or reluctant to react with electrophiles. Conversion to the more reactive dianion species seemed an obvious remedy. Second, some investigators hoped to develop an intermediate capable of reacting twice with a bifunctional electrophile leading to novel ring-constructing processes (e.g., geminal sulfone dianions; Section 2.2.4).

Because geminal dianion-based molecules require a specific atomic arrangement and a non-reactive functional group acting as a carbanion stabilizing moiety (to avoid deprotonation elsewhere), most of the examples of geminal dianions are simple hydrocarbons bearing one carbanion-stabilizing group. Four different types of carbanion-stabilizing groups have been most often used to direct formation of the geminal dianion or serve as a precursor for this intermediate; α-halocarbonyl, nitrile, nitro, and sulfonyl. Sequential deprotonation of these four molecule types leads to intermediates that can be depicted in various resonance forms as shown in Figure 2.2. The 1,1- or geminal dianion forms are shown for convenience, as the dianion likely is better stabilized as the vinylic, enolate-type structure.

In general, strong base combinations (tBuOK, BuLi) are required to access these putative intermediates, although other bases have been employed for the initial deprotonation (e.g., KH, LDA).

The nature and depiction of certain geminal dianions have been challenged. Meth-Cohn and coworkers studied the reaction between two equivalents of LiHMDS and benzylnitrile or diethyl benzylphosphonate.[10] Using the C-13 chemical shift and multiplicity as indicators, they suggest that no geminal dianion is formed since an initial change in chemical shift from 23.1 (t) to 32.1 (d) follows the addition of one equivalent of base, but virtually no further chemical shift is noted following the addition of a second equivalent. To account for the sole initial deprotonation reaction yet enhanced overall reactivity by the dianionic species, the authors suggest the formation of a monoanion complexed to the second equivalent of base, or possibly a bridged dimer structure (Figure 2.3).

Since formation of the dianion proceeds though initial α-deprotonation and heteroenolate formation, a significant change in chemical shift and multiplicity would be expected in the change in hybridization from sp^3 to sp^2. When subsequent vinylic deprotonation occurs, no further change in hybridization occurs and therefore the shift would not be expected to be dramatically altered, although the multiplicity would be affected. Further spectral work is needed to better clarify the nature of the benzylic type dianions, which can possibly equilibrate with deprotonation at the ortho position.

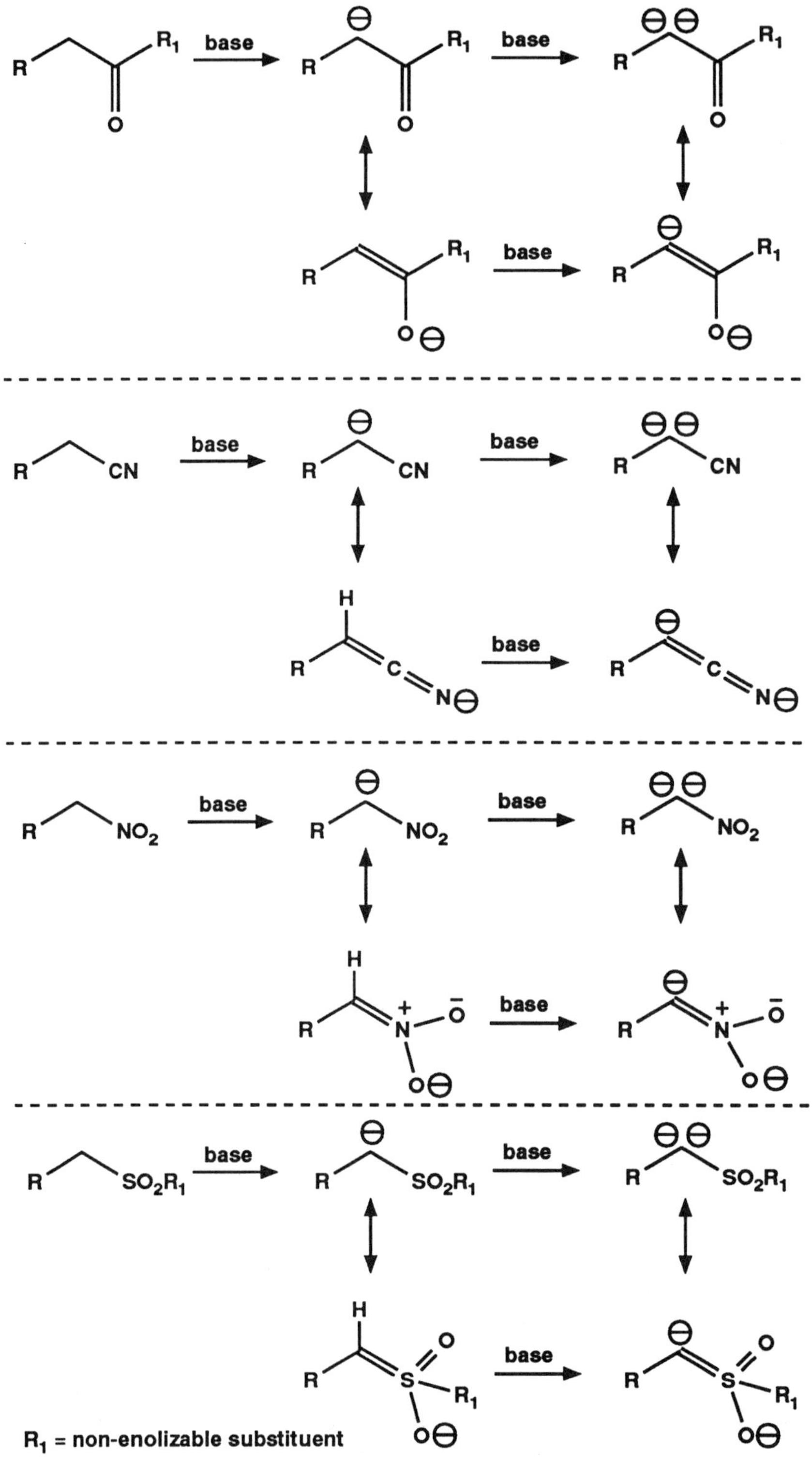

Figure 2.2. Representative geminal-(1,1)-dianions

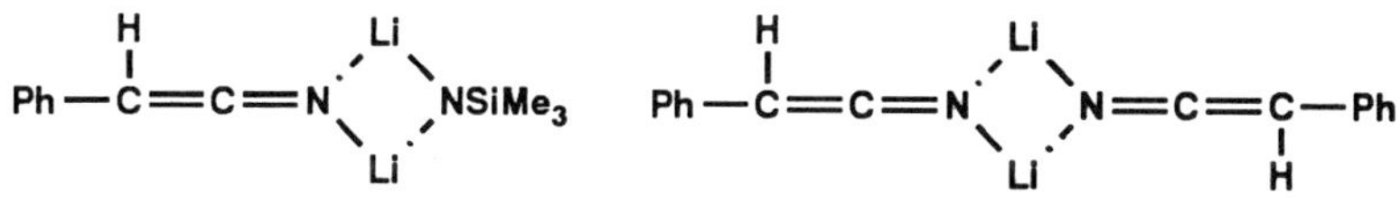

Figure 2.3. Possible chelated forms of benzylic geminal dianions

Although geminal dianions may be viewed as possessing a high degree of coulombic repulsion, reorganization to the 1,3-dianion, via the enolate or heteroenolate, results in a dianion system that places a suitable distance between charges and more likely accounts for their ease of generation. For convenience, the various geminal or 1,1-(C,C)-dianions will be classified alphabetically according to the carbanion-stabilizing group appended, namely, (halo)carbonyl-, nitrile-, nitro-, and sulfonyl-stabilized.

2.2.1 α-Halocarbonyl-stabilized 1,1-(C,C)-dianions

Although there was report of a ketone-stabilized geminal dianion in 1971,[11] it was not until Kowalski and coworkers initiated a series of investigations in 1980 that the full utility of these intermediates was realized.[12] Carbonyl-stabilized geminal dianions, unlike the others covered in this section, could suffer from possible attack by alkyllithium bases during preparation. This limitation led to the clever use of α-bromoketones as precursors to the 1,1-dianions.

Treatment of an α-bromoketone with amide bases yields the α-bromo lithioenolate (α-carbanion stabilized by both carbonyl and halogen), which is then subjected to metal-halogen exchange to yield the desired intermediate (Eq. 2.1).

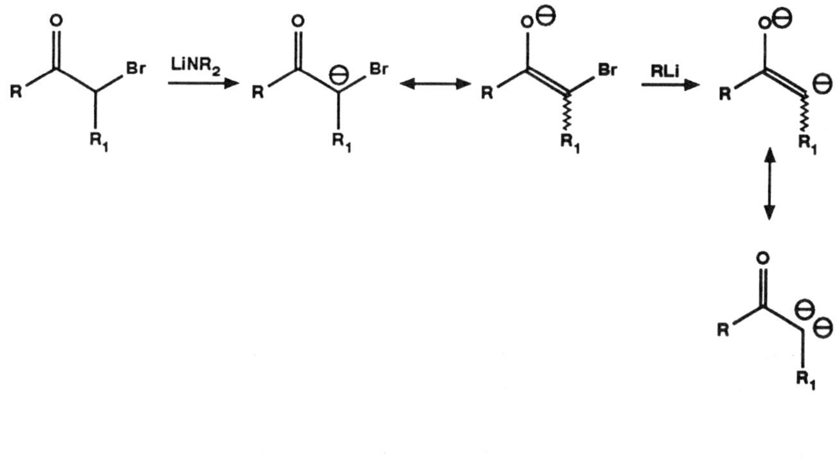

(2.1)

Preliminary studies reacted these intermediates with electrophiles showing that bis-α-deuteration and the bis-silylated enol ether supported the geminal dianion constitution. Both cyclic and acyclic α-haloketones were converted to geminal dianions. The initially formed α-bromoenolate also could be accessed by reaction of the α-bromoenol acetate with MeLi,[12] or by conjugate addition, and metal-halogen exchange (Eq. 2.2).[13]

1,1-Ketone dianions formed by this procedure undergo aldol-type reactions with aldehydes and ketones (Eq. 2.3).[14] In some cases, immediate dehydration to the enone was observed. Noteworthy was the reactivity of the dianion toward highly congested carbonyls including reaction with 2,2,6,6-tetramethylcyclohexanone.[14]

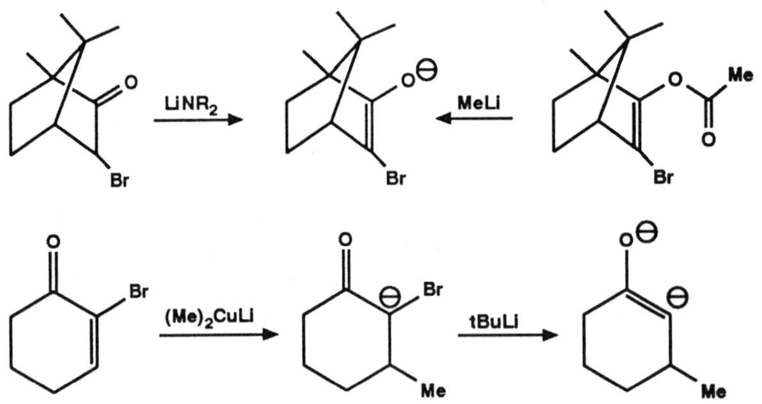

(2.2)

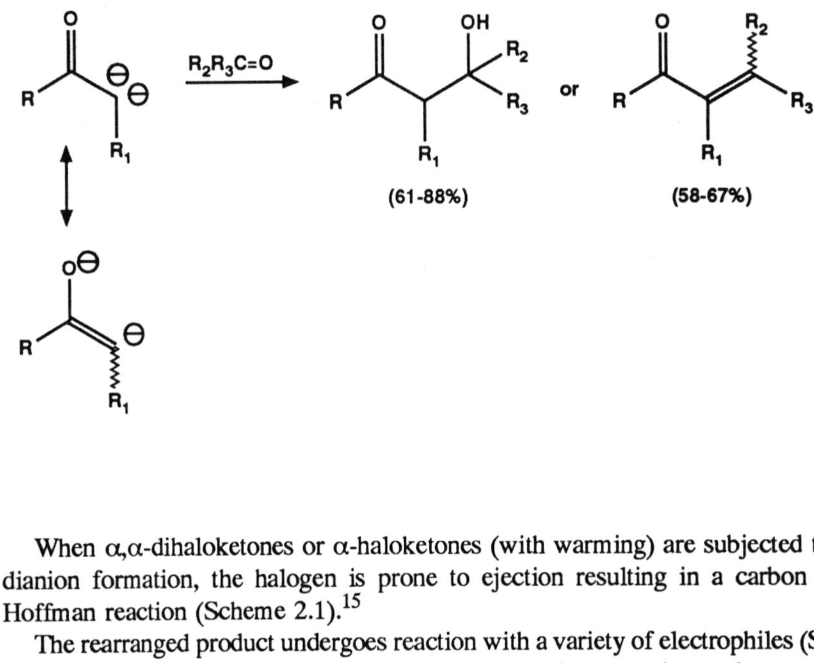

(2.3)

When α,α-dihaloketones or α-haloketones (with warming) are subjected to the geminal dianion formation, the halogen is prone to ejection resulting in a carbon analog to the Hoffman reaction (Scheme 2.1).[15]

The rearranged product undergoes reaction with a variety of electrophiles (Scheme 2.1).[15] The net conversion of a ketone to an ester prompted a more thorough examination of this process as a homologation sequence.[16,17] Ethyl esters reacted sequentially with dibromomethyl lithium and BuLi to afford the α-haloketone enolate dianion (Eq. 2.4), which upon quenching with ethanol gives the methylene homologated ester in 53-75% yield.[16,17]

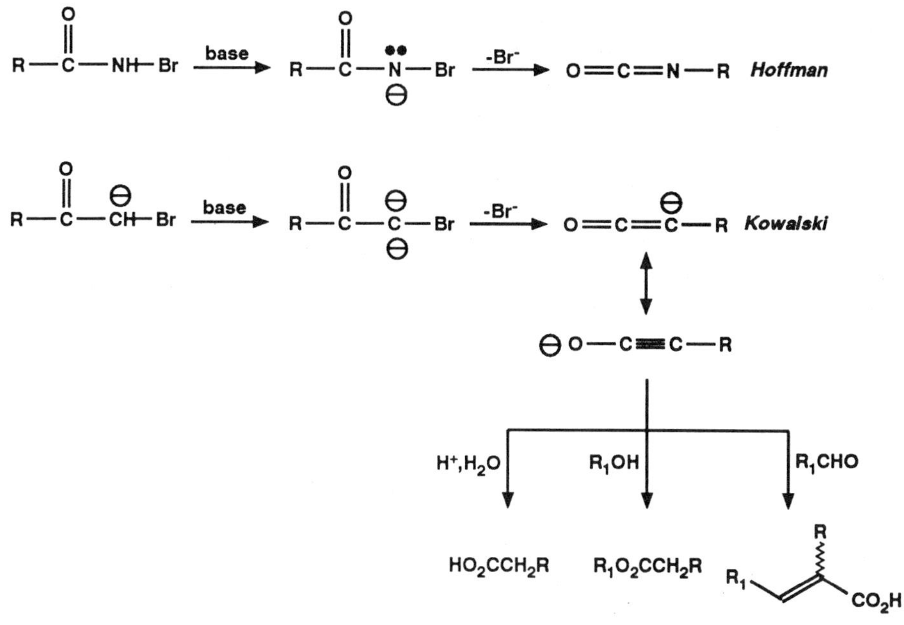

Scheme 2.1. The carbon analog of the Hoffman reaction

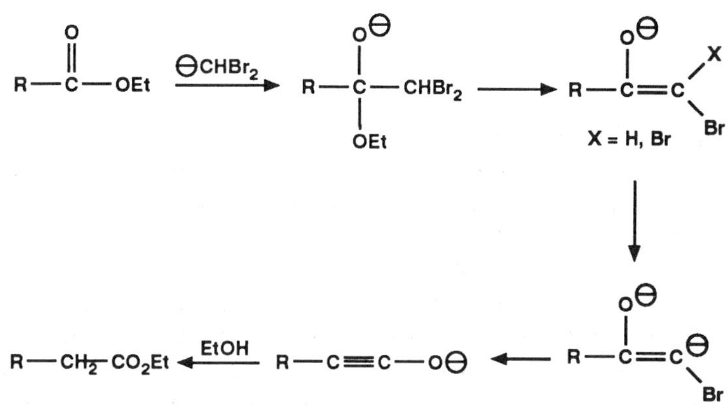

(2.4)

Wiemer and colleagues utilized this geminal dianion methodology to access β-ketophosphonates by quenching the intermediate dianions with $(RO)_2P(O)Cl$ at low temperature (< -78°C) (Eq. 2.5).[18] Simple enolates react with $(RO)_2P(O)Cl$ to give enol phosphates that must be rearranged to the α-phosphonyl compounds.

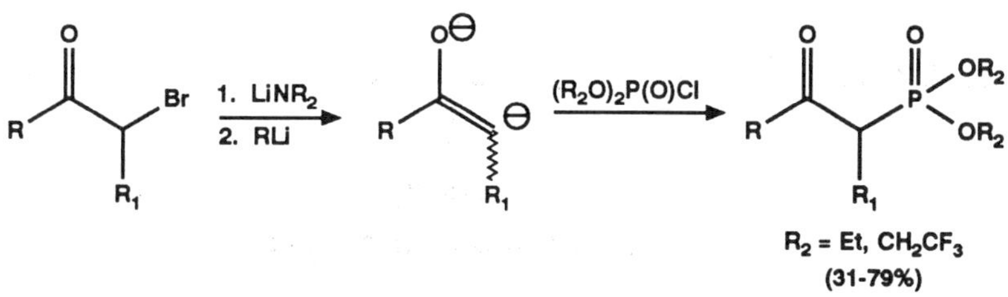

(2.5)

2.2.2 Nitrile-stabilized 1,1-(C,C)-dianions

Although only marginal research in this area has been conducted, the nitrile group can serve to stabilize 1,1-dianion formation (Figure 2.2). Reduced interest in this particular class of intermediates may result from their propensity to undergo bis-reaction, and/or mixed alkylation at the α-position and nitrogen atom.

Over twenty years ago, acetonitrile and α-trimethylsilylacetonitrile were converted to their dianions. Both dianion species underwent bis-reaction with aldehydes and TMS-Cl.[19]

Benzylnitrile[11,20] and its 3,4-dimethoxy analog[21] were converted to the geminal dianions, but again bis-alkylation predominated. This reactivity was used to advantage in the preparation of cyclopropanes, and could be suppressed by acylation, which forms a subsequent non-nucleophilic enolate (Eq. 2.6).

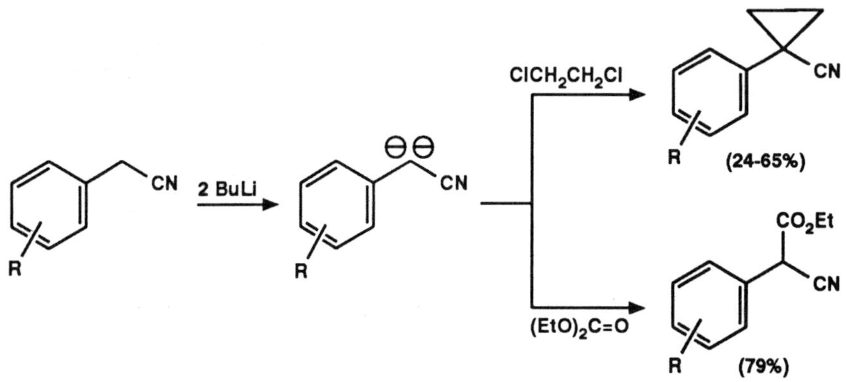

(2.6)

2.2.3 Nitro-stabilized 1,1-(C,C)-dianions

The nitro group imparts a very low pKa to an adjacent CH-containing group, which makes initial deprotonation quite facile. Unfortunately, this ease of proton abstraction is offset by a lowered nucleophilicity. Conversion to the dianion was sought as a method that would increase the nucleophilicity of the α-nitro position.

Several advantages to the use of the nitro moiety in dianion formation include its inability to undergo addition reactions with alkyllithiums, and possible transformation into amine (reduction) and ketone (Nef reaction) functional groups following use as a dianion stabilizing moiety. Seebach and coworkers were among the first to explore the enhanced reactivity of the nitro-stabilized geminal dianion, which were found to undergo high yielding reactions with a variety of electrophiles (Eq. 2.7)[22] Added carbanion-stabilizing groups (R = alkyl, PhS, Ar, MeO_2C) were also examined as precursor to the dianions, and similar outcomes were observed with the notable difference in the use of LDA as the base when the carboxyester was present.[23]

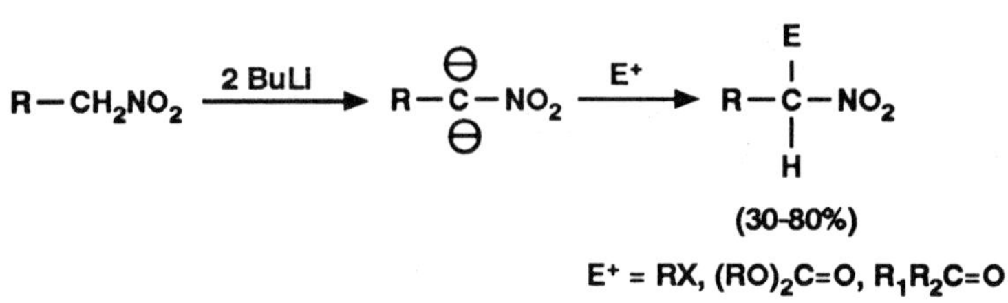

(2.7)

When alkene groups are present as part of the the R group, double deprotonation leads to resonance-stabilized dianions.[24] 3-Nitro-1-propene affords a geminal dianion that gives exclusive α-alkylation. By comparison, 4-nitro-1-butene gives a mixture of alkylation products at the 1- and 3-positions consistent with formation of a 1,2- and 1,4-dianion system (Scheme 2.2). Similar regioisomeric mixtures may occur when additional carbanion-stabilizing groups appear at the β-position.[23]

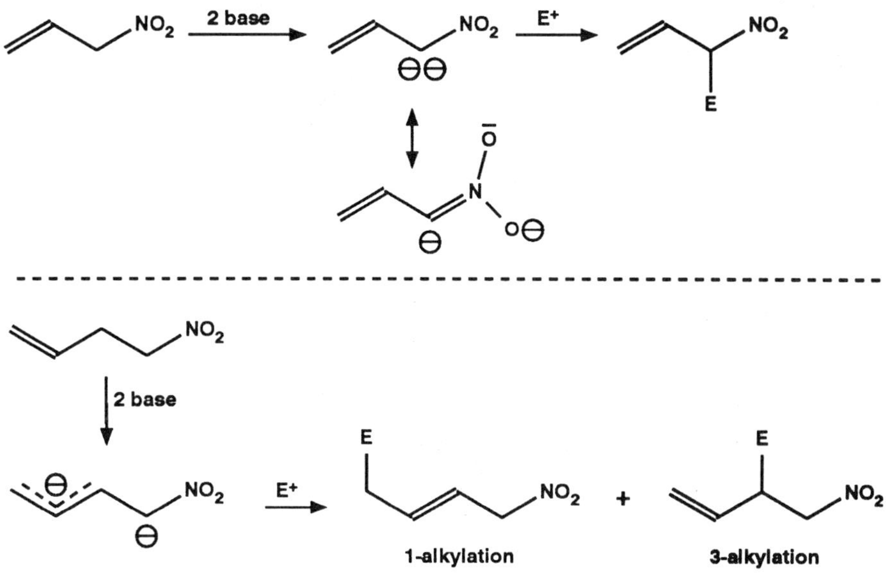

E+	1-alkylation		3-alkylation
RCHO	1	:	1
iPrX	3	:	97
		(40-82%)	

Scheme 2.2. Formation and reactions of 3- and 4-nitroalkene dianions

Yamada and coworkers found that the vicinal (1,2) dianion could be formed from alkylnitronates by simply changing the order of the addition of reagents.[25] Whereas Seebach added BuLi to a solution of the nitroalkane in THF/HMPA to afford the geminal dianion, Yamada found that addition of the nitroalkane to a preformed solution of the BuLi/HMPA complex in THF led to predominant generation of the vicinal dianion.

Seebach and Eyer explored the geminal nitro dianions bearing a β-heteroatom substituent.[26] The THP ether of 2-nitroethanol was converted to its 1,1-dianion, which was proposed to exist in the intramolecular chelate form (Fig. 2.4).

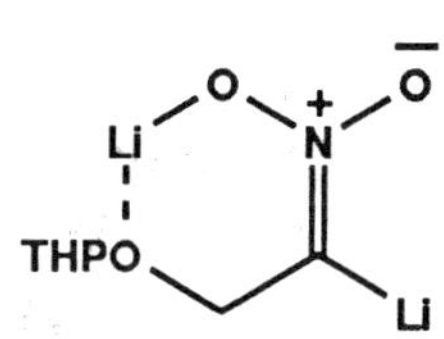

Figure 2.4. Proposed chelate form of a β-alkoxy nitronate dianion

β-Hydroxy or β-ether nitronate dianions condense with aldehydes and ketones to yield *threo*-(di)hydroxy nitronates in 75-95% yield.[26] Important to the use of these intermediates is the fact that virtually no β-elimination was observed.

Williams and Mosher used a geminal nitro dianion derived from glyceraldehyde to synthesize furanosides and pyranosides (Eq. 2.8).[27]

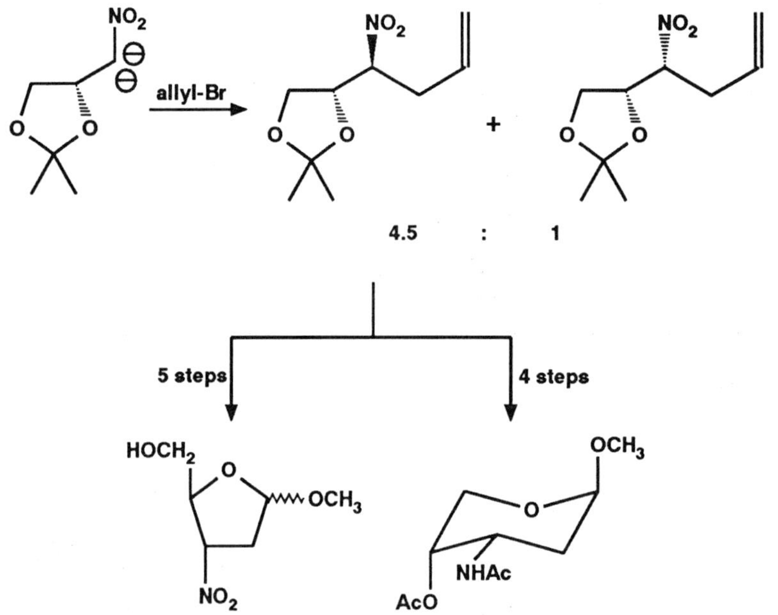

(2.8)

2.2.4 Sulfonyl-stabilized 1,1-(C,C)-dianions

Sulfones, like the nitro compounds, do not react directly with alkyllithium bases, and can aid 1,1-bis-deprotonations. Most of the pioneering work with these intermediates was conducted with tertiary sulfonamides.[20,28] Bis-reaction at the α-site was observed with simple electrophiles.

Benzyl phenylsulfone also forms the 1,1-dianion and bis-alkylates[20] but undergoes reaction with carbonyl compounds to afford vinyl sulfones (Eq. 2.9).[29] Further evidence for the 1,1-sulfonyl dianion was provided by Bosworth and Magnus who observed an interesting E1cB elimination following bis-deprotonation of a complex sulfone.[30]

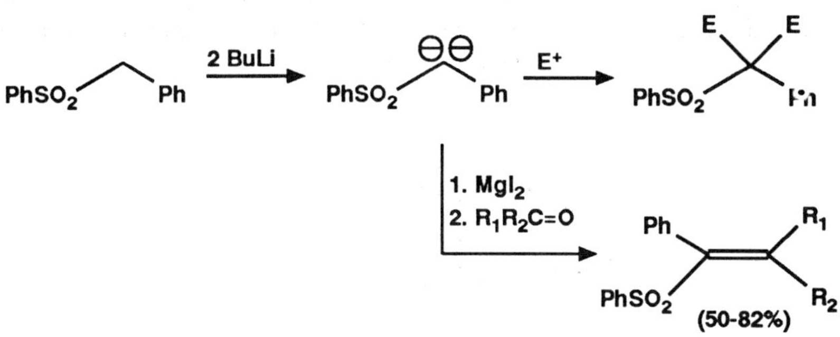

(2.9)

These preliminary investigations paved the way for more controlled use of the sulfonyl-stabilized geminal dianions. Sulfone-stabilized geminal dianions react with monosubstituted epoxides to give γ-hydroxy sulfones with good to excellent diastereoselectivity (Eq. 2.10).[31]

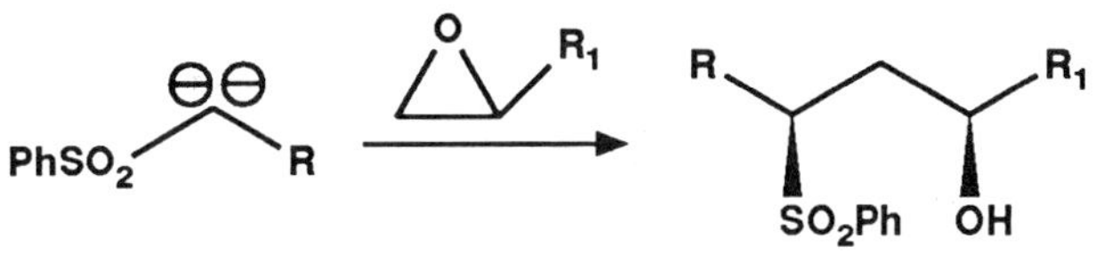

(2.10)

	Yield	de
R = -(Me)C=CH$_2$, R$_1$ = Ph	71	99:1
R = -(Me)C=CH$_2$, R$_1$ = 3-butene	90	95:5

Ethyl phenylsulfone forms the geminal dianion with 2 BuLi and reacts with *cis*-1,4-dichlorobut-2-ene to give 3-methyl-3-(phenylsulfonyl)cyclopentene through an intramolecular alkylation (Eq. 2.11).[32]

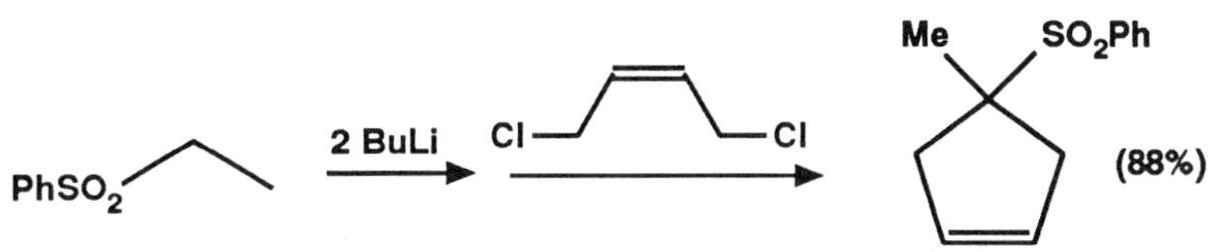

(2.11)

A variety of bifunctional electrophiles react with geminal sulfone dianions to give excellent outcomes of highly functionalized products (Scheme 2.3).[33-36]

A novel use of sulfonyl-stabilized geminal dianions was conducted by McCombie and coworkers who conducted an intramolecular cyclization of the dianion center onto an epoxide (Eq. 2.12).[37] In this instance, a β-carbonyl group aided bis-deprotonation of the α-site.

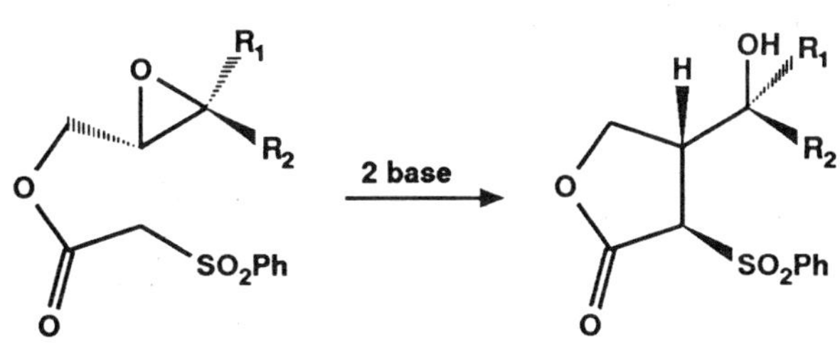

(2.12)

The dianion of allyl phenylsulfone also has been investigated.[38] Initial bis-deprotonation gives the 1, *ortho*-dianion, which, upon warming to 50°C, undergoes isomerization to the geminal 1,1-species (Eq. 2.13), which likely resides in resonance-delocalized form.

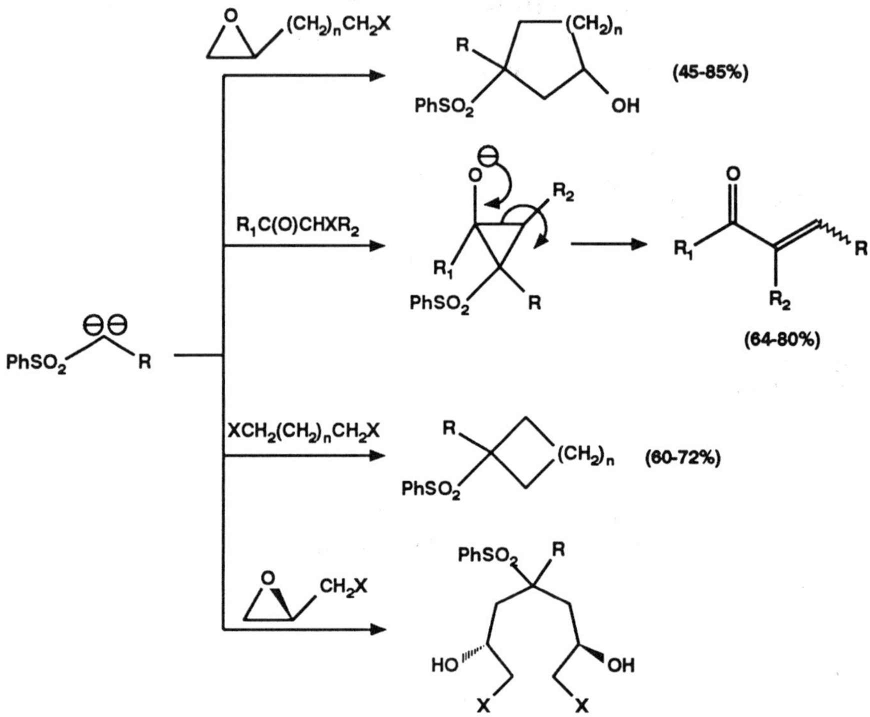

Scheme 2.3. Reactions of 1,1-(C,C)-sulfone-stabilized dianions

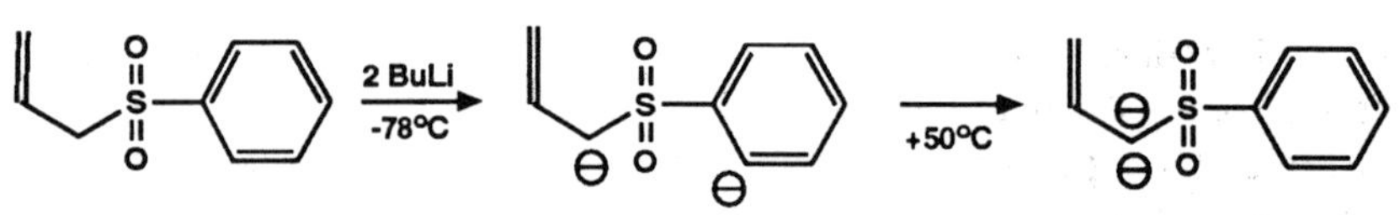

(2.13)

α-Heteroatoms aid geminal dianions formation, and makes possible the direct synthesis of carbonyl compounds[39] and vinylsulfones[40] (Scheme 2.4).

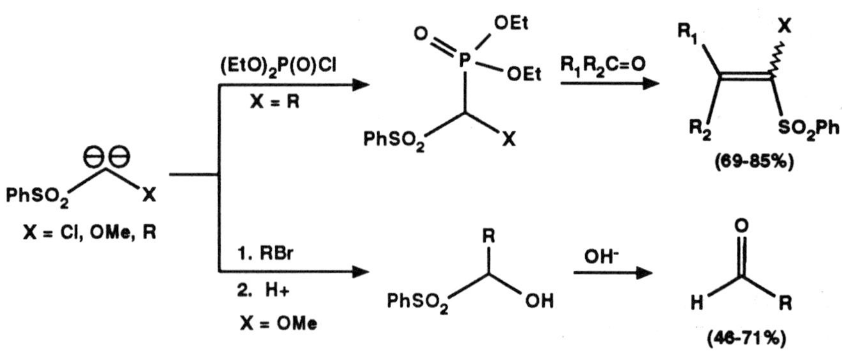

Scheme 2.4. Reactions of α-heteroatom-substituted, geminal sulfone dianions

2.3 1,2-(C,C)-DIANIONS (VICINAL DIANIONS)

Two suitably positioned carbanion-stabilizing groups (e.g., 1,2-diesters) can direct sequential deprotonations leading to vicinal- or 1,2-(C,C)-dianions. Vicinal dianions exist largely as the conjugated 1,3-diene-1,4-diolate when the stabilizing groups are carbonyl (Figure 2.5). The second carbanion stabilizating group is needed to produce a vicinal C,C-

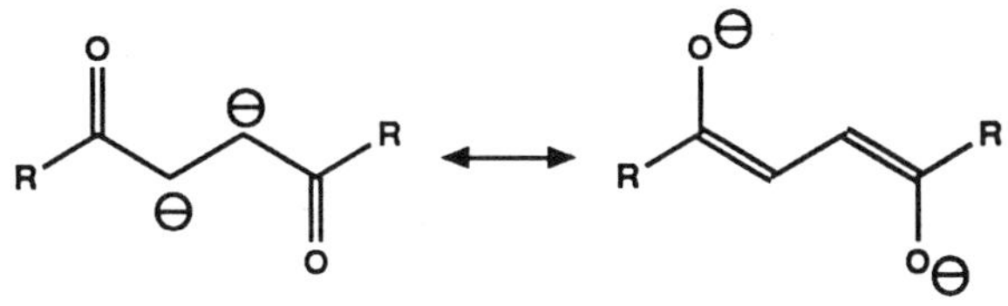

Figure 2.5. A vicinal (1,2)-dianion stabilized by carbonyl groups

dianion, as O,C-, N,C-, and S,C-vicinal dianion types are accessed by more direct methods, namely, first deprotonation of the heteroatom-H linkage, followed by CH-deprotonation at a vicinal site normally mediated by a sole carbanion stabilizing group (Eq. 2.14). Amides (Figure 2.5; X=NHC(O)R) of α-amino esters can be considered structurally akin to the 1,4-dicarbonyl precursors of vicinal C,C-dianions (see Chapter 4), and some uses in synthesis have revealed these likenesses.

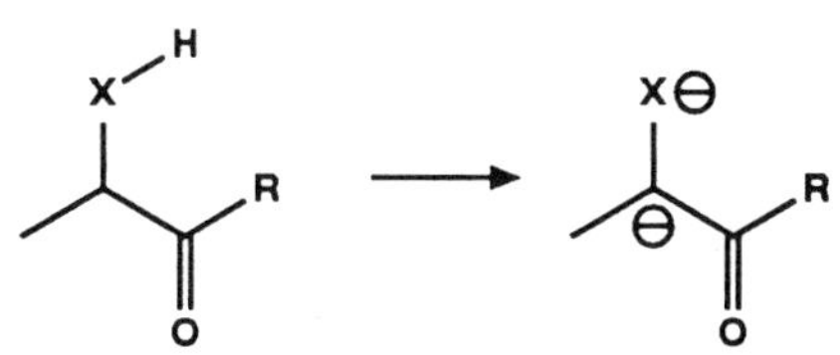

(2.14)

Several vicinal C,C-dianions have been generated solely for studies of theoretical and spectroscopic interest. The 1,1-dianions of stilbene,[41] triphenylethylene,[42,43] 2,3,3-triphenylpropionitrile,[44-46] dimethylcyclobut-3-ene-1,2-dicarboxylate,[47] and phenyl[tris(trimethylsilyl)]ethylene[48] (Figure 2.6) are examples that are not used in synthesis. Studies of these vicinal dianions are useful toward a better understanding of the reactivity profile and character of related vicinal dianions.

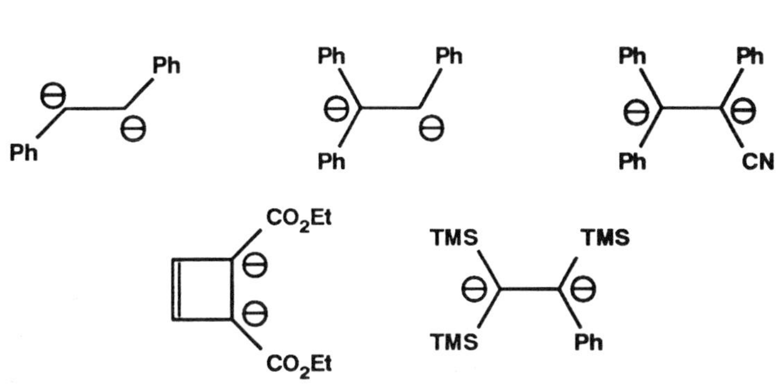

Figure 2.6. Vicinal (C,C)-dianions of theoretical and spectroscopic interest

Certain related vicinal C,C-dianions may find categorization elsewhere in this text owing to structural overlap with other dianions. For example, certain allylic and homoallylic nitro compounds are bis-deprotonated to give either/or a mixture of the geminal and vicinal (resonance-stabilized) dianions, respectively. These mixed "vicinal type" dianions are grouped with the geminal C,C-dianions to aid understanding of the specific characteristics of this class of dianion precursors.

The purpose of this section is to define and explore the synthetic uses of C,C-dianions that are formed through stepwise deprotonations of vicinal C-H groups. A majority of the vicinal C,C-dianions have been accessed from 1,4-dicarbonyl compounds, or their masked carbonyl counterparts. Two basic types of vicinal C,C-dianion have found utility in organic synthesis, acyclic and cyclic, the latter showing far more synthetic use to date.

2.3.1 Acyclic vicinal C,C-dianions

The dianion of diethylsuccinate undergoes mono-[49] or di-alkylation,[50] and condensation reactions[51] to give good yields of products. Condensation with aldehydes and ketones leads directly to paraconic esters (43-86%) following reaction with acid (Eq. 2.15).

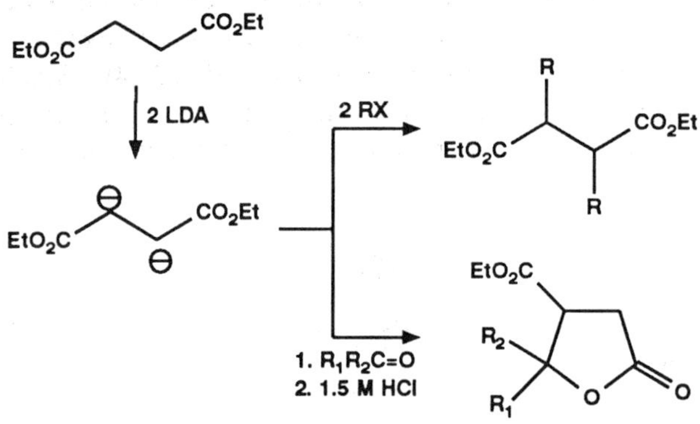

(2.15)

Yamamoto used esters and oxazoline-protected succinic acids to access vicinal dianions that served in ring-forming and ring-annulating procedures (Eq. 2.16) (Scheme 2.5).[51,52]

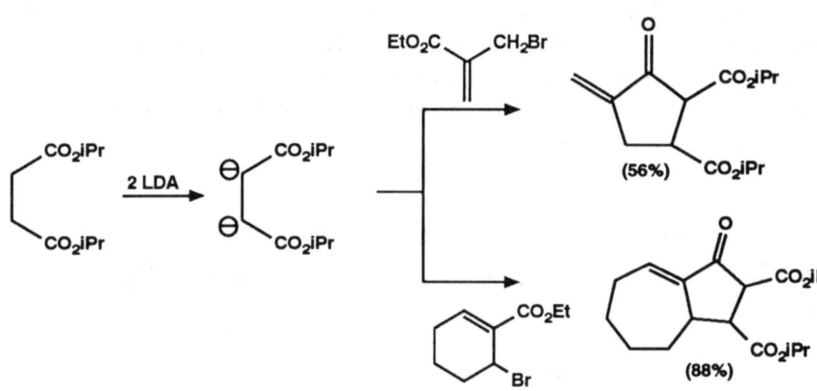

(2.16)

A somewhat unusual entry, which defies categorization as a true vicinal C,C-dianion type, is dianions accessed from disubstituted α-ketoamides.[53] Deprotonation of the α-keto position gives an enolate, which upon reaction with a second equivalent of base gives the initially formed vicinal dianion leading to the diene-1,2-diolate (Scheme 2.6). Alkyl halides reacted predominantly to give tertiary alcohols (α-alkylation via resonance structure III).

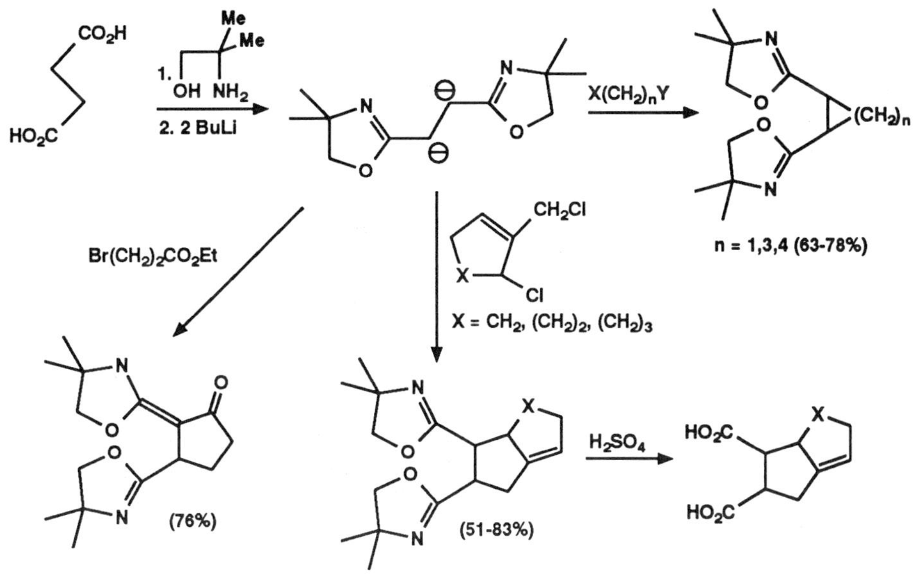

Scheme 2.5. Vicinal (C,C)-dianions prepared from protected carboxylic acids

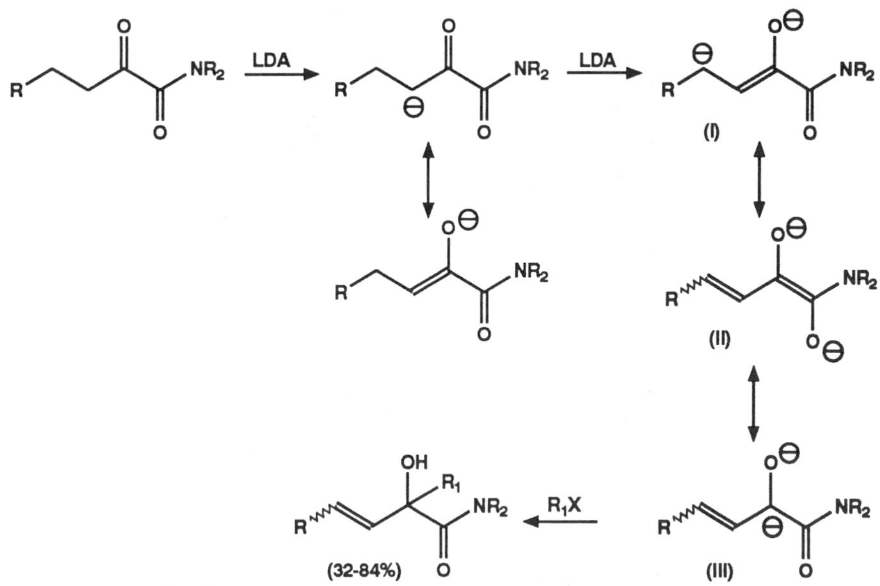

Scheme 2.6. α-Ketoamides leading to a 1,1-(C,C)-dianion (I) and O,C-dianion (III)

2.3.2 Cyclic vicinal C,C-dianions

Most of the activity in the area of cyclic vicinal C,C-dianions has been conducted by Garratt and coworkers who explored a variety of alkylation and ring-building reactions from cyclic 1,2-diesters. Their studies have been nicely complemented by related investigations of cyclic vicinal dianions that are stabilized by only one carbonyl group using an alternative functional group or moiety to stabilize the second deprotonation. The following discussion will first cover a ring size progression of cyclic 1,2-vicinal dianions derived from 1,2-diesters followed by a presentation of the alternative approaches to cyclic vicinal dianions.

Cyclobutane rings can be difficult to prepare and incorporate into molecules. Garratt and coworkers found a useful general method to circumvent these problems by exploiting the reactions of vicinal dianions formed from the readily available dimethylcyclobutane-1,2-dicarboxylate (Scheme 2.7).[54-56] The cyclobutyl 1,2-dianion reacts twice with alkyl halides[54] and aldehydes,[54] once with Michael acceptors,[54] and undergoes interesting cyclizations with dihalides[55] and diesters[55,56] (Scheme 2.7).

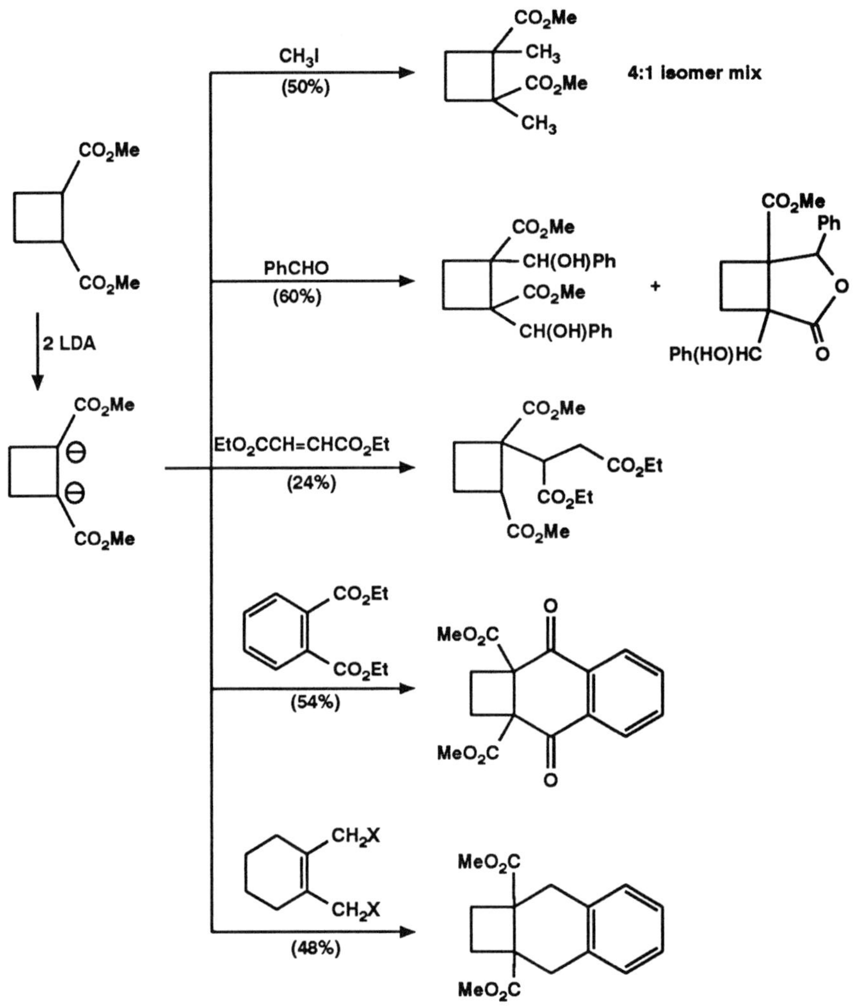

Scheme 2.7. Synthetic uses of the vicinal-(C,C)-dianion of cyclobutane-1,2-dicarboxylate

Cyclopentyl and cyclohexyl dicarboxylate vicinal 1,2-dianions also have been reacted with β-bromoethyloxirane to give annelated γ-lactones (Scheme 2.8).[57] The dianion of dimethylcyclohexane-1,2-dicarboxylate was used in a synthesis of (+/-)-valerane[58] and tricyclobuta[1,2:3,4:5,6]naphthalene.[55] An unsaturation in the [dianion containing] ring may be present, and offers the opportunity for further elaboration of the molecule following use as a ring annulating reagent. Dimethyl 4-cyclohexen-1,2-dicarboxylate was converted to its vicinal dianion and reacted successfully with β-bromoethyloxirane (Scheme 2.8),[57] dibenzylic dibromides,[59] ethyl-4-bromobutyrate,[60] 1,4-dihalides,[61] and alkynoates[62] (Scheme 2.9).

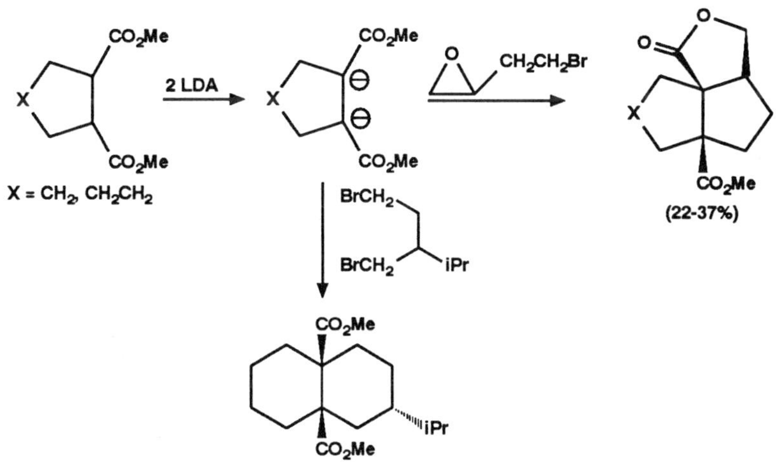

Scheme 2.8. Ring-forming reactions of cyclic vicinal dianions

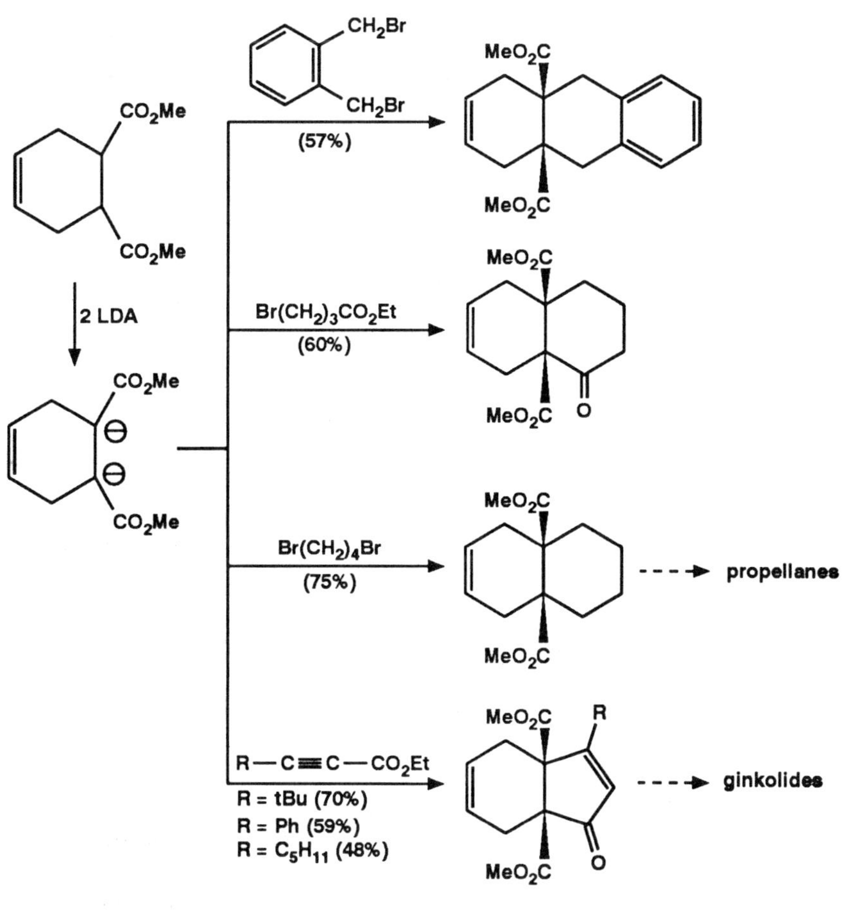

Scheme 2.9. Synthetic uses of the vicinal dianion of cyclohexene-3,4-dicarboxylate

As mentioned, several vicinal C,C-dianions have been explored that do not bear the 1,2-diester-stabilizing group combination. For example, the added stability imparted by aromaticity to the formation of cyclopentadiene carbanion proved attractive. Ide,[63] Koreeda,[64,65] and Trost[66] utilized a cleverly placed cyclopentane carbonyl group to form the initial carbanion that, through conversion to its enolate, forms the requisite cyclopentadienyl system that probably lowers the pKa of remaining methylene group and readily gives deprotonation upon reaction with the second equivalent of base (Scheme 2.10).

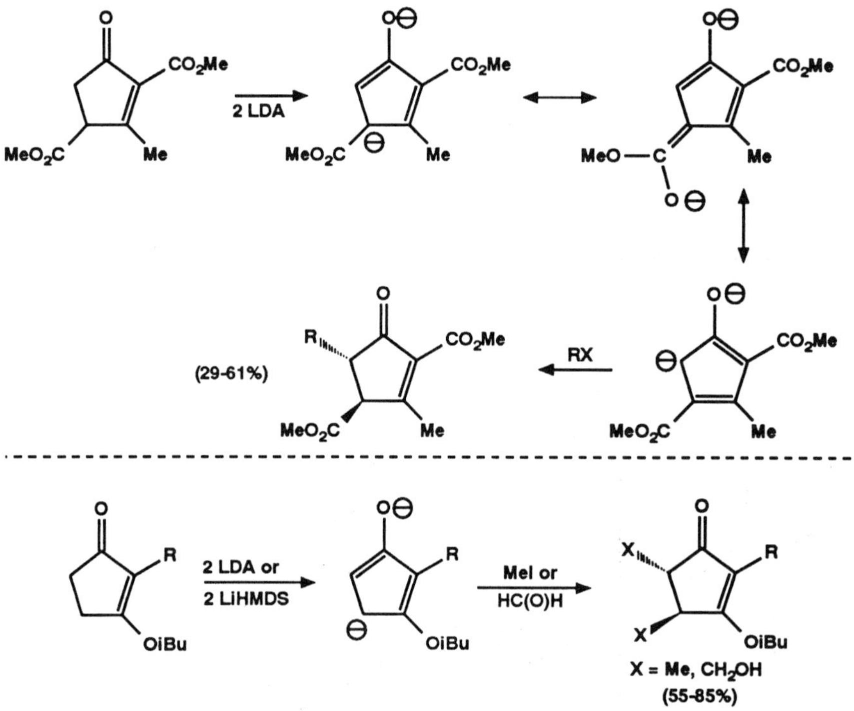

Scheme 2.10. Formation and utility of cyclopentadiene-based vicinal dianions

2,4-Dimethoxycarbonyl-3-methyl-2-cyclopentanone forms the cyclopentadienyl dianion system via vicinal deprotonation, and undergoes regio- and stereoselective alkylations (Scheme 2.10).[63] Alkylation with ethyl bromoacetate provided a route to the Corey lactones.

The isobutyl enol ether of 2-alkylcyclopentane-1,3-diones proved an excellent precursor for vicinal deprotonations in route to a cyclopentadienyl aromatic dianion system that underwent bis-alkylation and bis-condensation in good yield (Scheme 2.10).[64] Included in this study was an efficient intramolecular alkylation using 1,4-diiodobutane. This intramolecular strategy was extended to the synthesis of (dl)-coriolin (Eq. 2.17).[65]

The benzofused cyclopentadienyl system was formed through initial deprotonation of 1-indanone (Eq. 2.18).[66] Alkylation occurs almost exclusively at the benzylic position although some alkylation at the oxo-carbon was observed.

As final examples of vicinal dianions, a bicyclic succinimide dianion which, irrespective of the initial exo or endo stereochemistry, was found to alkylate predominantly from the exo face.[67] Beslin and Dlubala found that the sulfur functional groups of a β-thiosubstituted dithiopropanoate served to direct vicinal deprotonation.[68]

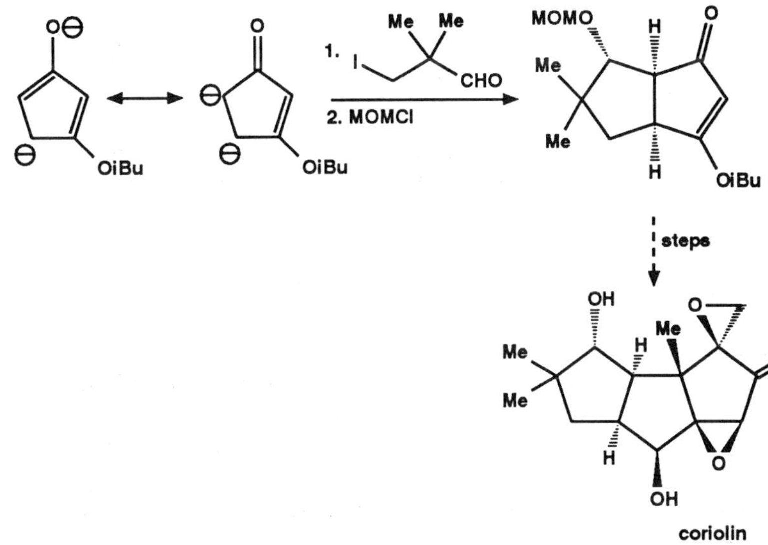

(2.17)

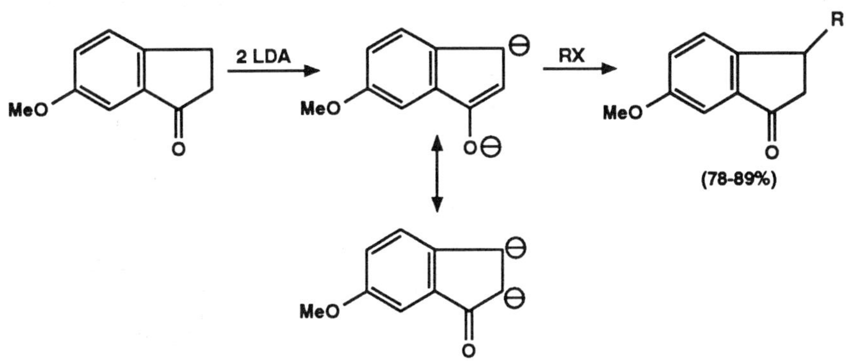

(2.18)

2.4 1,3-(C,C)-DIANIONS (β-DICARBONYL DIANIONS)

Ketones that bear a β-carbonyl moiety, for example an aldehyde, ester, or second ketone group, can be deprotonated stepwise between the carbonyl groups and then at the "external" ketone enolate site to give a 1,3-(C,C)-dianion (Figure 2.7). C,C-Dianions arising from the bis-deprotonation of β-dicarbonyl compounds have been the most widely used in organic synthesis owing to their ready access, predictable reactivity, and expanded reaction repertoire. Because of the widespread use, β-dicarbonyl dianions have been tagged with a variety of nomenclatures. Early work in this area termed the products of the bis-deprotonation of β-dicarbonyl as α,γ-dianions to indicate the position of the carbanion sites relative to the first carbonyl group. Several investigators prefer the α,α'-dianion designation of β-dicarbonyls to support the carbanion positions about a single carbonyl group. As mentioned in Section 2.1, we find the numerical designation as 1,3-(C,C)-dianion most easily recognizable and are used herein. The 1,3-dianion descriptor also more clearly distinguishes these dianions on the basis of deprotonation site rather than the nature of the functional groups (Figure 2.1). This is an important clarification since β-dicarbonyls also can give rise to 1,1-(C,C)- or remote dianions. Examples of 1,3-(C,C)-dianions are shown in Figure 2.7.

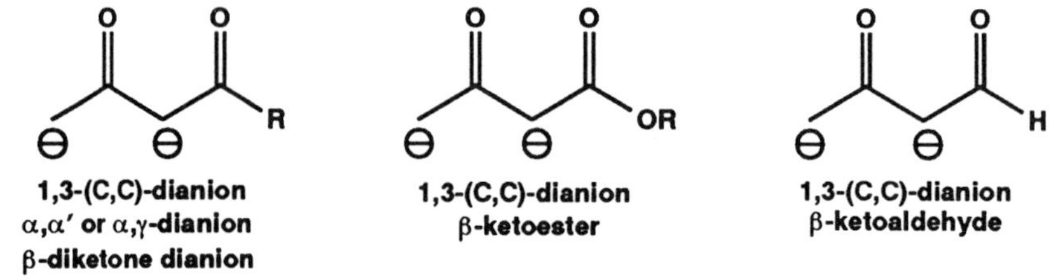

Figure 2.7 Representative β-dicarbonyl dianions

Several reviews have covered the preparation and reactivity of 1,3-dianions[3,5,8,9] including the landmark review by Harris and Harris.[2] Their use is both commonplace in total synthesis and methodology, and is looked upon as a reliable tool when three-, four-, or five-carbon chain extension is warranted. More recent variations in the use of 1,3-dianions derived from dicarbonyl compounds have permitted for even greater breadth of application including bis-deprotonations from cyclic dicarbonyls, intramolecular alkylations, and stereochemical advances.

The minimum requirement for entry to 1,3-dianions derived from β-dicarbonyls is the presence of a ketone group, which establishes two carbonyl-flanking positions available for deprotonation (as compared to malonic diesters). The second carbonyl-based group may be

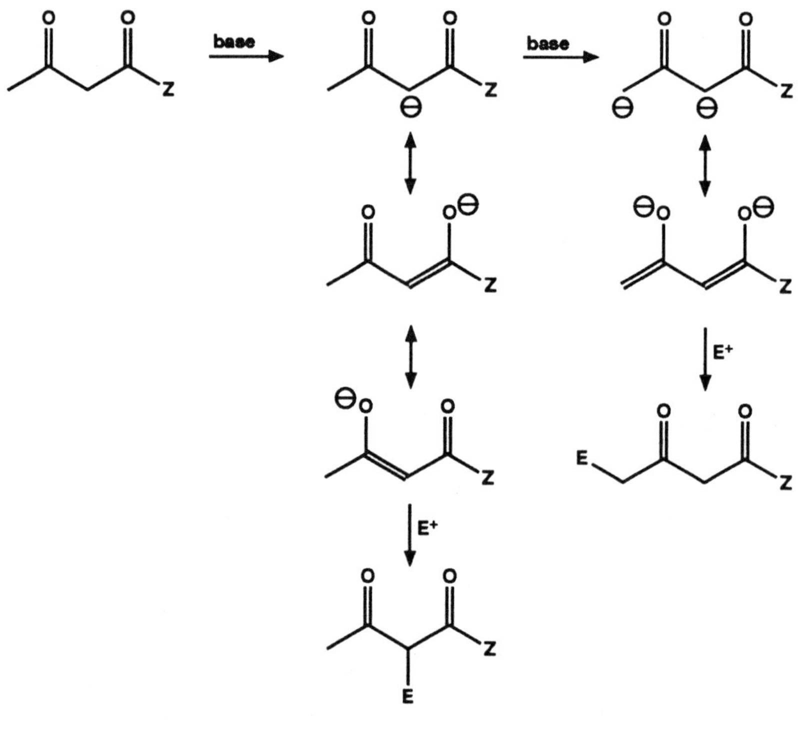

Scheme 2.11. Deprotonation of β-dicarbonyl compounds and possible reactions

an aldehyde, ester, or ketone, which may be present in acyclic or cyclic form. Owing to the reactive nature of carbonyl-containing carbanion stabilizing groups, alkyllithium or organomagnesium reagents are not recommended for the initial deprotonation, but may be used to conduct the second deprotonation. A few base combinations have emerged as the most reliable in terms of high yield of the dianion and minimum reaction of the carbonyl. Metal amides like $LiNH_2$, $NaNH_2$, and KNH_2 have been used with great success, although some reaction with the ester group of methyl acetoacetates has been described. Sterically hindered amide bases, especially LDA, have been used extensively to access 1,3-(C,C)-dianions derived from β-dicarbonyls including reactive β-ketoaldehydes. Another favorite among base choices has been the NaH/BuLi combination, which has great advantages in the ease of dianion production. Since the NaH is unreactive toward carbonyls and the BuLi is introduced with a known concentration, extensive experimental preparation is circumvented. Representative examples of the preparation of β-dicarbonyl 1,3-(C,C)-dianions are given in the experimental section.

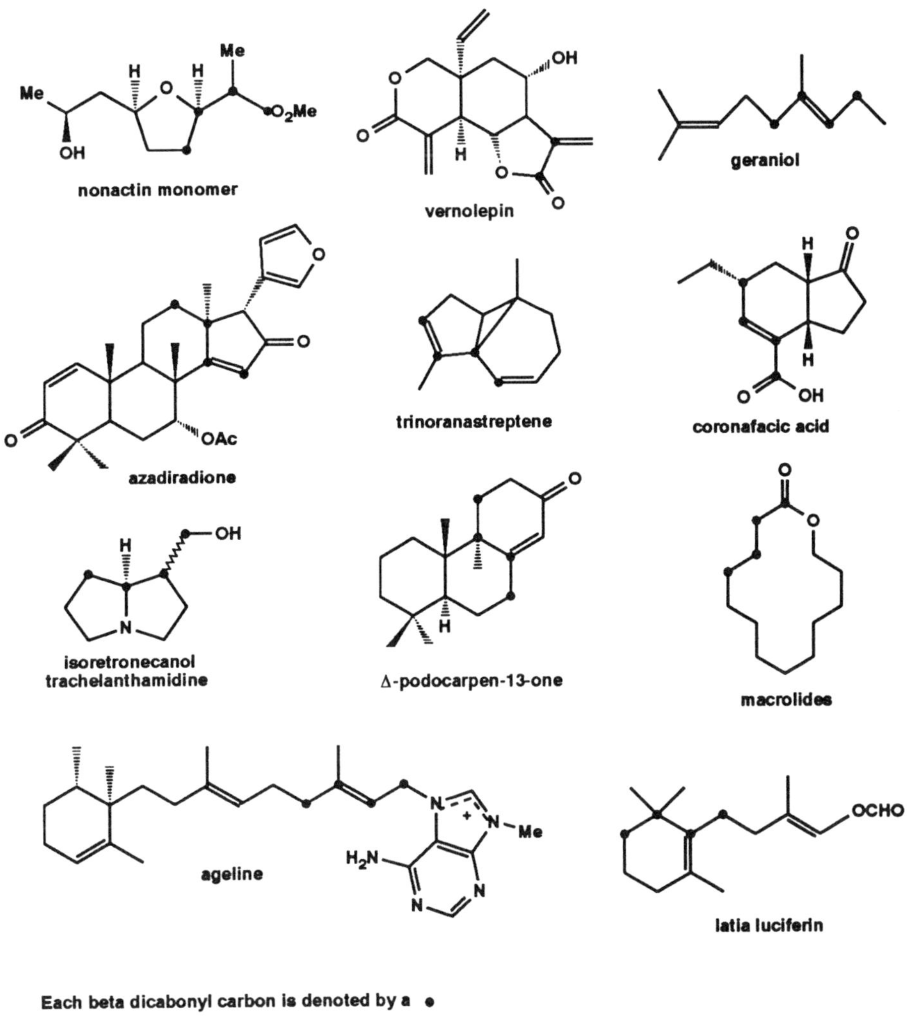

Figure 2.8. Synthetic products resulting from the use of 1,3-(C,C)-dianions derived from β-dicarbonyl compounds

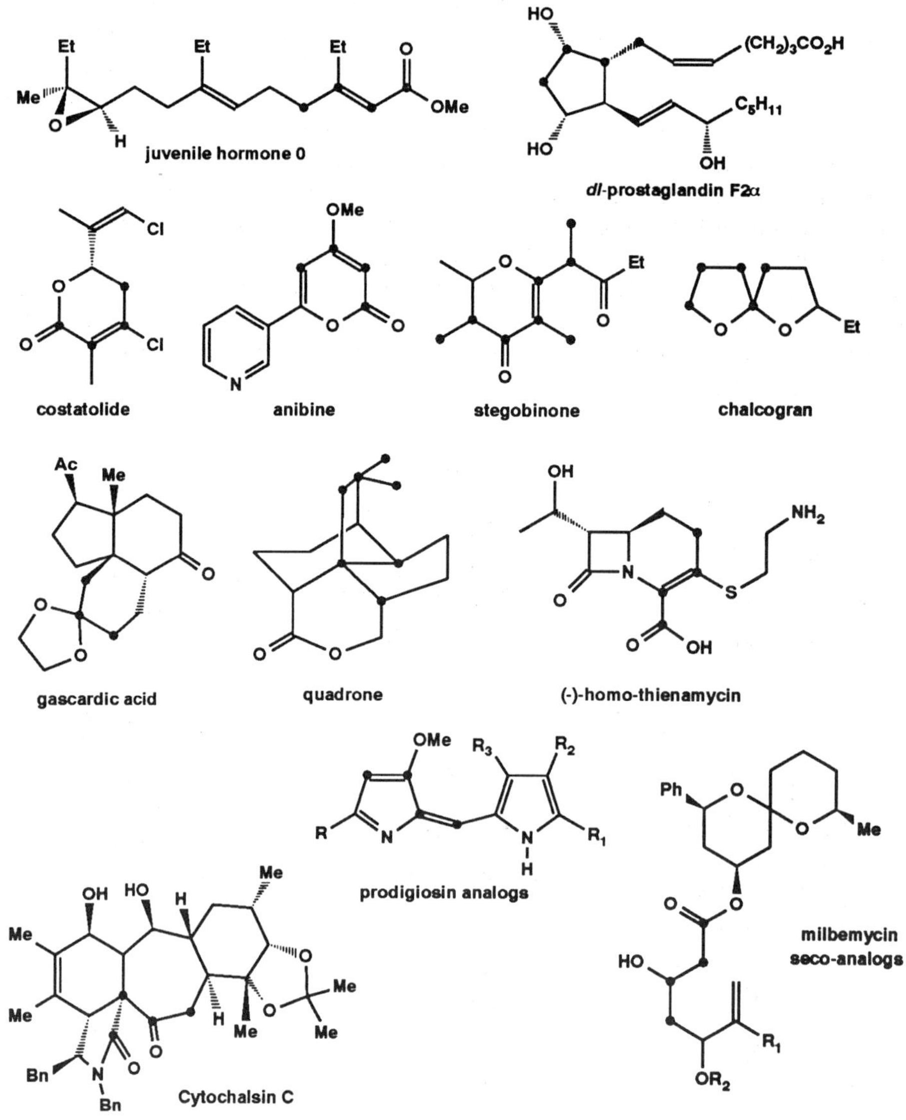

Each beta dicabonyl carbon is denoted by a •

Figure 2.8 (cont.). Synthetic products resulting from the use of 1,3-(C,C)-dianions derived from β-dicarbonyl compounds

When a β-dicarbonyl compound is first deprotonated at reduced temperature, a distabilized enolate is formed that is relatively non-nucleophilic. These monoanions do undergo the well known condensation reactions with aldehydes, ketones, and esters but usually require extended reaction times or elevated temperatures. A second deprotonation results in the formation of a highly reactive terminal carbanion site forcing formation of a conjugated dienolate intermediate that reacts regiospecifically with electrophiles. Following reaction of the dianion intermediate with one equivalent of an electrophile, the monoanion remains, which can be further reacted under certain conditions (Scheme 2.11). Alternatively, the dianion may be regenerated following addition of the electrophile, and a second reaction may be conducted at the 3-carbanionic position thereby difunctionalizing the β-dicarbonyl compound in one experiment.

Since dianions of β-dicarbonyl compounds have been so widely used in organic synthesis, it is preferable to organize their utility according to the reaction type rather than the compound type (ketoester, acetoacetone, cyclic or acyclic, etc.). In part, this is because dianions of β-ketocarbonyls undergo similar reactions leading to products that share a common point of attachment. As a result, this section will arrange the common synthetic uses of dianions of β-dicarbonyls according to the nature of the electrophile. This preference will permit certain tabular collections of common reactions to be made that should provide the reader with the needed depth of coverage and comparative reactivity. The organization of β-dicarbonyl dianions by reaction with electrophile type is consistent with that found in a prior review.[5] In this section, we shall cover alkylation (including alkyl halides, epoxides, etc.), condensation, acylation, and addition reactions. It is no surprise that investigators have made extensive use of these readily accessed dianions in synthesis. As an introduction to their utility, a pictoral survey of some natural and unnatural products prepared using dianions derived from β-dicarbonyl compounds is shown in Figure 2.8. The carbons comprising the chain of the β-dicarbonyl component are denoted by • to denote the nature of the carbon-carbon bond connectivity.

Following coverage of 1,3-(C,C)-dianions derived from β-dicarbonyl compounds, a brief survey of related 1,3-(C,C)-dianions prepared from ketones that bear a β-carbanion stabilizing group other than a carbonyl (e.g., nitro, cyano, etc.) will be conducted. 1,3-(C,C)-Dianions that bear a sulfur- or phosphorus-containing moiety as the β-stabilizing group are covered in Sections 2.5 and 2.6, respectively.

2.4.1 Alkylation of β-dicarbonyl dianions

Alkylation of β-dicarbonyl dianions with alkyl halides

The alkylation of β-dicarbonyl compounds with alkyl halides continues to be one of the most attractive pathways to prepare γ- or 3-substituted ketoesters and diketones (Eq. 2.19). The alkylation of these dianions has been reviewed regularly,[2,3,5] with the earlier coverage[2,5] focused more upon the dianions generated from metal amides (MNH_2). As mentioned, most experimenters now prefer 2 LDA or the base combination of NaH and BuLi to generate the dianion followed by capture with the alkylating agent. Regardless of the base choice, alkylations occur best with primary, allylic and benzylic halides, and to a lesser extent with secondary halides. Tertiary halides undergo exclusive elimination owing to the base strength of the dianion. A summary of the alkylation of β-dicarbonyl 1,3-(C,C)-dianions with alkyl halides is presented in Table 2.1.

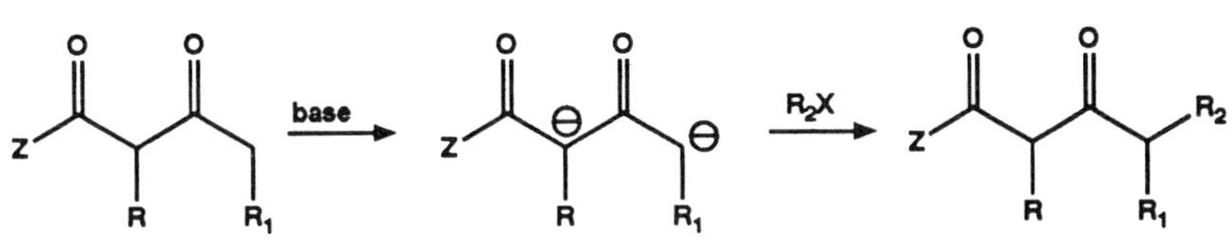

(2.19)

Table 2.1
Representative alkylations of β-dicarbonyl dianions (Eq. 2.19)

Z	R	R_1	Base	R_2X	Yield(%)	Ref.
Me	H	H	$NaNH_2$	MeI	59-65	69
Me	H	H	$NaNH_2$	EtBr	70	69
Me	H	H	$NaNH_2$	nPrBr	77	2
Me	H	H	$NaNH_2$	iPrBr	66	69
Me	H	H	KNH_2	iPrBr	27	70
Me	H	H	KNH_2	CH_2=$CHCH_2Br$	65	70
Me	H	H	$LiNH_2$	nBuBr	25	69
Me	H	H	$NaNH_2$	nBuBr	67-88	71
Me	H	H	KNH_2	nBuBr	68	70
Me	H	H	$NaNH_2$	iBuBr	78	69,72
Me	H	H	$NaNH_2$	sBuBr	78	69,72
Me	H	H	$NaNH_2$	$nC_6H_{13}Br$	69	73
Me	H	H	$LiNH_2$	BnCl	0	70
Me	H	H	$NaNH_2$	BnCl	69-73	69,70
Me	H	H	KNH_2	BnCl	60	1,70
Me	H	H	$NaNH_2$	LiO_2CCH_2Cl	82	74
Me	H	H	$NaNH_2$	$(MeO)_2CHCH_2Br$	38-49	75
Me	H	H	$NaNH_2$	$PhCH_2CH_2Br$	43	76
Et	H	H	$NaNH_2$	MeI	56-66	77
iPr	H	H	$NaNH_2$	MeI	56	77
iPr	H	Me	$NaNH_2$	MeI	58	77
$-(CH_2)_3-$		H	KNH_2	BnCl	62	78
$-(CH_2)_3-$		Me	$NaNH_2$	nBuBr	60	2
$-(CH_2)_4-$		H	KNH_2	BnCl	58	78
$-(CH_2)_4-$		Me	$NaNH_2$	nBuBr	82	2
C_5H_{11}	H	H	$NaNH_2$	EtBr	70	69
C_5H_{11}	H	H	$NaNH_2$	nBuBr	68	69
C_5H_{11}	H	H	$NaNH_2$	BnCl	52	70
Me	H	Ph	$NaNH_2$	nBuBr	77	72,79
Me	H	Ph	$NaNH_2$	BnCl	77	77
Me	H	Ph	$NaNH_2$	$Ph(CHCl)CH_3$	44-48	80
iPr	H	Me_2	$NaNH_2$	nBuBr	26	72
Ph	H	H	$NaNH_2$	EtBr	85	72
Ph	H	H	KNH_2	BnCl	77	1
Ph	H	H	LDA	EtI	-	81
Ph	H	H	$LiNH_2$	2-Cl-quinoline	71	82,83
Ph	H	Me	$NaNH_2$	EtBr	84	72
Me	Ph	H	KNH_2	BnCl	62	84
EtO	H	H	KNH_2	MeI	36-37	85
EtO	H	H	KNH_2	EtBr	27-29	85
EtO	H	H	$NaNH_2$	BnCl	41	2

Table 2.1 (cont.)
Representative alkylations of β-dicarbonyl dianions (Eq. 2.19)

Z	R	R_1	Base	R_2X	Yield(%)	Ref.
MeO	H	H	LDA	$R_3SiOC(R_1R_2)CH_2CH_2I$	25	86
MeO	H	H	NaH/BuLi	MeI	81	87,88
MeO	H	H	NaH/BuLi	EtBr	84	87,88
MeO	H	H	NaH/BuLi	iPrI	73	87,88
MeO	H	H	NaH/BuLi	nBuBr	72	87,88
MeO	H	H	NaH/BuLi	CH_2=$CHCH_2Br$	83	87,88
MeO	H	H	NaH/BuLi	BnCl	81	87,88
MeO	H	nBu	NaH/BuLi	MeI	81	87,88
MeO	H	nBu	NaH/BuLi	CH_2=$CHCH_2Br$	77	87,88
MeO	H	nBu	NaH/BuLi	BnCl	63	87
MeO	H	Me	NaH/BuLi	BnCl	76	87
MeO	H	Bn	NaH/BuLi	MeI	86	87
EtO	H	H	NaH/BuLi	CH_2=$CHCH_2Br$	77	87
EtO	Me	H	NaH/BuLi	EtBr	78	87
MeO	-$(CH_2)_3$-		NaH/BuLi	CH_2=$CHCH_2Br$	67	87
MeO	-CH_2CH_2-		NaH/BuLi	CH_2=$CHCH_2CH_2CH_2I$	89	89
MeO	-CH_2CMe_2-		NaH/BuLi	CH_2=CHCH=$CHCH_2CH_2I$	-	90
MeO	H	H	NaH/BuLi	$Br(CH_2)_4Br$	26	91
MeO	H	H	NaH/BuLi	$Br(CH_2)_5Br$	25	91
MeO	H	H	NaH/BuLi	$Br(CH_2)_{10}Br$	39	91
MeO	H	H	NaH/BuLi	$ClCH_2CH$=$CHCH_2Cl$	18	91
MeO	H	H	NaH/BuLi	$MeOCH_2Cl$	65	92
MeO	H	H	NaH/BuLi	MeOCH(Cl)Me	89	92
MeO	H	H	NaH/BuLi	EtOCH(Cl)Me	73	92
MeO	H	Me	NaH/BuLi	$MeOCH_2Cl$	87	92
MeO	H	H	NaH/BuLi	(Z)-Me_2C=$CHCH_2Br$	79-85	92-95
MeO	H	H	NaH/BuLi	1-$(BrCH_2)C_6H_{11}$	70	96
MeO	H	H	NaH/BuLi	geranyl-Br	61	97
tBuO	H	H	NaH/BuLi	geranyl-Br	-	98
MeO	H	H	NaH/BuLi	geranyl-Cl	82	99
EtO	H	H	NaH/BuLi	allyl-Br	97	100
MeO	H	H	NaH/BuLi	CH_2=CHCH=$CHCH_2Br$	70	101
MeO	H	H	NaH/BuLi	CH_2=CHCH=$C(Me)CH_2Br$	50	101
MeO	H	H	NaH/BuLi	C_3H_5CH=CMe-CH_2Br	53	101
MeO	H	H	NaH/BuLi	farnesyl-Br	≥80	102
MeO	H	H	NaH/BuLi	(E)-Br(Me)C=$CHCH_2Br$	-	103
a	H	H	LDA	-	43	104
b	H	H	LDA	-	45	104
c	H	H	LDA	-	49	104
MeO	H	H	NaH/BuLi	$TsOCH_2$cyclopentene	-	105
tBuO	H	H	BuLi	4-iodomethyl-β-lactam	84	106
Me_2N	H	H	LDA	BnCl	92	107
tBuS	H	H	NaH/BuLi	MeI	78	108
tBuS	H	H	NaH/BuLi	EtI	69	108
tBuS	H	H	NaH/BuLi	CH_2=$CHCH_2Br$	70	108
tBuS	H	H	NaH/BuLi	$THPOCH_2CH_2I$	75	108

Table 2.1 (cont.)
Representative alkylations of β-dicarbonyl dianions (Eq. 2.19)

(note substituent order change)

R	R_1 Z	Base	R_2X	Yield(%)	Ref.
H	$-CH_2CH_2-$	LDA	MeI	41	109
H	$-CH_2CH_2-$	LDA	$CH_2=CHCH_2Br$	56	109
H	$-CH_2CH_2-$	LDA	BnBr	67	109
H	$-C(Me)_2CH_2-$	LDA	MeI	42	109
H	$-C(Me)_2CH_2-$	LDA	$CH_2=CHCH_2Br$	57	109
H	$-C(Me)_2CH_2-$	LDA	nPrBr	32	109
H	$-C(Me)_2CH_2-$	LDA	nPrI	62	109
H	$-C(Me)_2CH_2-$	LDA	iPrBr	75	109
H	$-C(Me)_2CH_2-$	LDA	iPrI	78	109
H	$-C(Me)_2CH_2-$	LDA	cyclohexyl-Br	16	109
H	$-C(Me)_2CH_2-$	LDA	cyclohexyl-I	54	109
H	$-C(Me)_2CH_2-$	LDA	BuBr	70	109
d	$-CH_2CH_2-$	LDA	MeI	80	110
d	$-CH_2CH_2-$	LDA	nPrI	69	110
d	$-CH_2CH_2-$	LDA	iPrI	67	110
Me	$-CH_2-$	BuLi	$CH_2=CHCH_2Br$	50	111
Me	$-CH_2-$	BuLi	$n-C_5H_{11}I$	71	111

a = $-O(CH_2)_9Br$, b = $-O(CH_2)_{10}Br$, c = $-O(CH_2)_{11}Br$ and are intramolecular alkylations
d = $-CH_2CH_2OEt$

As seen in Table 2.1, straightforward alkylation of acetoacetone and acetoacetate dianions and their substituted derivatives may be conducted with relative ease.[2,3,5] Alkyl bromides and iodides have been most widely used regardless of the base employed.

Two types of cyclic β-dicarbonyl compounds have been converted to the corresponding dianion and reacted with alkyl halides; those bearing one carbonyl on the ring and one exocyclic, and those bearing both carbonyl groups as part of the cyclic system. Both cyclic dianion types may be generated using the same base conditions noted for the acyclic congeners and reacted with alkyl halides (Eq. 2.20 and 2.21).

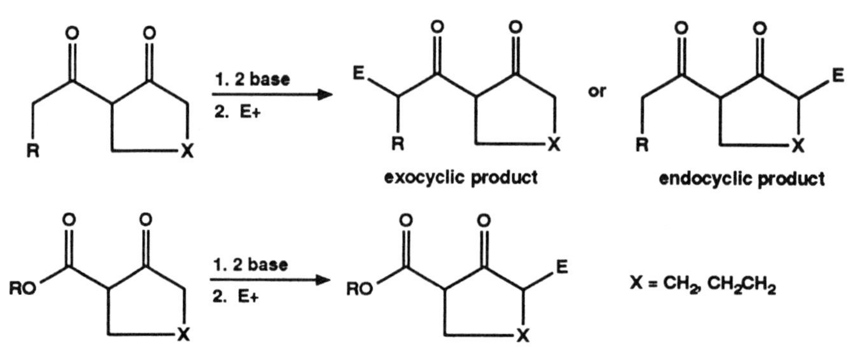

(2.20)

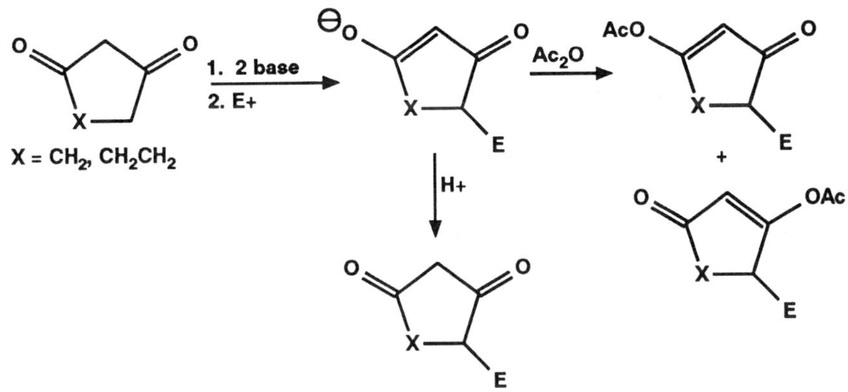

(2.21)

Over thirty years ago, Harris and Hauser[78] showed that simple 2-acetyl- or 2-propionyl cyclopentanones and cyclohexanones (R = H, Me; X = CH_2, CH_2CH_2) undergo regioselective exocyclic alkylation. When the exocyclic carbonyl group is an ester moiety, alkylation can occur only at the endocyclic position.[87,89,90] This cyclic β-dicarbonyl 1,3-dianion strategy was used in the synthesis of inside-outside stereoisomers[89] and quadrone.[90]

Harwood and coworkers[109-110] found that cyclohexanediones undergo clean alkylation following dianion formation with 2 LDA/HMPT (Eq. 2.21). The alkylated products could be isolated as a mixture of enol acetates after reaction with acetic anhydride. 4-Substituted cyclopentanediones can be accessed through similar methodology.[111] In this latter system, deprotonation may be conducted with 2 BuLi with no apparent addition to the carbonyl.

Heteroatom-containing cyclic β-dicarbonyl dianions also have been generated and reacted with alkylating agents. Wachter and Harris reacted triacetic lactone with alkali amides to form the resonance-delocalized dianion that alkylates at the 5- and 6α-positions (Eq. 2.22).[112]

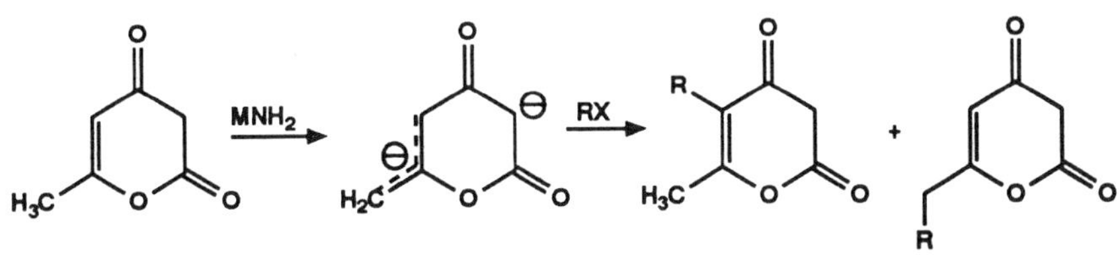

(2.22)

4-Carboalkoxy-pyrrolidin-3-ones react twice with LDA to give the corresponding dianion that reacts in 51-73% yield with a variety of alkyl halides and in better yield with aldehydes (Eq. 2.23).[113]

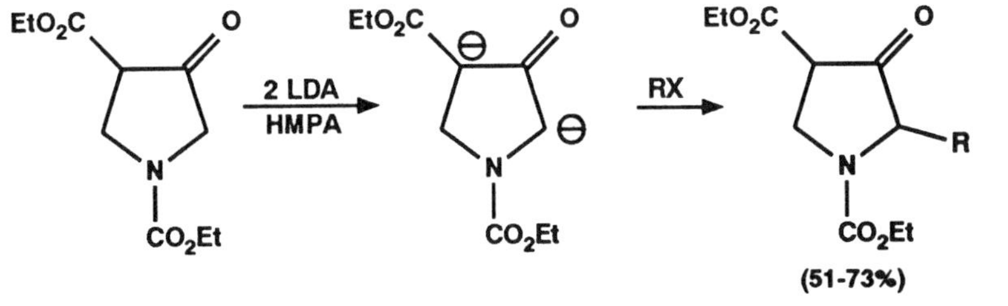

(2.23)

Intramolecular alkylation also has been a successful variation conducted with β-dicarbonyl dianions. Weiler and coworkers continued in their exploration for new uses of β-ketoesters dianions by using an intramolecular alkylation[91] in the synthesis of macrocyclic lactones.[114] Although the method was accompanied by elimination products, the 25- and 27-membered β-ketolactones could be prepared using this technique (Eq. 2.24).

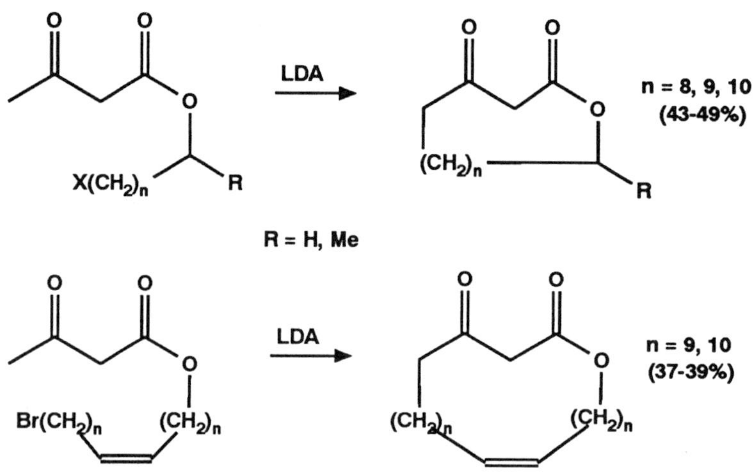

(2.24)

Alkylation of β-dicarbonyl dianions with epoxides

Epoxides serve as excellent alkylating agents for β-dicarbonyl 1,3-(C,C)-dianions. Although any 1,3-dicarbonyl dianion would be expected to undergo this alkylation reaction, most studies to date have been conducted with the dianions of β-ketoesters.[115-120] Clear advantage for the use of β-ketoesters in this reaction sequence include cyclization of the resultant oxyanion onto the ester moiety to form tetrahydrofuran derivatives. In all instances, attack on the epoxide occurs regioselectively at the least substituted carbon atom.

Schlessinger[115] and Mori[16] were first to explore the reaction between epoxides and acetoacetate dianions. Substituted cyclohexene oxides underwent regioselective ring opening with the dianion derived from the t-butyl ester of acetoacetate in 90% yield.[115] The success of this reaction was then applied to the synthesis of vernolepin. The cyclic version of an acetoacetate reacted cleanly with enantioenriched (R)-1-butene oxide to give an intermediate product that was used in the preparation of chalcogran (Eq. 2.25).[116] The preparation of the chalogran stereoisomers proceeds through the aforementioned cyclization.

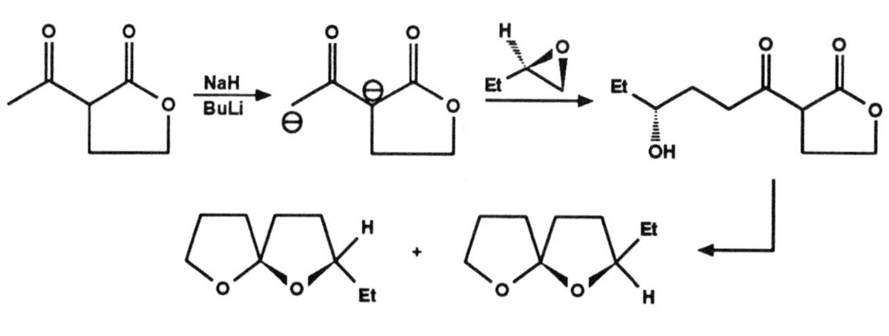

(2.25)

Yamaguchi and Hirao showed that markedly improved yields of alkylation products could be obtained with the methyl acetoacetate dianion when the epoxide was pretreated with BF_3-Et_2O.[117] The influence of the Lewis acid was most prominent when polysubstituted epoxides were used. As a notable extension to this work, 2-furylidenes could be prepared by reacting the crude 6-hydroxy-3-oxoalkanoate with TsOH in refluxing benzene (Scheme 2.12).

Lygo and coworkers reacted dianions of β-ketoesters with epoxides to synthesize tetrahydrofurans, methyl nonactate and homononactates (Scheme 2.12).[118-120]

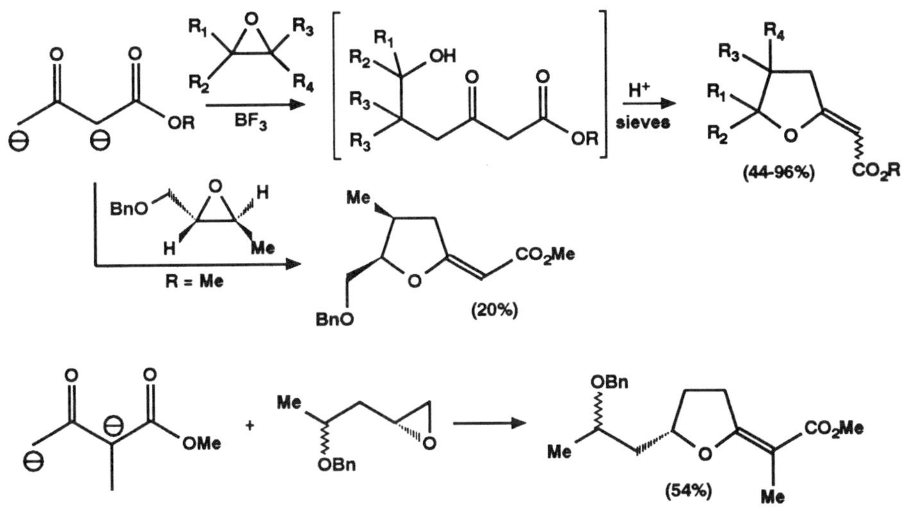

Scheme 2.12. Reaction of acetoacetate dianions with epoxides

2.4.2 Condensation reactions of β-dicarbonyl dianions

The aldol reaction with dianions of β-dicarbonyl dianions has been reviewed.[3,5] Although this reaction has been explored to a lesser degree than alkylation, the procedure forms a regiochemically precise C-C bond giving a product rich in functionality. The aldol products may spontaneously cyclize to the hemiketals or lactones depending upon the nature of the dianion and this property has been exploited in the preparation of novel oxygen-containing heterocycles. Alternatively, the aldol product may be dehydrated with acid to furnish α,β-dicarbonyls. Regardless of the pathway chosen, the procedure results in a five-carbon fragment with alternating carbons bearing oxygen functional groups. Such fertile fragments are highly sought intermediates in natural product synthesis.

Dianions of diketones and ketoesters have been condensed with aldehydes and ketones to give the aldol products in good yield (Eq. 2.26) (Table 2.2). Again, much of the early work in this area used metal amide bases to form the dianion. The preference to a NaH/BuLi combination to generate the dianion followed the report by Huckin and Weiler,[121] who showed that the mixed metal dianion enolate enters into high yielding aldol reactions.

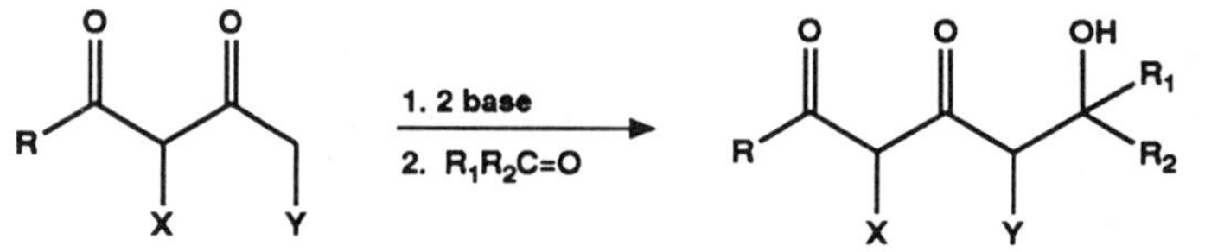

(2.26)

Table 2.2
Aldol reaction of β-dicarbonyl dianions (Eq. 2.26)

R	X	Y	R_1	R_2	Base	Yield(%)	Ref.
H	H	H	Ph	Ph	KNH_2	90	122
Me	H	H	Me	H	2 LDA	-	123
Me	H	H	Ph	Ph	KNH_2	73	124,125
Me	H	H	Ph	H	$LiNH_2$	40	124
Ph	H	H	PhCH=CH	Ph	KNH_2	77	126
$ArCH_2CH_2$	H	H	Ph	Ph	KNH_2	41	127
$PhC(O)CH_2$	H	H	Ph	Ph	$NaNH_2$	62	128
MeO	H	H	Ph	Ph	NaH/BuLi	93	121
MeO	H	H	Me	H	NaH/BuLi	26	121
MeO	H	H	Et	H	NaH/BuLi	73	121
MeO	H	H	nBu	H	NaH/BuLi	36	121
MeO	H	H	tBu	H	NaH/BuLi	82	121
MeO	H	H	Ph	H	NaH/BuLi	89-91	121,129
MeO	H	H	Ar	H	NaH/BuLi	68-73	121
MeO	H	H	2-furyl	H	NaH/BuLi	68	121
MeO	H	H	Me	Me	NaH/BuLi	70	121
MeO	H	H	$-(CH_2)_5-$		NaH/BuLi	63	121,129
MeO	H	H	$-(CH_2)_4-$		NaH/BuLi	25	121
MeO	H	H	CH_2=CH-	H	NaH/BuLi	63	129
MeO	H	H	2-thiophene	H	NaH/BuLi	93	130
MeO	H	H	$C_5H_{11}(CH=CH)_2-$	H	NaH/BuLi	100	131
EtO	H	H	C_5H_{11}	H	NaH/BuLi	53	132
EtO	H	H	PhSCH=CH-	H	NaH/BuLi	83	133
MeO	H	H	CH_2=C(Br)-	H	NaH/BuLi	75	134
Me	Me	H	ClCH=C(Me)-	H	NaH/BuLi	66	135
MeO	H	Bn	Me	H	NaH/BuLi	94	136
MeO	H	Bn	Pr	H	NaH/BuLi	96	136
MeO	H	Bn	C_7H_{15}	H	NaH/BuLi	82	136
MeO	H	Bn	Ph	H	NaH/BuLi	78	136
MeO	H	allyl	C_6H_{11}	H	NaH/BuLi	84	136
MeO	H	Bu	Ph	H	NaH/BuLi	83	136
tBuS	H	H	CH_2=CH-	H	NaH/BuLi	83	137
tBuS	H	H	Et	H	NaH/BuLi	84	137
tBuS	H	H	CH_2=C(Me)-	H	NaH/BuLi	88	137
NMe_2	H	H	Ph	Ph	NaH/BuLi	79	107

Table 2.2 shows that dianions of acetoacetates have been used most in condensation reactions. As testimony of the reliability of these dianions, many have been used early in a lengthy synthetic sequence toward natural products, which would necessarily require both large scale and repetitive reactions. Some examples of the use of methyl acetoacetate dianion in total synthesis include the preparation of 4-methylprodigiosin,[130] prostaglandin $F_{2\alpha}$,[131] phototoxin analogs,[132] costatolide,[135] dihydroxy vitamin D_3,[133] and gascardic acid[138]

(see product structures in Figure 2.8). The ethyl acetoacetate dianion was condensed with a 4-oxo-tetrahydroindole to afford the corresponding aldol product.[139] Dehydration led to the 4-substituted indole product.

Clarke and Ley found that the substitution of a thioester led to a dianion reagent that, following condensation, could be transesterified.[137] The thioacetoacetate dianion reagent then became pivotal in synthetic routes to milbemycin.

As mentioned, the aldol products from these dianions can be induced to undergo elimination or used to modify or create a heterocycle system. Surprisingly, the hydroxy ketoesters did not cleanly form the alkene upon treatment with $POCl_3$/pyridine or pTsOH in refluxing benzene. However, anhydrous HCl in $CHCl_3$ worked well.[121]

The conversion of the aldol products into pyranones has been of particular interest. Recent studies by Peterson[129] and Sakai[136] are representative of the efficacy of this transformation (Eqs. 2.27 and 2.28). When acidic conditions are applied to the aldol product from the 2,4-pentanedione dianion, 2-pyranones result whereas basic conditions lead to formation of the 4-pyranone from the acetoacetate aldol product (Eq. 2.27).[129] Sakai and coworkers found that acid-promoted cyclization (pH = 6) of the aldol products gave predominantly *the trans*-geometry and good yields of 4-oxo-2-pyrones (Eq. 2.28).[136]

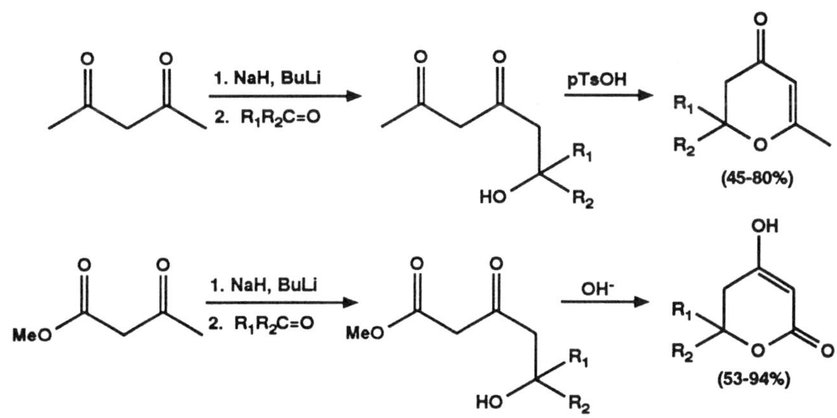

(2.27)

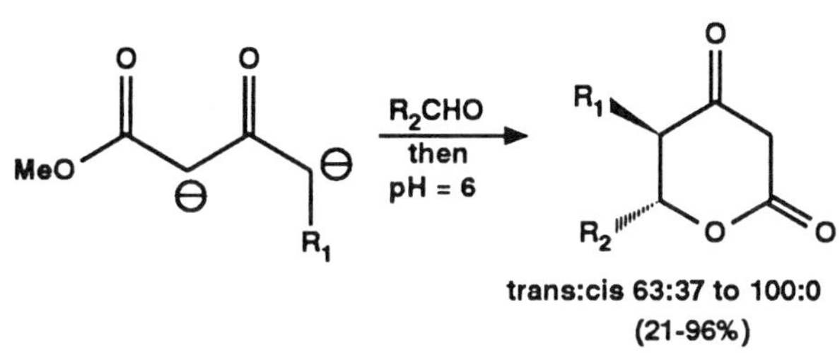

(2.28)

An interesting cyclic variant to the β-dicarbonyl dianion aldol condensation was investigated by Honda and coworkers.[140] Tetronic acid, which in keto form exists as a cyclic β-ketoester, was deprotonated with 2 LDA and reacted with various ketones to give predominant formation of the *erythro*-adduct presumably via a six-membered chelate-controlled transition state (Scheme 2.13). Exchange of the lithium for zinc and magnesium enhanced the stereoselectivity. Like dianions leading to the aromatic cyclopentadiene monoanion structure, tetronic acid dianions result in the aromatic furan resonance structure through the enolate contributor.

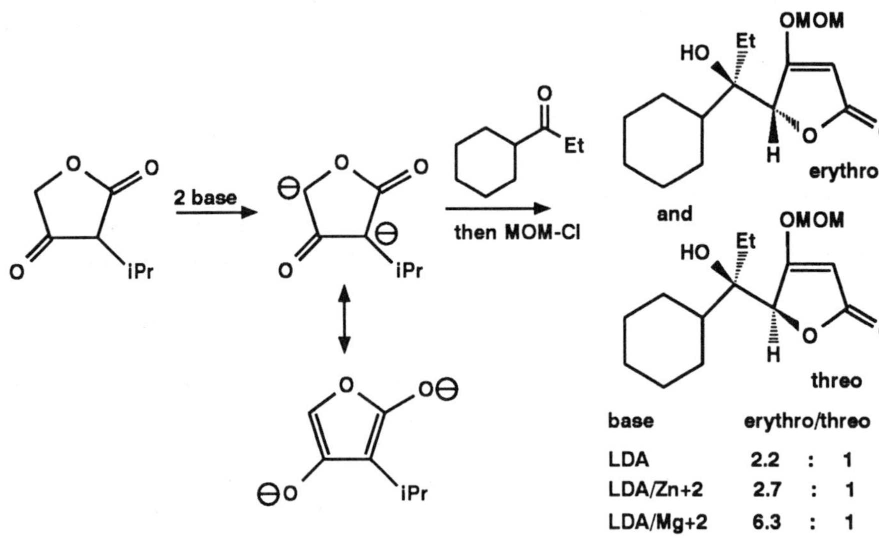

Scheme 2.13. Reaction of the tetronic acid dianion with ketones

2.4.3 Acylation of β-dicarbonyl dianions

Dianions of various β-dicarbonyls react with acylating agents to afford tricarbonyl compounds, which subsequently enolize and/or cyclize. During the reaction, the monoanion of the tricarbonyl product has a more acidic proton than the monoanion of the β-ketoester and proton transfer should occur. As a result, the maximum conversion for these reactions should be 50%. Huckin and Weiler,[141] who were among the first to study the acylation reactions of β-dicarbonyl dianions, examined this limitation in some detail and found that alternate addition of base improved the yield. For example, addition of one-half of an equivalent of the ester to the dianion, followed by regeneration of the dianion increased the yield to over 50%. Most acylation reactions now use a large excess of dianion to drive the reaction to completion.

Yamaguchi and coworkers have made extensive use of the acylation of β-dicarbonyl dianions in route to the synthesis of polyketides,[142] phenols,[143] and aryl C-glycosides (Scheme 2.14).[144] Although reaction with glutarate esters were successful, dimethylamides, previously treated with boron trifluoride, gave superior yields of acylated products.

Several modifications to the reaction of β-dicarbonyl dianion with acylating agents have been performed. For example, the dianion of ethyl acetoacetate (from LDA) reacted with ethyl benzoate to give the acylated product in 35% yield. However, the addition of 0.02 M TMEDA increased the yield to 85%.[145] Hanamoto and Hiyama found that N-methoxy-N-methyl amides react with the dianions of acetoacetates in good yields offering an alternative to ester and dialkylamide as acylating agents.[146]

The dianions of simple β-diketones have been acylated and used in the synthesis of stegobinone[147] and xanthones.[148]

2.4.4 Addition reactions of β-dicarbonyl dianions

Dianions of β-dicarbonyl compounds have been shown to react with carbon dioxide, nitriles, imines and nitroolefins among others. Most of these additions were covered in prior reviews,[3,5] and representative reactions are shown in Scheme 2.15.

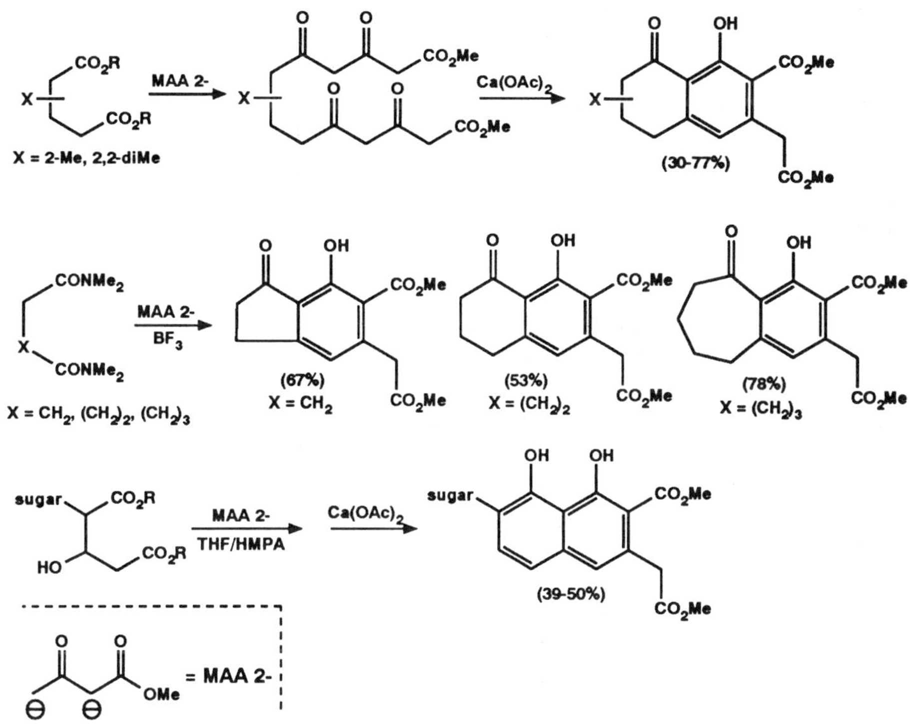

Scheme 2.14. Polyketide synthesis via the acylation of β-dicarbonyl dianions

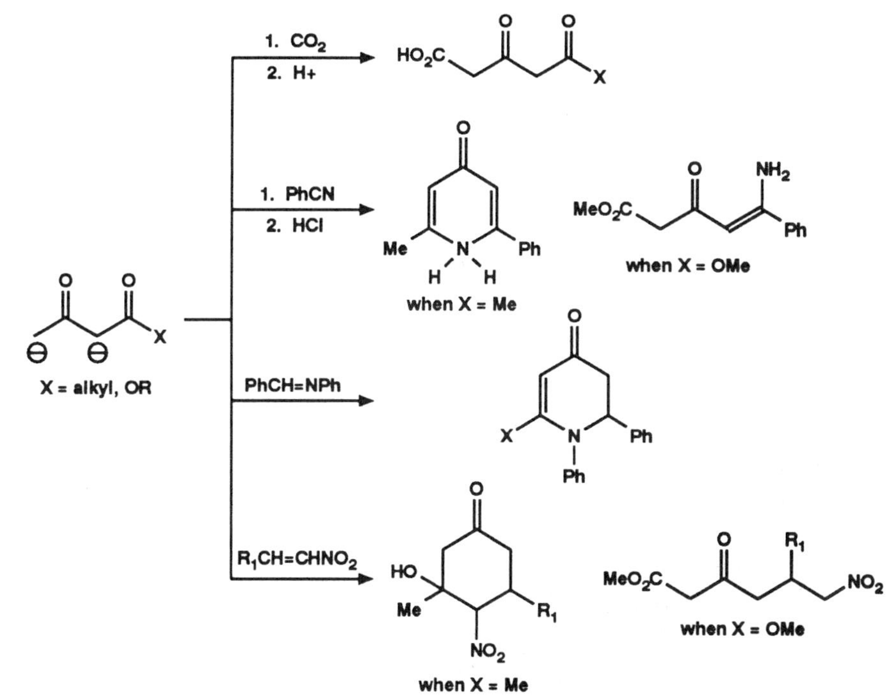

Scheme 2.15. Miscellaneous addition reactions of β-dicarbonyl dianions

Nitroolefins were shown to be excellent Michael acceptors for a host of dianions including those derived from 1,3-(N,C)-carboxamides, diketones, ketoaldehydes, and ketoesters.[149] Depending upon the workup conditions, 6-nitro-1,3-diketones or 3-hydroxy-4-nitrocyclohexanones can be isolated (Scheme 2.15).

Dianions of β-dicarbonyl compounds also undergo Michael reaction with α,β-unsaturated sulfones. Pyne and coworkers installed this clever usage of a β-dicarbonyl dianion in route to a synthesis of cytochalasin C.[150] Optically active β,γ-diketo-p-tolylsulfoxides have recently been prepared from β-diketone dianions.[151]

2.4.5 Miscellaneous 1,3-(C,C)-dianions derived from ketones other than β-dicarbonyl

As shown in the prior sections, ketones normally receive assistance in the deprotonations from a second carbonyl group or, as will be seen in Sections 2.5 and 2.6, from sulfur- or phosphorus-containing carbanion stabilizing groups. In this section, we will briefly cover some related 1,3-(C,C)-dianions that bear a β-stabilizing group other than a carbonyl or no second carbanion stabilization group. In addition, by virtue of the central carbonyl, certain ketones can be bis-deprotonated at both enolate positions to give a 1,3-(C,C)-dianion without the aid of a second β-positioned group. Such dianions have also been termed α,α′- or α,γ-type in the literature.

The most simple system, acetone, may be bis-deprotonated and reacted with bifunctional esters to form β-polycarbonyl products through a one-pot, double Claisen condensation (Eq. 2.29).[152,153] The use of the dianion rather than enolate monoanion obviated the formation of subsequent cyclization side products, which would take place after the first addition.

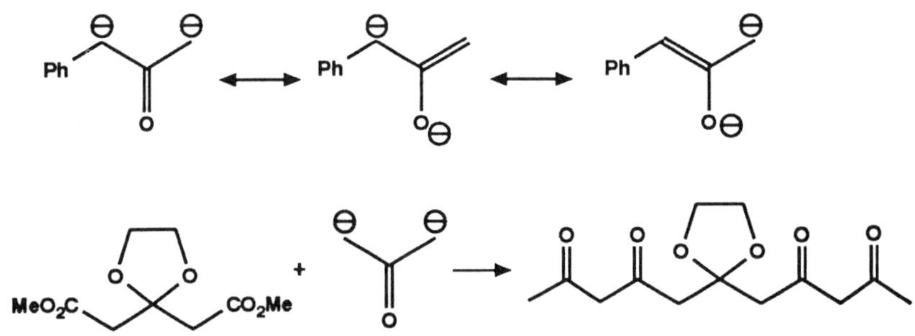

(2.29)

Hubbard and Harris[154] were among the first to explore the general attributes of acyclic and cyclic aliphatic ketones dianions. They found that these dianions reacted with alkyl halides, aldehydes, ketones, and esters to form the α-functionalized product in excellent yields (47-92%), which in many cases was a marked improvement when compared against the enolate monoanion.

A second electrophile may be added directly to the reaction to give difunctionalized products although the ensuing regiochemistry depends upon the nature of the intermediate product. Specifically, following reaction with the first electrophile, the remaining enolate may equilibrate to the thermodynamically favored form, which would lead to 1,1-functionalization. Equilibration to the thermodynamic form is markedly accelerated in the case when the first reaction is an acylation, which leads to a β-dicarbonyl and proton transfer is rapid.

Ketone dianions that bear a phenyl group as a carbanion stabilizer have been explored by Trimitsis and Bays and coworkers.[155-157] Interestingly, the phenylacetone dianion was found to react predominantly at the benzylic position with most alkyl halides; methyl iodide gave a 1:1 mixture. Since the benzylic position would have been expected to be deprotonated first, the terminal carbanion was anticipated as the most reactive. However, the phenylacetone dianion, generated from 2 KNH_2, reacts via a Michael reaction.[158]

The nitrile group has served as an interesting β-substituent to influence the direction and 1,3-(C,C)-dianion formation.[159,160] Two examples will serve to illustrate their utility. α-Cyanoacetone forms the 1,3-dianion and reacts with lactim ethers to form the intermediate β-ketoimines that, upon treatment with methanol, undergo immediate cyclization to form bicyclic pyridinones (Eq. 2.30).[159] A highly interesting route to α-cyanoketone dianions was devised by Vinick and Gschwend.[160] Methyl isoxazoles ring open upon reaction with two equivalents of LDA to form the requisite 1,3-dianion. Alkylation reactions proceeded in good yield and regioselectivity (Eq. 2.31).

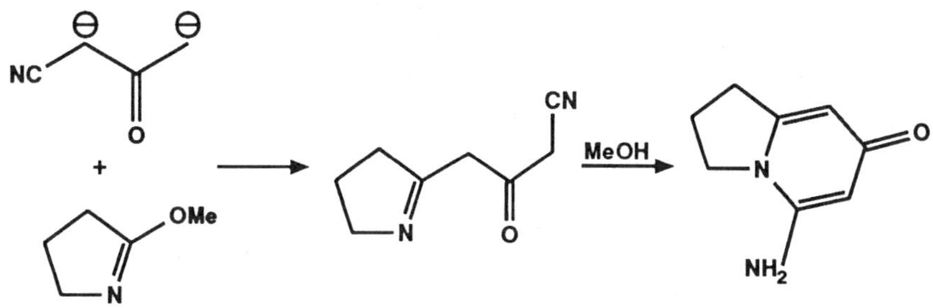

(2.30)

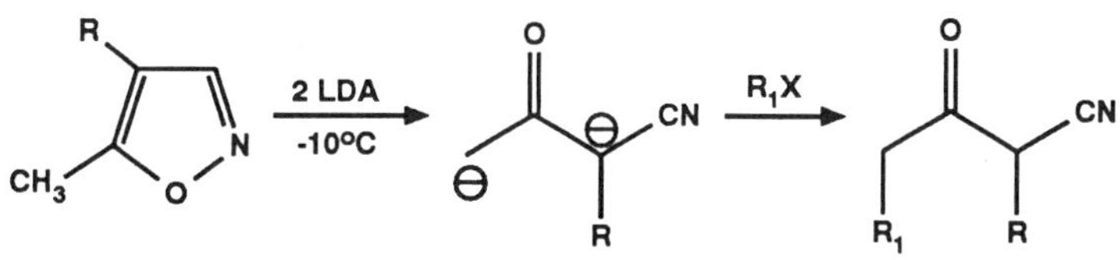

(2.31)

β-Ketoimines can form the 1,3-(C,C)-dianion in a system appearing as an aza-analog to the β-diketone dianion. Boatman and Hauser showed that the 2-N-phenyl imine of 2,4-pentanedione could be bis-deprotonated with metal amides, which reacted with alkyl halides and carbonyls at the 5-position.[161]

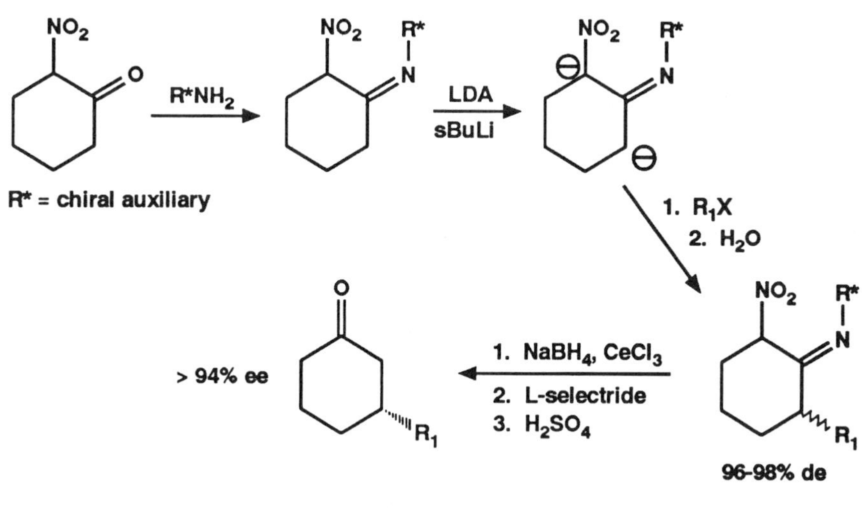

Scheme 2.16. Use of a chiral auxiliary in the alkylation of 1,3-(C,C)-dianions

Denmark and coworkers made elegant use of the imine-based dianion system by appending a chiral auxiliary to the N-substituent derived from certain α-nitroketones (Scheme 2.16).[162] The dianion reacted with electrophiles to give high diastereoselectivities. Subsequent manipulation gave enantioenriched 3-substituted cyclohexanones, a substituent pattern difficult to access by other methods.

Recently, Bartoli and coworkers showed that α-nitroketones form the expected 1,3-dianion and react at the 3-position with electrophiles.[163]

2.5 SULFINYL AND SULFONYL-STABILIZED 1,3-(C,C)-DIANIONS

The sulfinyl (sulfoxide) and sulfonyl (sulfone) groups furnish excellent stabilization of neighboring carbanionic centers, and therefore extension to form 1,3-(C,C)-dianionic species is possible through dual stabilization provided by the sulfur moiety or, in the case of β-ketosulf(o)inyl compounds, assisted deprotonation of the α-CH. The generation and reactions of sulfoxide and sulfone-stabilized 1,3-dianions has been covered.[5,6,36] Prior to a discussion of the specific synthetic uses of these dianions, it is important to first identify and distinguish among the various dianion types that may form (Figure 2.9). When C-H deprotonation sites are available at both the 1- and 3-positions (α,α'), the 1,3-dimetalated species predominates. When R_2 (Figure 2.9) is aryl or does not bear an acidic α-CH, 1,1-bis-deprotonation may occur to form the geminal dianion, which is discussed in Section 2.2.

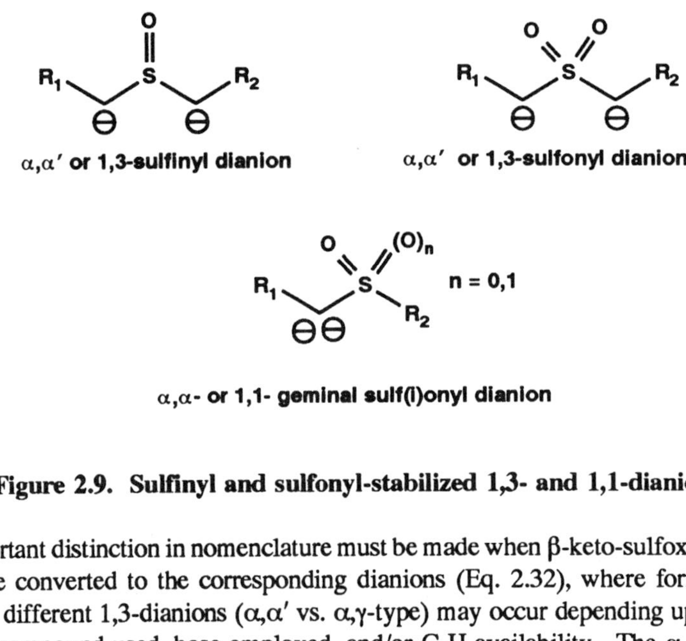

Figure 2.9. Sulfinyl and sulfonyl-stabilized 1,3- and 1,1-dianions

An important distinction in nomenclature must be made when β-keto-sulfoxides or β-keto-sulfones are converted to the corresponding dianions (Eq. 2.32), where formation of two structurally different 1,3-dianions (α,α' vs. α,γ-type) may occur depending upon the sulfur-containing compound used, base employed, and/or C-H availability. The α,α'-designation for the $S(O)_n$- juxtaposed sulf(o)inyl dianion was chosen to stay consistent with

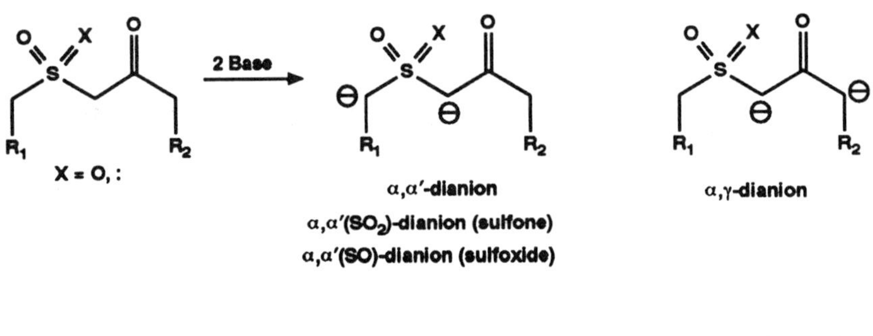

(2.32)

the prior assignments for simple sulfones and sulfoxide dianions, and also because early literature reports carry this descriptor. Unfortunately, this nomenclature overlaps with the assignment for simple ketone 1,3-dianions (Section 2.4), which also bear the α,α′ designation. To better clarify and focus these differences, both sulfone and sulfoxide α,α′-dianions will heretofore be assigned α,α′(SO_2)-dianions and α,α′(SO)-dianions, respectively, to indicate their position of the carbanion, namely, surrounding the sulfur functional group, and to make their nomenclature distinct from the α,γ-dianion type.

In all cases, β-ketosulf(o)inyl compounds undergo initial deprotonation at the "shared" methylene group with the second deprotonation site (α′ or γ) regulated by the structural type. When α′- and γ-sites have available C-H's, the second deprotonation may result in competition and/or equilibration leading to a mixture of dianions. Most investigations to date, however, circumvent this problem through the use of non-enolizable constituents at either the α′- or γ-positions. Because addition to the carbonyl group also is possible, the use of sodium hydride or an amide base is suggested for the first deprotonation. This section will cover the major uses of sulfone- and sulfoxide-stabilized 1,3-dianions. For a comprehensive account of sulfur-stabilized dianions, see the excellent review by Tanaka and Kaji.[6]

Simple acyclic and cyclic sulfones have been bis-deprotonated to form the α,α′(SO_2)-dianion and reacted with an array of electrophiles. (Eq. 2.33) (Table 2.3).[164-172] In most instances, the electrophile reacts twice with these 1,3-dianions. More complex bicyclic[173] and tricyclic[174] sulfone-stabilized α,α′-dianions have also been bis-alkylated.

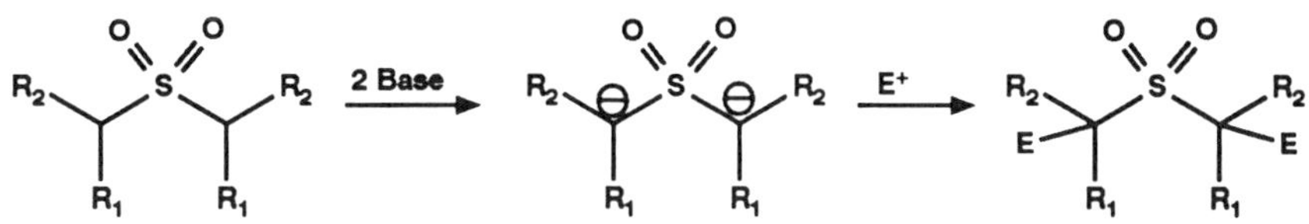

(2.33)

Table 2.3
Bis-reaction of α,α′-Sulfonyl Dianions with Electrophiles (Eq. 2.33)

R_1	R_2	Base	E	Yield(%)	Ref.
H	H	2 EtMgBr	$Me_2C(OH)$	15	164
H	H	2 $LiNH_2$	$Ph_2C(OH)$	80	165,166
H	H	2 $LiNH_2$	Bu	55	165,166
H	H	2 NaH	PhC(O)	87	167
H	$-CH_2CH_2-$	2 $NaNH_2$	$Ph_2C(OH)$	48	166
H	$-CH_2CH_2-$	2 BuLi	$Ph_2C(OH)$	80	168
Ph	H	2 $NaNH_2$	Bn	77	169,170
Ph	Me	2 BuLi	D	-	171

One highly interesting usage of α,α′-(SO_2)-1,3-dianions has been conversion into alkenes following treatment with copper (II) salts (Eqns. 2.34-2.37).[175] For example, sulfones bearing a β-carbonyl group may be bis-deprotonated to form either α,α′-(SO_2)-dianions or α,γ-dianions.

2.5.1 β-Ketosulfone 1,3-dianions

β-Ketosulfones can be bis-deprotonated in a manner analogous to β-dicarbonyl compounds to yield intermediates that were regiospecifically functionalized at the terminus. Early work exploring the 1,3-dianions of β-ketosulfones were aimed to avoid regiochemical ambiguities through use of the arylsulfone[167,176,177] or benzoyl[167,178,179,180] constituents (Eqn. 2.38-2.40).

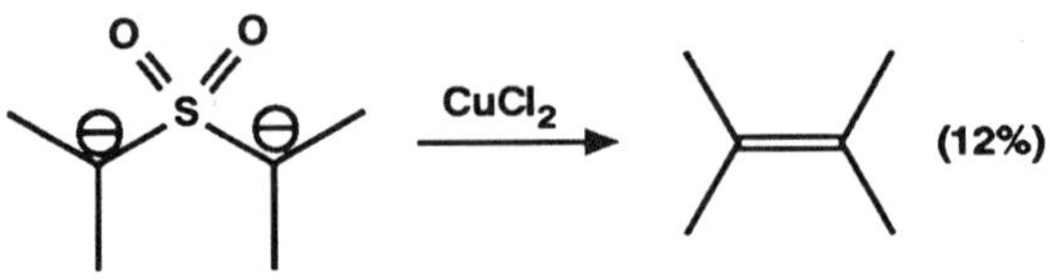

(2.34)

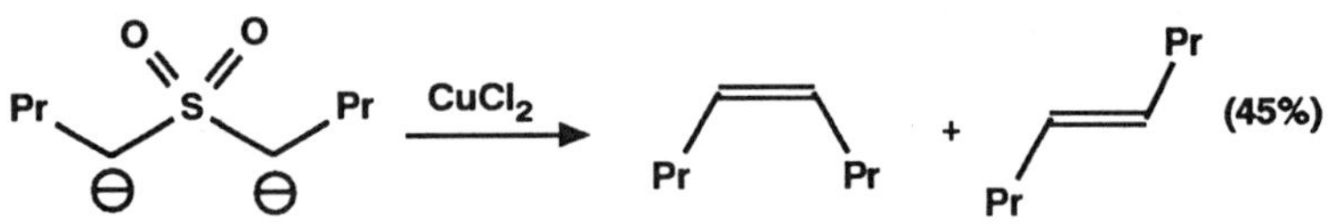

(2.35)

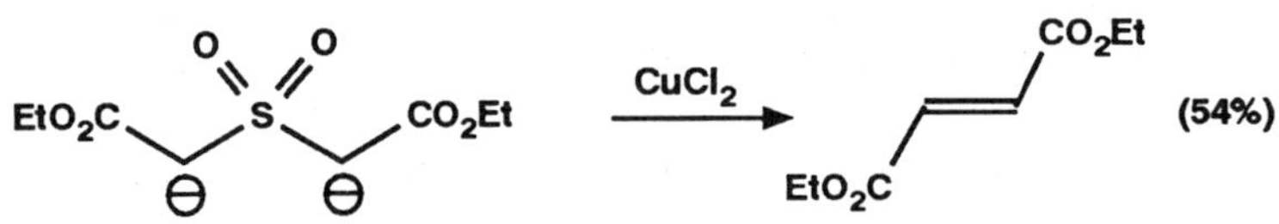

(2.36)

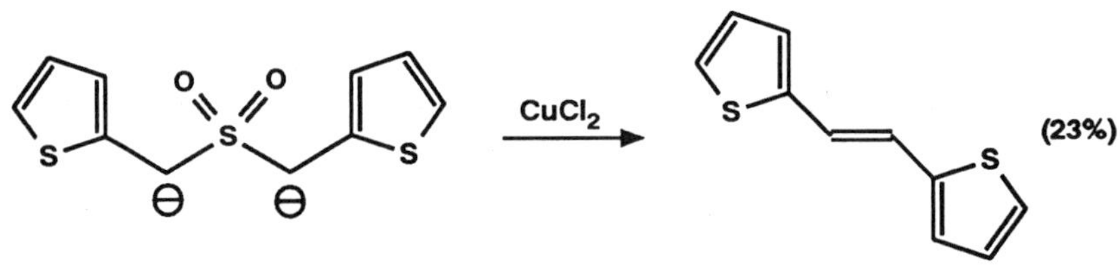

(2.37)

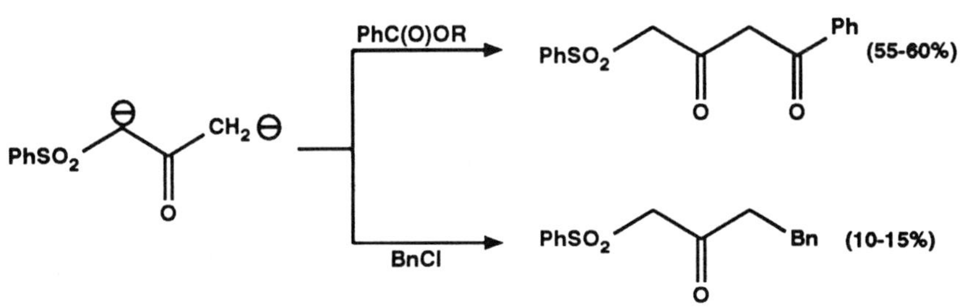

(2.38)

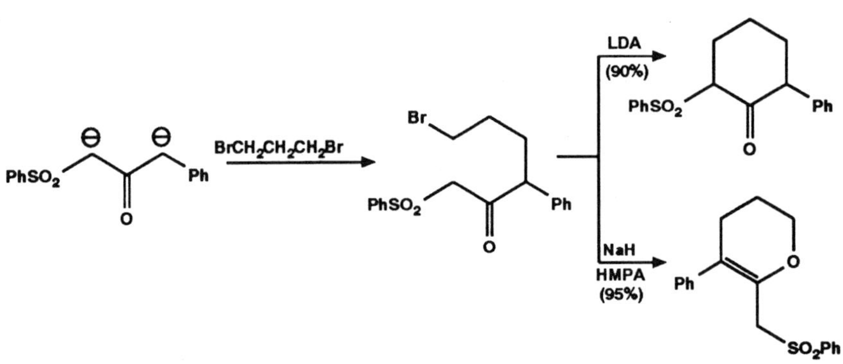

(2.39)

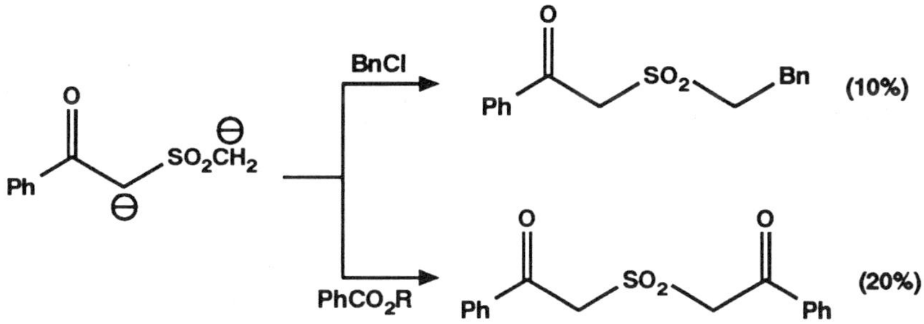

(2.40)

Cooke and Magnus[177] converted 1-phenyl-3-phenylsulfonyl-2-propanone to its crimson red α,γ-dianion using 2 LDA or sequential treatment with NaH and BuLi, and reacted it with 1,3-dibromopropane to give the dialkylated product in 60-75% yield. The product was cyclized to give predominant C-alkylation when LDA was the base, and O-alkylation when a NaH/HMPA was used in the subsequent ring closure. A similar strategy was used to form cyclic sulfones.[179] Lygo and O'Connor found that α,γ-dianions of certain β-ketosulfones react with epoxides to give yields of chain-extended products, although the reaction sequence was accompanied by tetrahydrofuran cyclization byproducts (Scheme 2.17).[181] One of the many useful advantages of sulfones was further exploited in this procedure. Subsequent protection of the resultant alcohol and alkylation at the α-position, followed by base-induced β-elimination of the sulfone moiety, furnished the unsaturated ketoester.

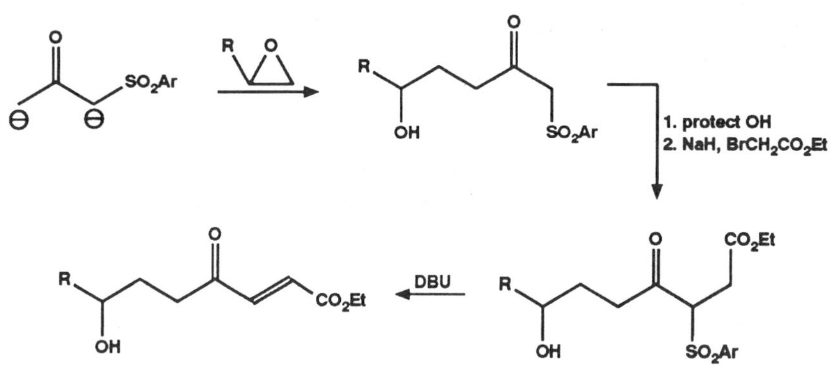

Scheme 2.17. Reaction of sulfonyl-stabilized 1,3-dianions with oxiranes

Sulfone-stabilized 1,3-dianions have been used to construct biabolenes (Eq. 2.41),[182] and have been oxidatively coupled (Eq. 2.42).[183]

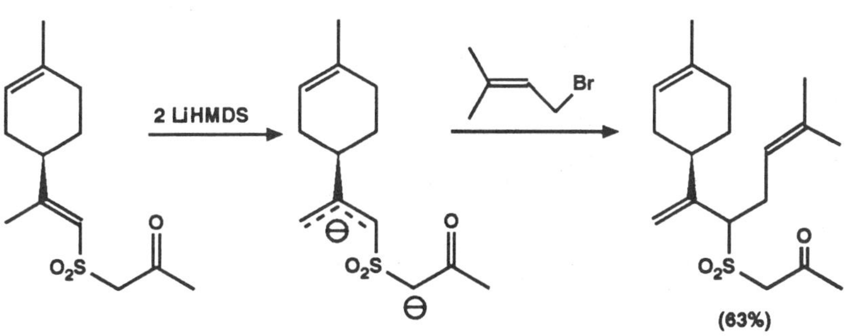

(2.41)

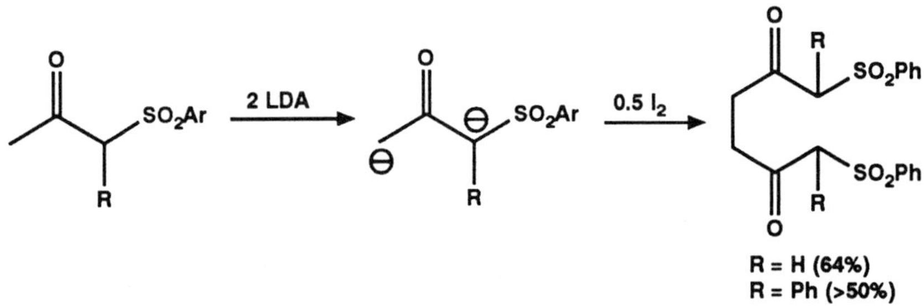

(2.42)

2.5.2 1,3-Sulfoxide dianions

Simple alkyl sulfoxides may be converted to their α,α′(SO)-dianions and condensed with electrophiles,[184] although bis-alkylation again predominates similar to the alkylsulfones.

2.5.3 β-Ketosulfoxide 1,3-dianions

Dianions of β-ketosulfoxides, like their sulfone counterparts, form either the α,α′(SO)- or α,γ-dianion types following reaction with 2 equivalents of base. Much of the work to date has studied reactions of the α,γ-type, which undergoes alkylation and condensation reactions to give moderate to good product yields (Eq. 2.43) (Table 2.4).[185-190]

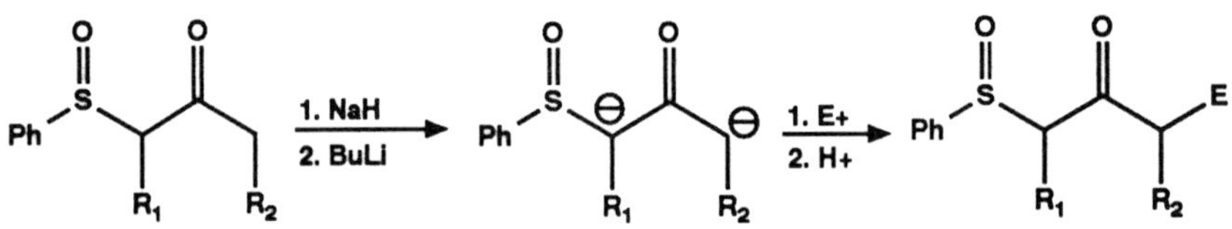

(2.43)

Table 2.4
Bis-reaction of α,α′-Sulfonyl Dianions with Electrophiles

R_1	R_2	E	Yield(%)	Ref.
H	H	Bu	78	185
H	H	iPr	73	185
H	H	Bn	63	185
H	H	CH_2=$CHCH_2CH_2$-	65	185
H	H	Me_2C=$CHCH_2CH_2$-	61	185
H	Bu	Bn	72	185
Me	H	Me	65	186
Me	H	Bu	84	186
Me	H	Bn	72	186
H	H	$Ph_2C(OH)$	91	187
H	H	$nC_5H_{11}CHOH$	80	187
H	H	Bn	73	187
H	H	CH_2=$CHCH_2$	64	187
H	H	nC_6H_{13}	79	187
H	H	$PhCH_2CH_2CHOH$	77	187
Me	H	R	63-70	189,190

R = 3-[2-(1,3-dioxolane)]propyl

Experimenters should be cognizant that the sulfoxide sulfur atom is asymmetric in these examples, and diasteromeric mixtures of products result. Fortunately, a major pathway of the β-ketosulfoxide dianion alkylation uses thermal elimination of the sulfoxide moiety to give substituted α,β-unsaturated ketones thereby removing much of this concern (Eq. 2.44)[186,188-190] Alternatively, Na-Hg or Al-Hg amalgam reduction furnishes the corresponding hydrocarbon.[191]

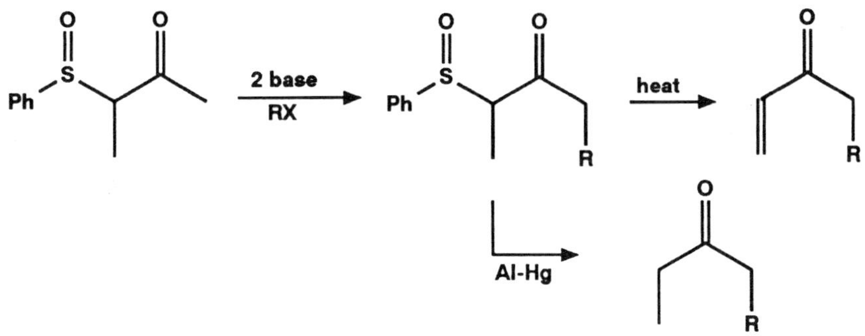

(2.44)

β-Ketosulfoxide α,γ-dianions undergo 1,2-additions with α,β-unsaturated ketones, but react by a Michael addition with α,β-unsaturated esters.[187]

Cyclic variations of the β-ketosulfoxide dianions have been explored by Grieco and Pognowski.[192] α-(Phenylsulfinyl)cyclopentanone and its cyclohexanone analog were reacted with NaH/BuLi to give low conversion to the dianion. With this base combination, significant transmetalation occurs to give the n-butylsulfoxide (Scheme 2.18), which forms a dianion that alkylates at the α′-site following treatment with alkyl halides. Use of 2.2 LDA gave high conversion to the desired α,γ-dianion. Pyrolysis of the product β-ketosulfoxides (75°C, CCl_4) gave the cyclic enones in 43-60% yield.

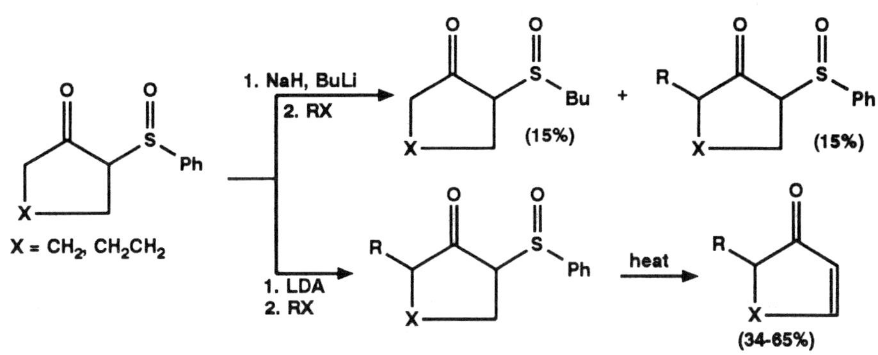

Scheme 2.18. Formation and reactions of cyclic β-ketosulfoxide dianions

Tamura and coworkers were able to cleanly access the α,α′-(S=O)-dianion of β-ketosulfoxides, which undergo alkylation, acylation, and condensation reactions in good yield (Scheme 2.19).[193]

The α,α′-(SO)-dianion reacts with β-monosubstituted α,β-unsaturated esters in a sequential Michael addition-Claisen condensation to give thian-3-one-1-oxides.[194] Alkylation at the alpha position may be conducted through the monoanion generated with NaH.

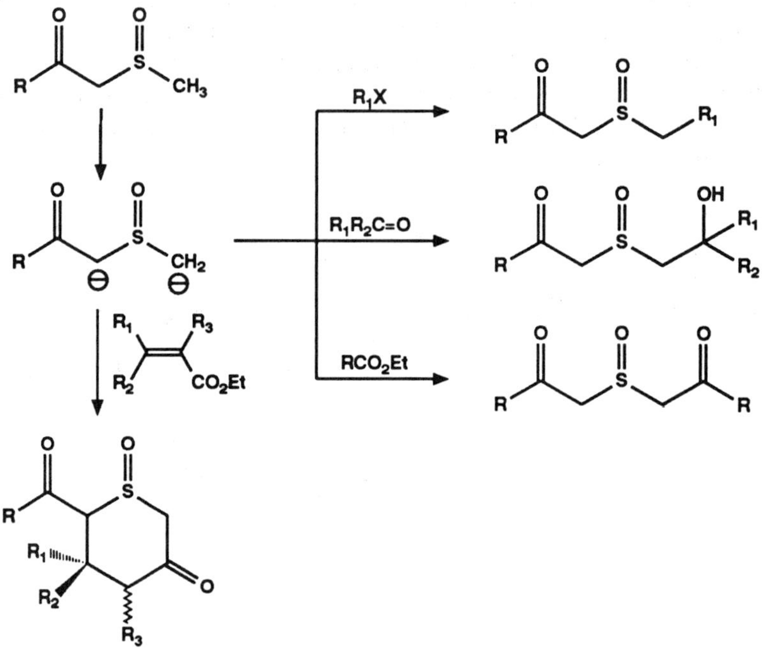

Scheme 2.19. Reactions of acyclic β-ketosulfoxide α,α′(SO)-dianions

Recently, an asymmetric sulfoxide moiety was exploited through initial use as a dianion (Eq. 2.45), followed by a tin-mediated cyclization.[178]

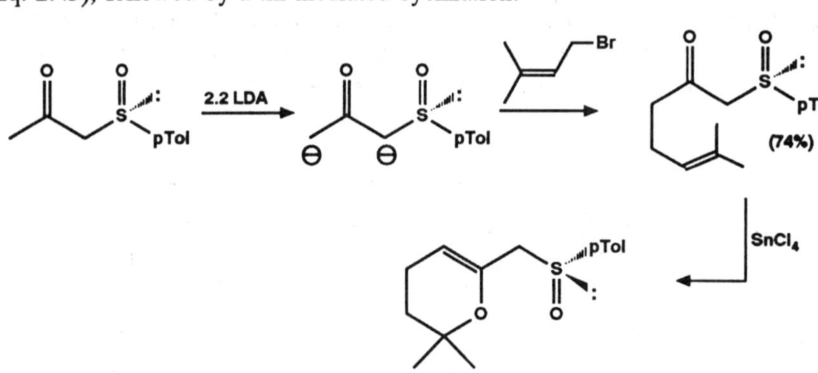

(2.45)

Last, there also has been a report of a resonance-delocalized α,α′-(S)-dianion bearing a thioether linkage, which alkylates at the α′-site (Eq. 2.46).[196]

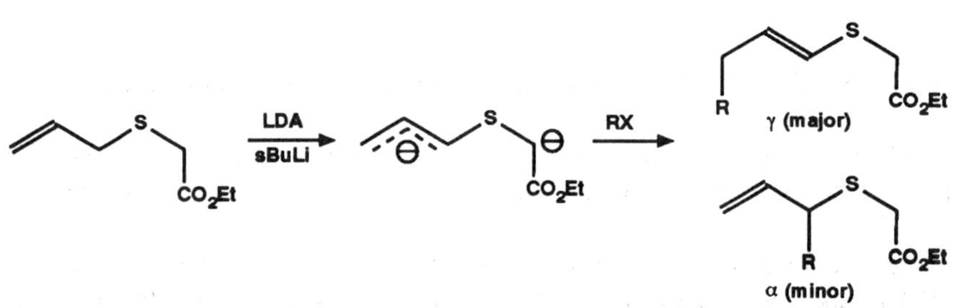

(2.46)

2.6 PHOSPHORUS-STABILIZED 1,3-(C,C)-DIANIONS

Carbonyl compounds, particularly ketones, bearing a β-phosphorus moiety may be converted to the corresponding 1,3-dianion (α,α′) yielding a β-keto ylid dianion. Two types of phosphorus-containing groups have been used to access these dianions: phosphines and phosphonates. Because deprotonation occurs at the α-phosphono-CH first, eventual reaction at the α′- or 3-position occurs with electrophiles retaining the ylid or phosphonate portion of the molecule used in future transformations.

β-Keto-phosphonium salts, prepared from triphenylphosphine and an α-haloketone, are readily converted into the ylid-based 1,3-dianion by stepwise treatment with KOH (followed by isolation) and n-BuLi (Eq. 2.47).[197] The ylid-based dianions (R = H) undergo reaction

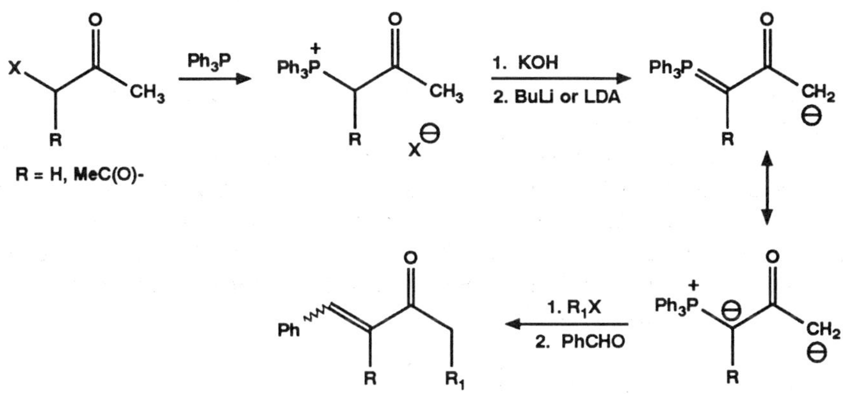

(2.47)

with BnBr (44%), allyl-Br (52%), and benzophenone (65%) at the α′-site, Subsequent reaction with benzaldehyde gives the anticipated α,β-unsaturated product (40% overall), the phosphonium group serving as a temporary regiocontrol element.[197] The corresponding red dianion of the diacetyl phosphonium (R = C(O)CH_3; Eq. 2.47) was prepared with LDA, and also enters into reaction with electrophiles.[198]

The corresponding β-ketophosphonates have attracted far more attention as dianion reagents presumably owing to greater ease of isolation of the intermediate products as compared to the phosphonium salts. Grieco and coworkers showed that β-ketophosphonates may be bis-deprotonated to form the α,α′-dianion (Eq. 2.48).[199,200] For example, dimethyl-

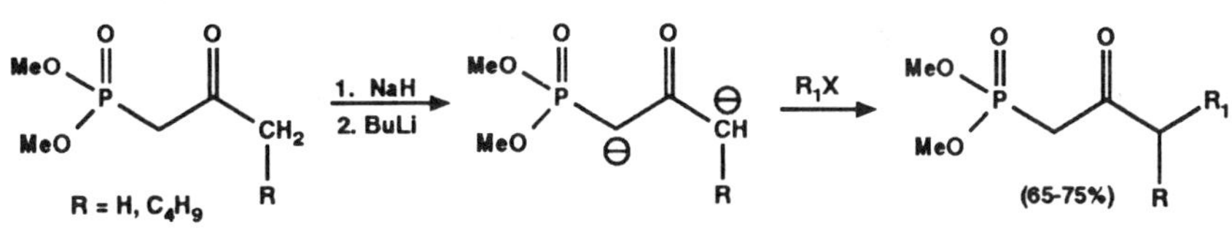

(2.48)

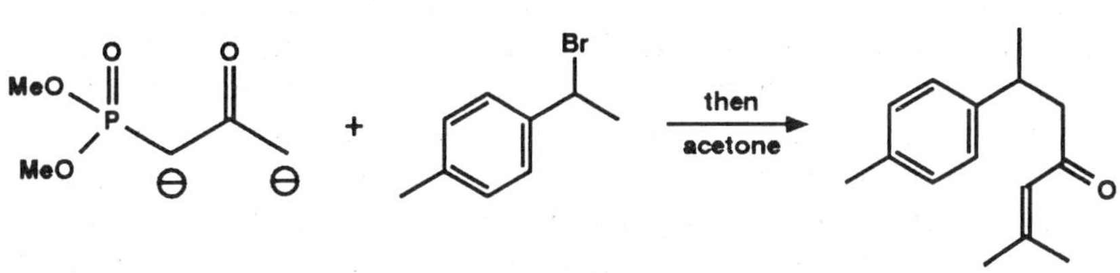

(2.49)

2-oxopropyl- or 2-oxoheptylphosphonate is converted to the dianion with NaH and BuLi, which reacts in good yield with alkylating agents (Eq. 2.48). No O-alkylation was observed. This reagent was used in the synthesis of the sequisterpene (+/-)-ar-tumerone (Eq. 2.49).[200]

The 2,4-dioxoalkylphosphonates may prepared in good yield by Claisen condensation of the diethyl-2-oxoalkylphosphonate dianion with esters or acyl chlorides (Eq. 2.50).[210]

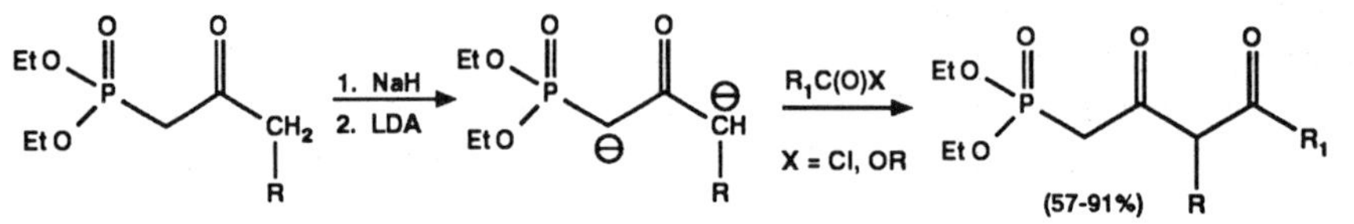

(2.50)

The product [substituted] dioxophosphonates also have been used as precursors to phosphorus-stabilized dianions (*vide infra*).

An interesting use of these dianions was explored by Wada, Kanemasa, and Tsuge.[202] Alkylation of the dianion of diethyl-2-oxopropylphosphonate with phenylthiomethyl iodide gave the alkylated product in 58% yield. Oxidation and elimination affords the 2-oxo-3-alkenylphosphonate, which undergoes cycloaddition as a heterodiene with certain enol ethers to give 2-alkoxydihydropyrans in satisfactory yield. Subsequent acid treatment leads to ring-opening and recyclization (Scheme 2.20).

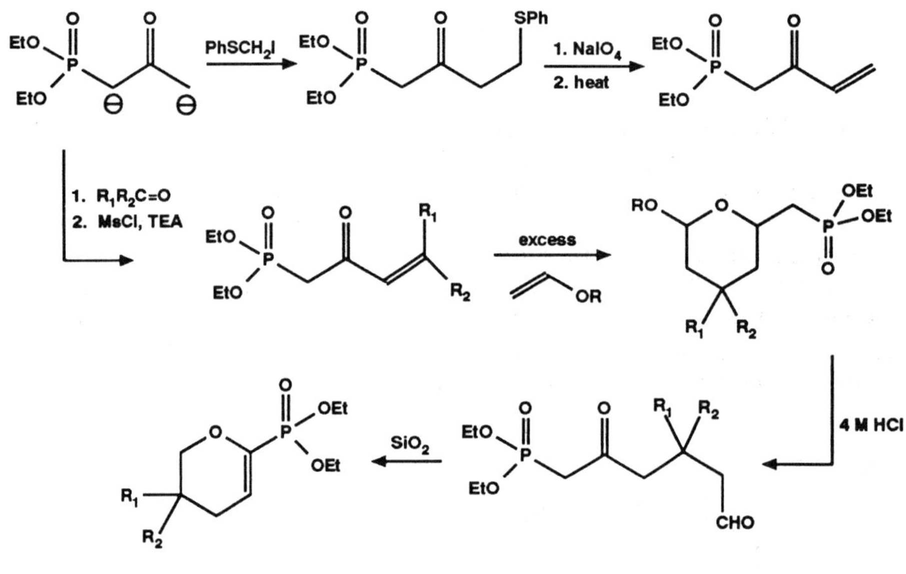

Scheme 2.20. Synthetic sequences using β-phosphonyl 1,3-dianions

The dianions generated from 2,4-dioxophos(o)inyl acetoacetate, where the terminal carbonyl moiety is an ester group (X = MeO, EtO), have found extensive use in synthesis. The dioxophosphonate has been prepared by acylation of the 2-oxophosphinyl dianion,[201] and by reaction of the corresponding halo carbonyl with ethyl diphenylphosphinate[203] or sodium diethylphosphite.[204] Dianions of 2,4-dioxophosph(o)inates undergo traditional ylid chemistry to form α,β-unsaturated carbonyl compounds in good yield (Eq. 2.51) (Table 2.5).[203-206] This outcome would seem to indicate that C-H groups experience greater acidity when juxtaposed between two carbonyl groups than between a carbonyl and a phosphonyl.

However, when 4-phosphinyl-3-oxobutanoate was treated with one molar equivalent of NaH or KH condensation, reactions were found to occur at both α- and α′-sites.

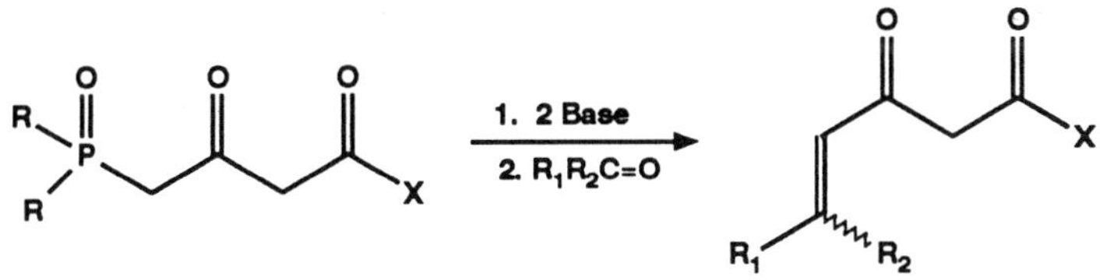

(2.51)

Table 2.5.
Reaction of β-Phosphonylacetoacetate Dianions with Carbonyl Compounds (Eq. 2.51)

R	X	Base	R_1	R_2	Yield(%)	Ref.
Ph	OEt	2 NaH	H	H	52	203
Ph	OEt	2 NaH	Me	Me	81	203
Ph	OEt	2 NaH	nC_7H_{15}	H	73	203
Ph	OEt	2 NaH	p-XPh	Me	80-90	203
Ph	OEt	2 NaH	Me	Me	72	203
Ph	OEt	2 NaH	Me	Pr	65	203
Ph	OEt	2 NaH	$-(CH_2)_5-$		85	203
Ph	OEt	2 NaH	Ph	Me	65	203
EtO	OEt	2 NaH	Me	H	89	204
EtO	OEt	2 NaH	iPr	H	96	204
EtO	OEt	2 NaH	CH_2=CH	H	91	204
EtO	OEt	2 NaH	Ph	H	84	204
EtO	OEt	2 NaH	Me	Me	70	204
EtO	OEt	2 NaH	$-(CH_2)_4-$		65	204
EtO	OEt	2 NaH	Ph	Me (E/Z; 3:1)	60	204
EtO	OEt	2 KH	p-XPh	p-XPh	65-74	205
EtO	OEt	2 KH	Ph	H	72	205
EtO	OEt	2 KH	PhCH=CH	H	15	205
EtO	OEt	2 NaH	Me_2C=CH	H	53	206
MeO	OEt	2 NaH	Me_2C=CH	H	47	206
MeO	OEt	NaH/BuLi	Me_2C=CH	H	69	206

X = H, Me, Cl, MeO, etc.

Minor reaction variations using C,C-dianions stabilized by phosphorus groups have been reported. Goswami generated a "remote" phosphorus-stabilized 1,4-dianion, which reacts with a variety of electrophiles in good yield (Eq. 2.52).[207] The inclusion of this dianion here rather than in the remote C,C-dianion sections illustrate that no rearrangement to the more stable α,α′-dianion was observed. McClard and coworkers generated the dianion of phosphonylphosphinyls, which undergoes alkylation in good yield (Eq. 2.53).[208] Use of the isopropoxy phosphodiester groups permits the use of BuLi as base, which generates a higher portion of the dianion without addition to the phosphorus group.

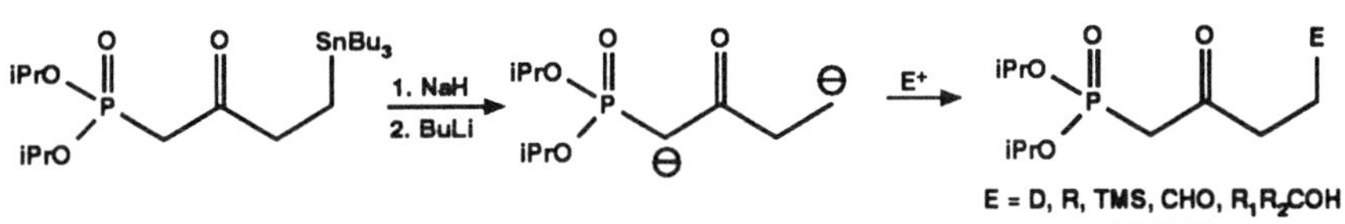

(2.52)

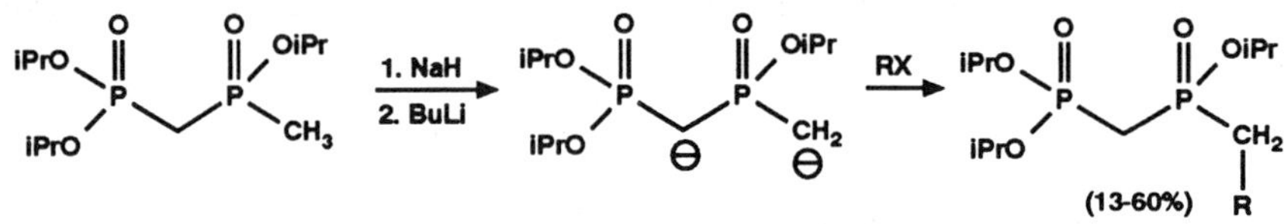

(2.53)

2.7 REMOTE C,C-DIANIONS

Whenever successive deprotonations are generated on a molecule that are at least one intervening carbon- or heteroatom- containing group apart, the dianion thus formed can be termed "remote." The remote designation specifically differentiates this class from the geminal and vicinal C,C-dianions (numerically represented as 1,1- and 1,2-(C,C)-dianions respectively), but does overlap somewhat with the nomenclature assigned to the popular α,α′- or 1,3-(C,C)-dianions. As a result of this overlap, certain nomenclature distinctions must be made. The prior coverage of 1,3-(C,C)-dianions (Sections 2.4, 2.5, and 2.6) focused upon bis-deprotonations that were conducted adjacent to a central carbanion-stabilizing group, for example, a carbonyl, sulfoxide or sulfone (Figure 2.10).

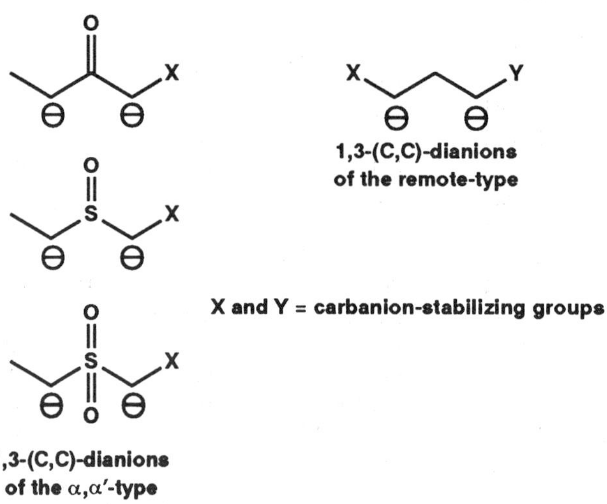

Figure 2.10. Comparison of α,α′-1,3-dianions and remote 1,3-dianions

Certain remote 1,3-(C,C)-dianions differ from these aforementioned specific α,α′-dianions in that no guiding central-stabilizing group is present (Figure 2.10). Moreover, the central atom of remote 1,3-dianions frequently plays an inconsequential role in the guidance of the carbanion position. Although the difference in designation is subtle, the reader will quickly recognize the distinctive dianion-forming strategy used by researchers of remote 1,3- vs. α,α′-dianions. The various forms of remote C,C-dianions are then easily categorized by extending the numerical system to accomodate the atom chain distance appearing between the carbanionic sites (Figure 2.11). As such, remote C,C-dianions are classified according to their intra-atomic spacing, namely, 1,4-(C,C)-dianions, 1,5-(C,C)-dianions, and so forth. Since there are only limited examples of C,C-dianions having four or greater than four atom spacers (1,6-dianions), these dianions will be grouped together as 1,Z-(C,C)-dianions. Remote dianions also may be generated with unsaturations in the three-carbon system, for example, alkene- and alkyne-based remote dianions.

This section will cover a broad array of molecule types arranged by the distance appearing between the two carbanion centers. In many cases, there will be very little structural resemblance between dianions falling in the same category, which is in contrast to the preceding sections. However, the reader will gain an understanding of the various functional groups used to stabilize remote carbanionic centers and the resultant compatibility of the [non-participating] functional groups needed for remote dianion preparation.

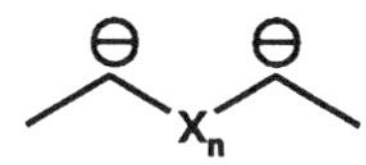

X = CH_2, O, S, etc.

n = 1; 1,3-(C,C)-dianion

n = 2; 1,4-(C,C)-dianion

n = 3; 1,5-(C,C)-dianion

n = 4; 1,6-(C,C)-dianion

Figure 2.11. Designations of remote dianions

Remote C,C-dianions have been formed using a wide variety of base combinations. In most cases, the selection of base is dependent both upon the pKa of the carbanion center to be formed, the reactivity of the appended functional groups, and the reaction intended for the dianion. In most cases, the [desired] reaction occurs at the second deprotonation site, and several reactions use both carbanion centers in the scheme to prepare cyclic compounds.

2.7.1 Remote 1,3-(C,C)-dianions

Three major types of remote C,C-dianions resulting from 1,3-deprotonations at carbon have been identified. These 1,3-(C,C)-dianions are separated into categories that describe a common stabilization group for the carbanion sites, namely, alkene-1,3-systems, alkyne-1,3-systems, and (exo)-1,3-dianions (Figure 2.12). Alkene and alkyne 1,3-dianions result

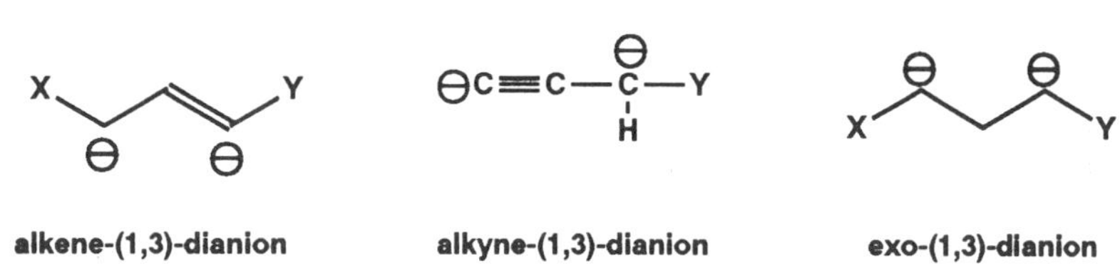

Figure 2.12. Representative remote 1,3-dianion types

from bis-deprotonations involving the pi-system and propargylic sites, respectively. The term (exo)-1,3-dianion is derived from the position of the carbanion stabilizing groups, which appear *outside* of the spacer group (1,3) system.

Certain araalkyl-based benzylic deprotonations, when followed by ortho ring metalation, form 1,3-dianions, which technically can be designated as 1,3-(C,C)-dianions. However, of these, toluene-based dianion systems undergo unique reactions and warrant a separate discussion (Section 2.8).

Alkene-based 1,3-(C,C)-dianions

Brenner and Bovete showed that 3-cyanopropene could be bis-deprotonated to form the resonance-stabilized 1,3-dianion, which undergoes reaction with excess TMS-Cl to give an 80% yield of the bis-silylated product (Eq. 2.54). The electron withdrawing cyano group controls the alkene regiochemistry.[209]

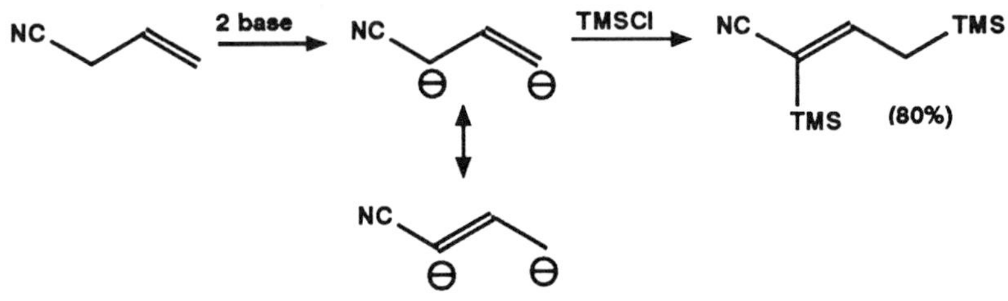

(2.54)

A more elaborate carbanion-stabilizing, alkene-based 1,3-dianion system was devised by Padwa, McCombie and coworkers.[210] Reaction of (E)-2-methoxy-2-(phenylsulfonyl)propene with BuLi gave abstraction of the α-sulfonylvinyl proton. Addition of a second equivalent of BuLi led to deprotonation at the allylic site, which reacts with MeI to give the expected alkylated product (Eq. 2.55). In addition to serving as a 1,3-dianion, this study shows the powerful directing influence of the sulfone moiety even toward a vinylic carbanion.

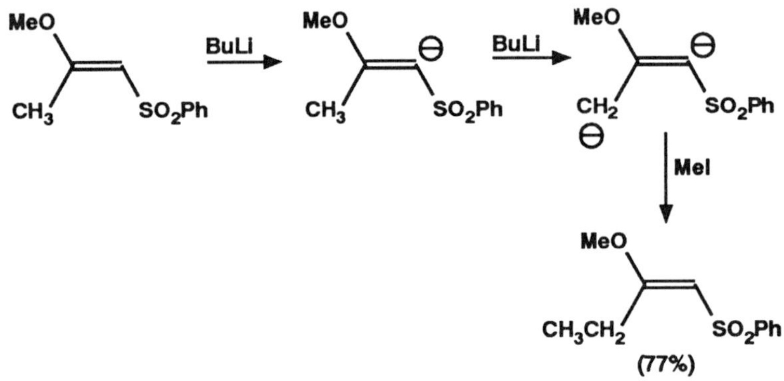

(2.55)

Alkyne-based 1,3-(C,C)-dianions

Terminal alkynes have served as precursors for the production of remote 1,3-dianions. Propyne may be bis-deprotonated by 2 BuLi to give the 1,3-dianion, which undergoes reaction with electrophiles to give the 1,3-disubstituted products.[211] The additives TMEDA or DABCO were needed to ensure clean alkylation. Sequential addition of different electrophiles afforded a new, one-pot 1,3-disubstitution procedure (Eq. 2.56).

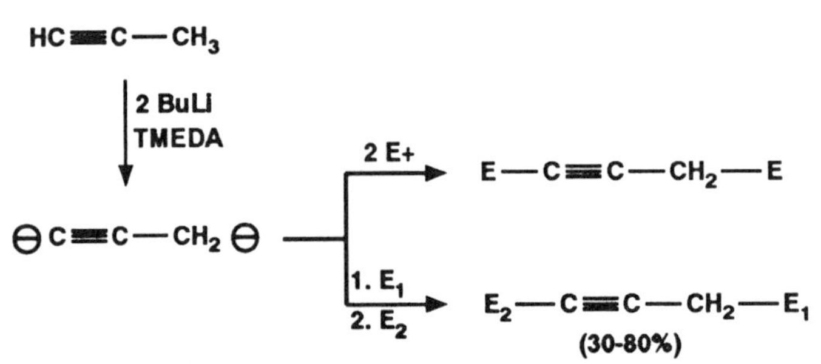

(2.56)

3-Thiapropyne 1,3-dianions have been investigated (Eqs. 2.57 and 2.58).[212-214] Alkylation occurs exclusively at the propargylic position, and no allene products were observed. The dianions of the 2-thiothiazoline derivatives[212,213] were applied to a squalene synthesis. Desulfurization led to a mixture of alkyne and allene products.[212]

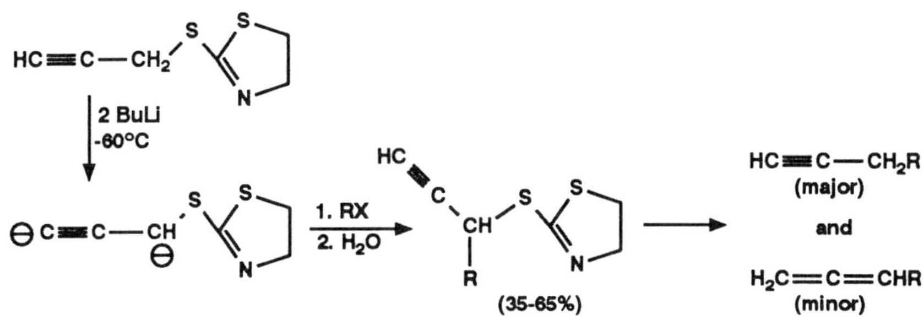

(2.57)

(2.58)

(Exo)-distabilized-1,3-(C,C)-dianions

Carbanion-stabilizing groups that are positioned with three atom spacer groups between them can direct bis-deprotonations toward the 1- and 3-positions.[215,216] Dimethyl-2-(methoxycarbonyl)glutarate forms the 1,3-(C,C)-dianion that undergoes reaction with 1,5,5-trimethyl-3-tosyl-1-imidazolium iodide to afford the imidazolium 2-substituted product (Eq. 2.59), which was further elaborated to a folate cofactor analog.[215]

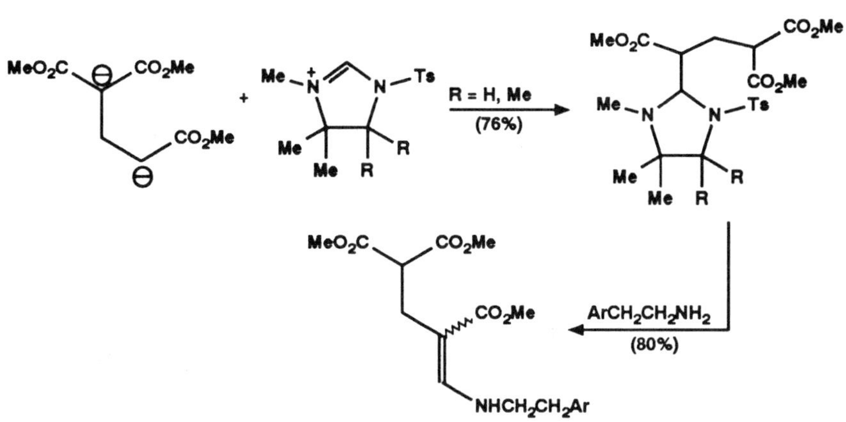

(2.59)

1,1-Bis(nitromethyl)cyclopropane serves as an interesting precursor to a remote 1,3-dianion. Reaction of the dinitro compound with two equivalents of dimsyl sodium [$CH_3S(O)CH_2$-Na+] gives the 1,3-dianion that cyclizes to the spiropentane upon reaction with I_2 (Eq. 2.60).[216]

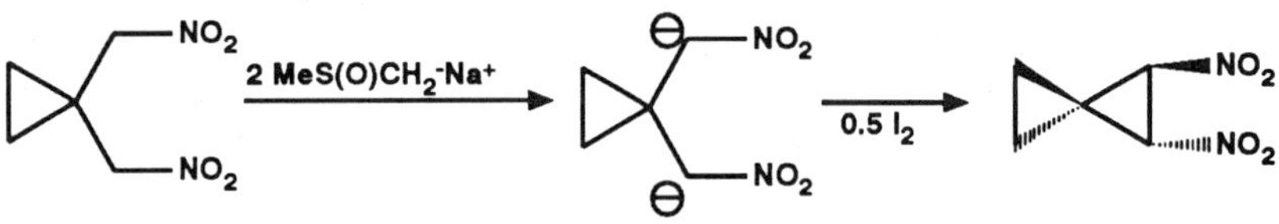

(2.60)

2.7.2 Remote 1,4-(C,C)-dianions

There has been significant activity in the preparation and reactions of 1,4-(C,C)-dianions. Most evident in this category is the interesting diversity of precursor molecules and selection of the carbanion stabilizing groups capable of introducing an electrophile at a unique site that was otherwise unaccessible. In most cases, the position of the second deprotonation is guided by the molecule design using precisely installed stabilizing moieties to direct the position of the carbanions. Several investigators have achieved surprising regiochemical selectivity in their design, particularly with those molecules that undergo secondary deprotonation by ortho-metalation.

In general, three types of 1,4-(C,C)-dianions have been explored (Figure 2.13). First, (endo)-stabilized-1,4-(C,C)-dianions, where at least one carbanion-stabilizing group appears within the domain of the spacer group, for example, at the 2- or 3-position. Second, (exo)-1,4-(C,C)-dianions represent intermediates that contain both carbanion stabilizing groups "outside" of the 1- and 4-positions, and therefore are not part of the 4-atom spacer group. Third, aromatic ring metalation (via metalation or halogen-metal exchange) also can furnish 1,4-(C,C)-dianions where an aromatic ring comprises part of the spacer group.

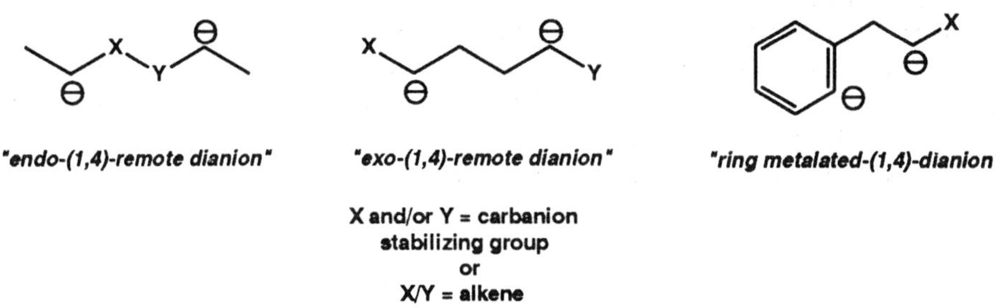

Figure 2.13. Representative examples of 1,4-(C,C)-dianions

(Endo)-1,4-(C,C)-dianions

Remote 1,4-dianion that contain carbanion-stabilizing groups between the carbanion centers are appended the prefix *endo* to indicate the general position of the directing group.

An internal dithioester served to direct 1,4-bis-deprotonation which, following alkylation, leads to a thia-Claisen precursor. Warming rearranges the molecule to afford a γ,δ-unsaturated ester after copper-mediated ethanolysis (Eq. 2.61).[217]

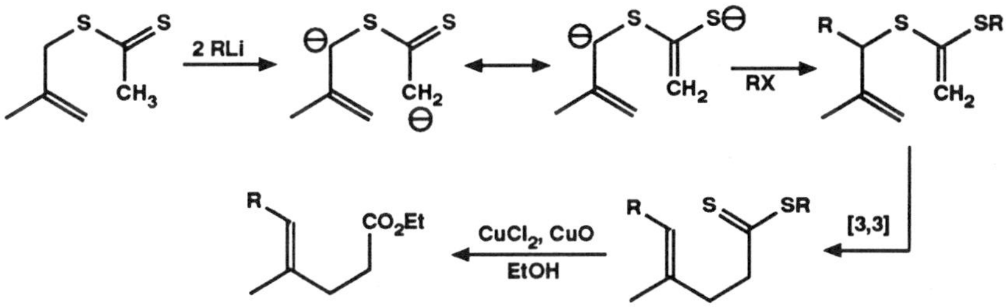

(2.61)

Kende and DeVita devised a clever endo-1,4-dianion derived from a 2-amido malonate that undergoes cyclocondensation when esters are used as the electrophile (Eq. 2.62).[218] The strategy was extended to the preparation of the lactam/lactone terminus of neooxazolomycin.

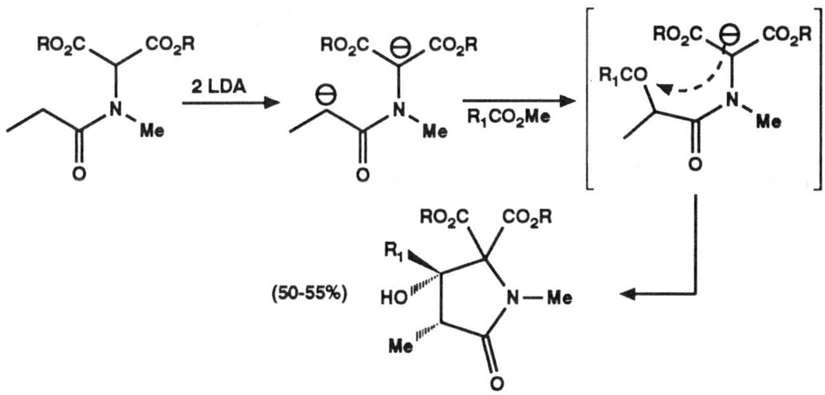

(2.62)

3-Substituted-1,2-cyclohexanediones furnish cyclic versions of the endo-1,4-dianion, which gives *threo*-selective aldol condensation products in good yield (Eq. 2.63).[219]

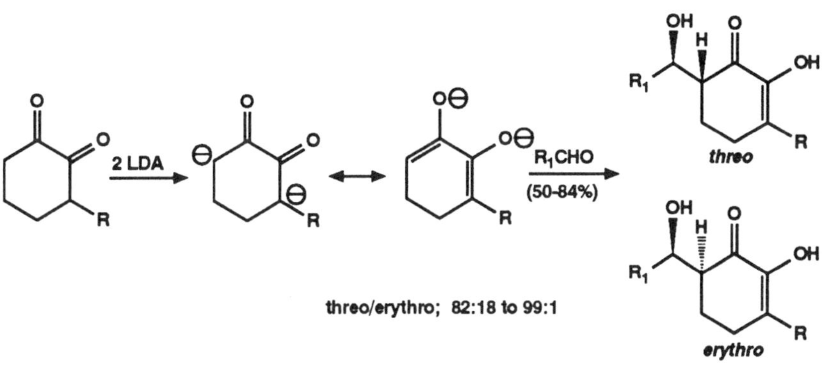

(2.63)

Hiyama and coworkers took advantage of certain N-substituted tertiary amides as the endo stabilizing group (in addition to an exo-stabilizing moiety (Z)) to give β-lactams following 1,4-bis-deprotonation and intramolecular coupling using N-iodosuccinimide (NIS) or copper (II) acetate (Eq. 2.64).[220,221] The procedure was amenable to an array of substituted R groups (H, Et, MeO, Bn_2N), and a high degree of asymmetric induction could be achieved when the amide substituent (R_1) was (R)-1-phenylethylamine.

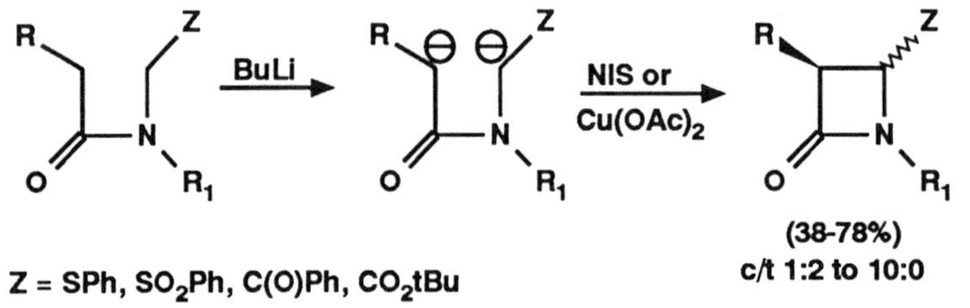

(2.64)

(Exo)-1,4-(C,C)-dianions

Remote 1,4-(C,C)-dianions can be generated when one or more carbanion-stabilizing groups are appended external to the carabnions, and used successfully in synthesis. As opposed to the endo descriptor, the exo prefix is added to indicate that the directing group or groups are not contained within the spacer group, and are generally attached directly to the 1- and 4-carbon-positions.

A straightforward first example of exo-stabilized 1,4-dianions is provided in the report by Sudweeks and Broadbent[222] and Lansbury and coworkers,[223] who used 1,4-dithianyl and 1,4-disulfones, respectively, to direct the bis-deprotonations (Eqs. 2.65 and 2.66).

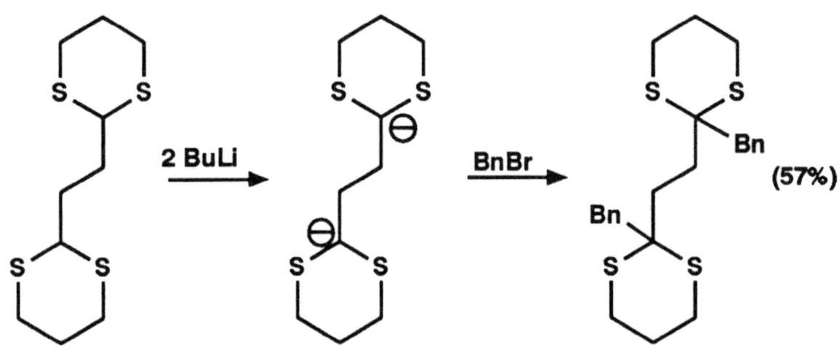

(2.65)

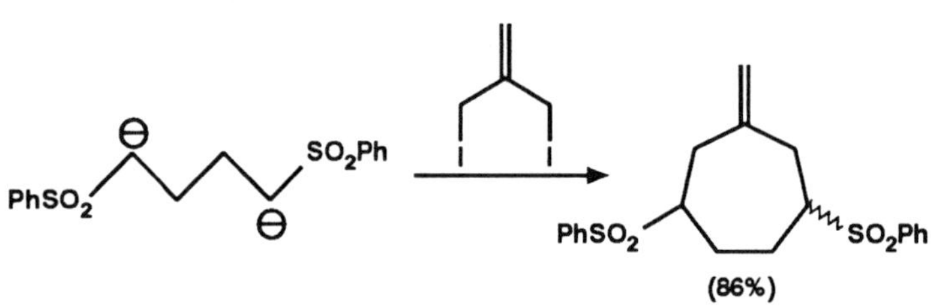

(2.66)

Disulfone-based dianions with progressive five (1,5) and six (1,6) methylene spacer groups also reacted with the dihaloalkene illustrated in Equation 2.66 to give good yields of cyclic products.[223] A 1,4-dianion stabilized by exo ketone and sulfone groups reacted with 1,3-dihalides to give the bicylic products shown in Scheme 2.21.

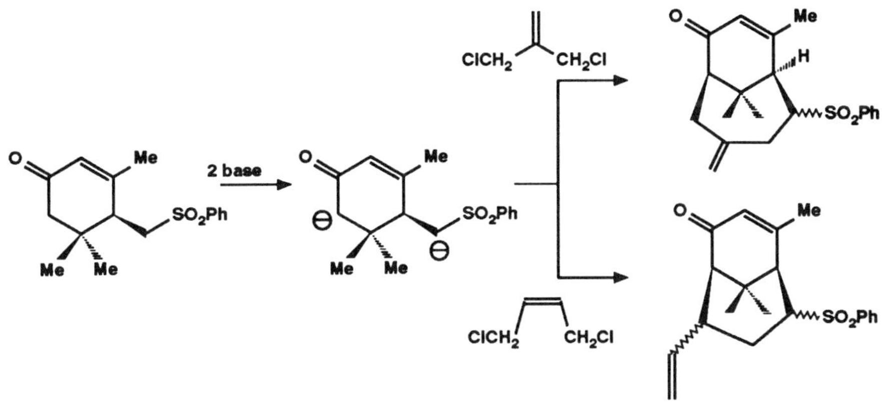

Scheme 2.21. Reaction of an exo-1,4-(C,C)-dianion with dihalides

Seebach and coworkers generated 1,4-dianions of γ,δ-unsaturated acetophenones that take advantage of the resultant resonance-stabilization to prepare terminal-substituted products.[224,225] The violet-red dianion of allyl acetophenone, prepared by sequential addition of KH (THF, 45 min, 0°C) and 2 sec-BuLi (TMEDA, 4h, -78°C → 0°C) reacts with most electrophiles at the 4-position to give products bearing exclusive cis-orientation (20-78%).[224] Interestingly, aromatic carbonyl-based electrophiles (e.g., benzaldehyde) give the sole trans adduct (Eq. 2.67). Coulombic repulsion and/or intramolecular chelation may be responsible, in part, for preferential formation and reaction by the terminal carbon position.

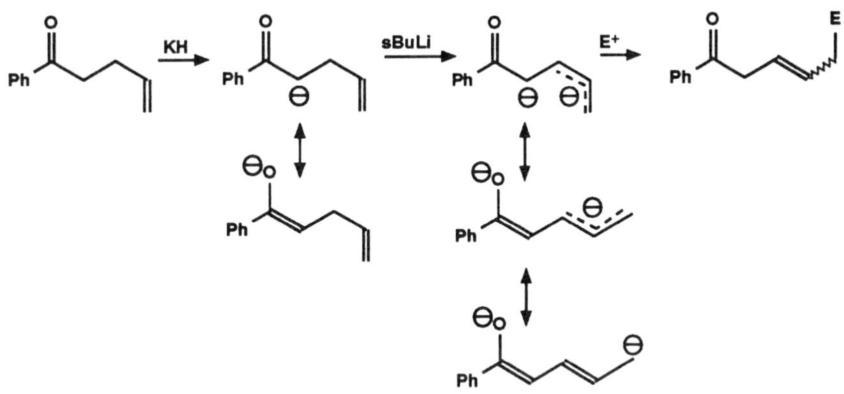

(2.67)

In a later report, the exo-stabilizing group was varied as acetyl, pivavoyl, and as a dithioester, all of which alkylate at the putative 4-carbanionic position.[225]

Cohen and coworkers have explored the synthetic utility and mechanistic basis of the 1,4-dianions prepared from bis-diphenylthioacetals.[226] They showed that the corresponding 1,4-dianion can be generated and cyclized to form the 1,2-bis(phenylthio)cyclobutene (Eq. 2.68).

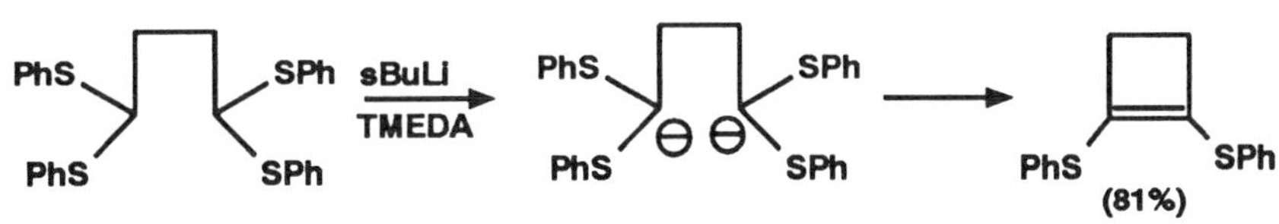

(2.68)

Ring metalated 1,4-(C,C)-dianions

Several substituted aromatic compounds undergo initial deprotonation at a side chain CH-position followed by ortho-metalation to give remote-1,4-(C,C)-dianions.

Diphenylacetylene undergoes addition by an alkyllithium reagent to furnish the α-substituted β-vinylcarbanion. Subsequent reaction with a second equivalent of RLi gives the ortho-metalated 1,4-dianion. The regiochemistry was established by quenching studies with deuterium and carbon dioxide.[227]

Phenylacetylene gives stepwise deprotonations at the alkyne CH and ortho position when the BuLi/t-BuOK base combination is used (Eq. 2.69).[228] Halogens may be introduced at the ortho position only after the cations were exchanged using $MgBr_2$.

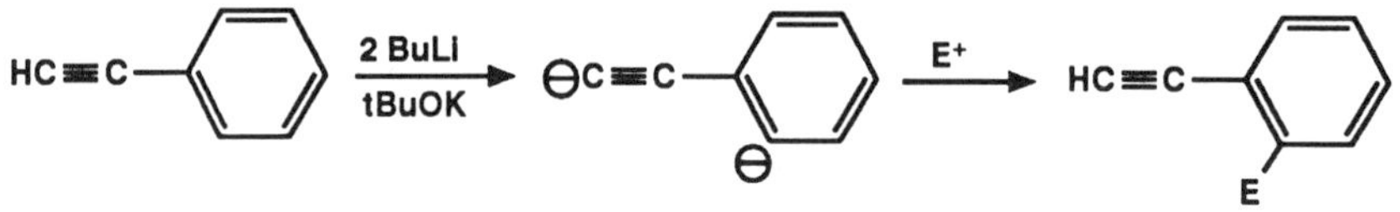

(2.69)

The direct metalation of aromatic ethers was studied.[229] 1,4-Dimethoxybenzene undergoes bis-deprotonation to form the para- or 1,4-dianion, which can be captured with excess TMS-Cl to give predominant formation of the bis-TMS derivative accompanied by some mono-TMS adduct (Eq. 2.70).

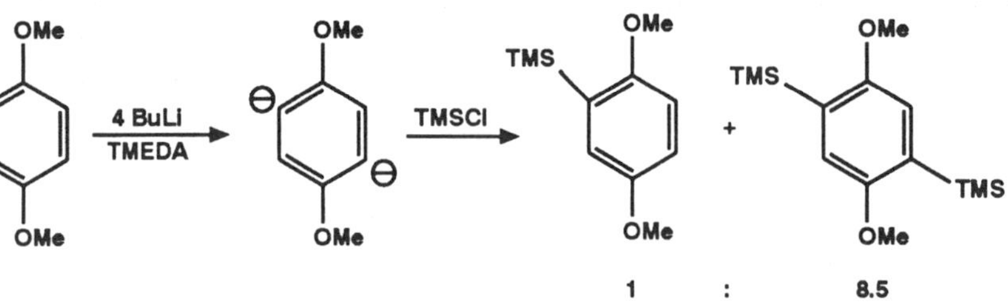

(2.70)

A large number of linked bromothiophenes have been converted into remote 1,4-dianions and, for example, oxidatively coupled (Eq. 2.71).[5,230,231]

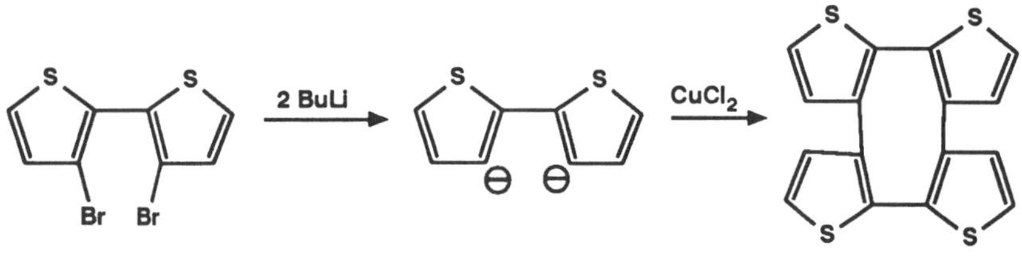

(2.71)

2.7.3 Remote 1,5-(C,C)-dianions

Several interesting remote 1,5-dianions have been used in synthesis, including straight-chain and aromatic-ring containing analogs. The straight-chain species may be formed either using endo- or exo-1,5-(C,C)-dianion types, or the synchronous use of both endo- and exo stabilizing groups. Overall, less 1,5-species have been reported than 1,4-dianions and therefore, no further categorization will be conducted in this section. Each example will be individually identified.

Both endo and exo carbanion-stabilizing groups were used to access 1,5-(C,C)-dianions by Jones and coworkers (Eq. 2.72),[232,233] Tanaka (Eq. 2.73),[234] and Nakamura (Eq. 2.74).[235]

The allylthioacetate dianions[234] alkylates predominantly at the α-position, and therefore, can be cross-listed as a 1,3-(C,C)-dianion (Eq. 2.73).

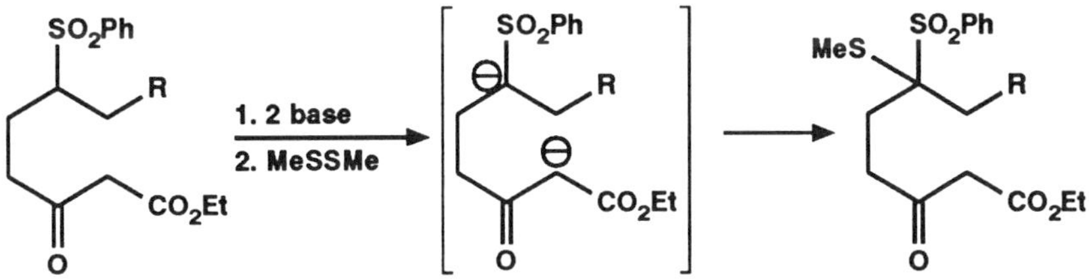

(2.72)

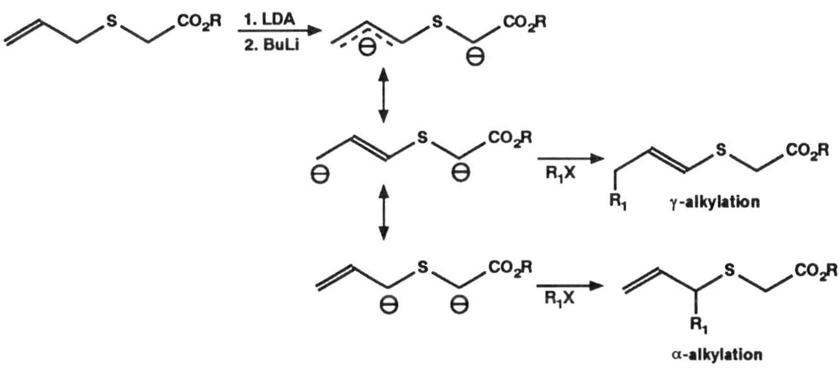

(2.73)

Fischer carbene complexes also have been converted to endo/exo 1,5-dianions, and reacted with aldehydes (Eq. 2.74).[235]

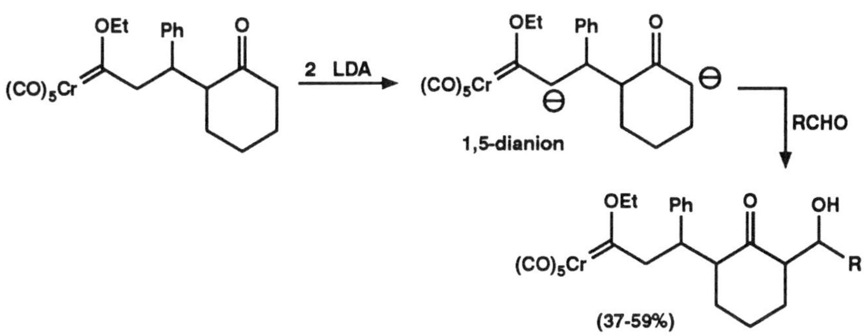

(2.74)

Sulfone-linked aromatic moieties can be bis-deprotonated to give the 1,5-dianions shown in Figure 2.14.[236]

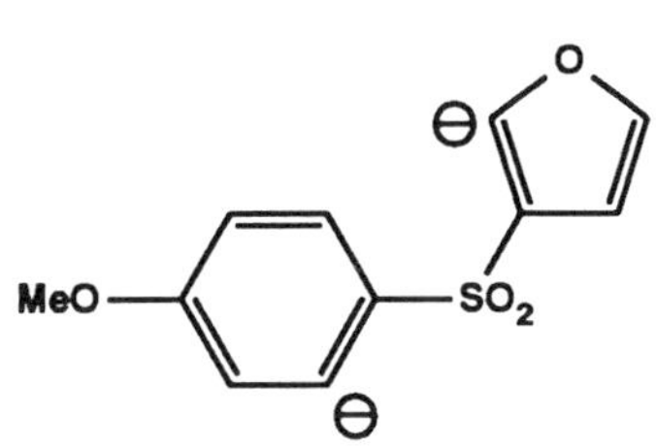

Figure 2.14. A 1,5-remote dianion accessed through bis-metalation

Several reports that incorporate an aromatic nucleus as the 2,3,4-spacer group have been converted to remote 1,5-(C,C)-dianions.[237-239] Levine and coworkers used 1,4-diazines to lead directly to these dianions,[237,238] whereas the corresponding benzene analog was examined as the bis-dithianyl derivative (Figure 2.15).[239] A more extensive coverage of toluene-based dianions can be found in Section 2.8.

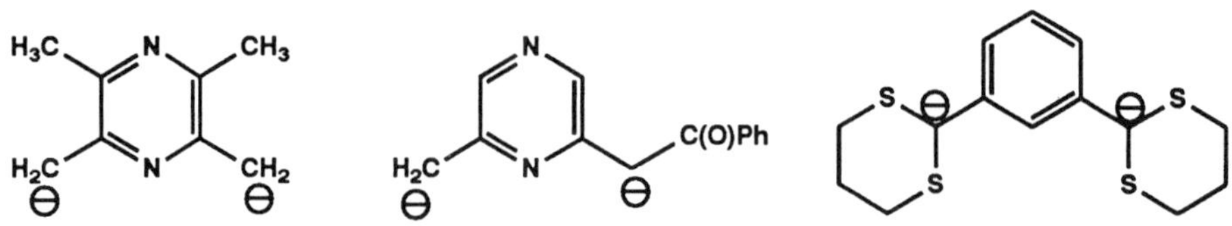

Figure 2.15. 1,5-Remote dianions containing an aromatic spacer moiety

2.7.4 Remote 1,Z-(C,C)-dianions

Distal C,C-dianions also have been used in synthesis although explorations of these intermediates are more scarce. Select examples that increase in the length of the spacer group of 1,Z-dianions are included here.

Remote 1,6-(C,C)-dianions have been prepared from azines,[240] 2,2′-[1,2-ethanediylbis(thio)]-bis-1,3-dithiolane,[241], 1-cyano-6-thiophenyl substrates,[242] 1,6-protected bis-cyanohydrins,[243] and [ethyl]bridged cyclopentadienes (Scheme 2.22).[244]

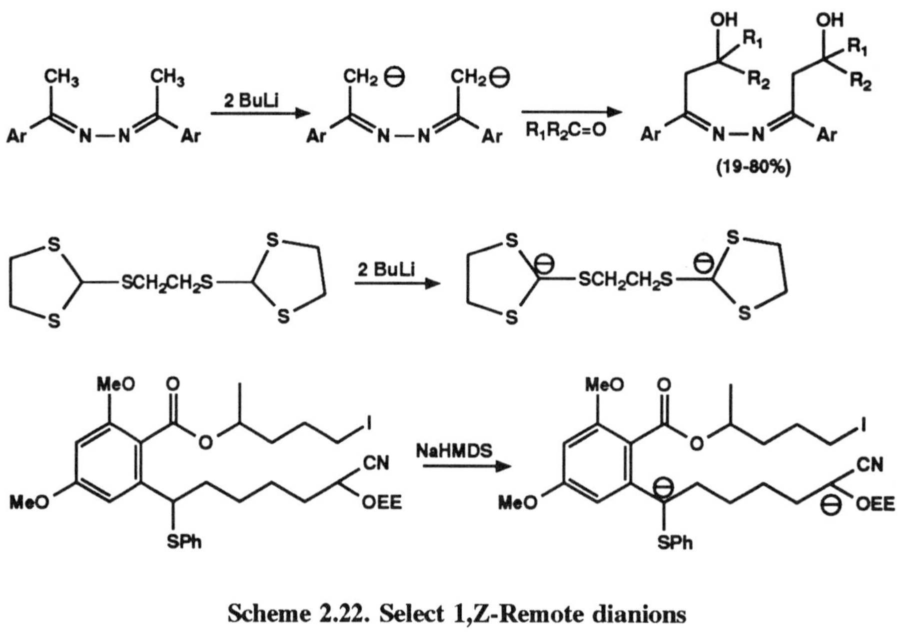

Scheme 2.22. Select 1,Z-Remote dianions

A remote 1,Z-dianion cyclization strategy was devised for the construction of large-molecule rings.[243] In this study, detailed information regarding the requirements of alkylating agent and carbanion-stabilizing group was provided.

The reaction of 1,2-bis(3-methyl-2-pyrazyl)ethane with LDA gives a 1,8-dianion that undergoes a low yielding cyclization upon treatment with 1,2-dibromoethylene (Eq. 2.75).[245]

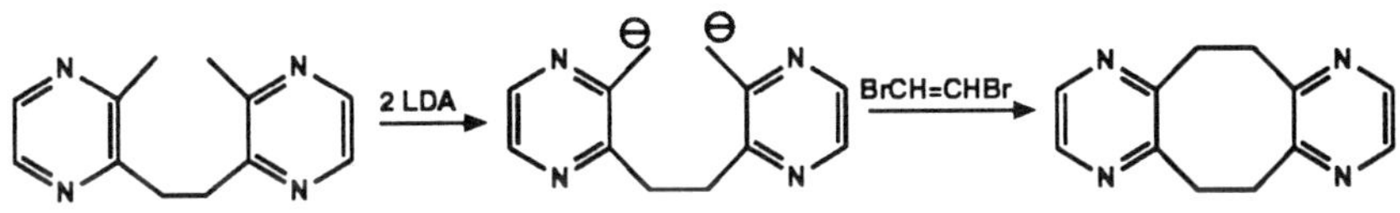

(2.75)

2.8 DELOCALIZED C,C-DIANIONS

Prior sections showed that allylic- and benzylic-positions can be deprotonated in route to C,C-dianions. However, most of those studies and, in coming sections that exploited these "π-delocalized" carbanion systems also have had their reactivity influenced to some extent by additional carbanion-stabilizing groups (e.g., carbonyl, sulfonyl, etc.). Therefore, the purpose of this section is to review C,C-dianion systems that result from successive deprotonations of allylic or benzylic CH-groups within molecules that do not contain additional carbanion stabilizing groups. The resultant intermediates are best described as "delocalized" or "resonance-stabilized" carbanions or dianions as termed in reviews by Bates and coworkers.[246,247] Brief coverage of these resonance extended dianion intermediates has also appeared to some degree in more general reviews,[3,5] and in a related review of the electronic structure of polycyclic conjugated dianions.[4]

It is widely appreciated that the deprotonation leading to monoanions of allylic and benzylic positions occurs with somewhat greater ease than deprotonation of an alkane-CH or an alkene vinyl-CH.[9,248] The enhanced acidity may be due, in part, to delocalization of the resultant anion through the π-system as illustrated in simplified form for propene and toluene (Figure 2.16). The allylic delocalization phenomena can cause regioisomeric mixtures of products (α- vs. γ-alkylation) following reaction with electrophiles when the π-system is unsymmetrically substituted.

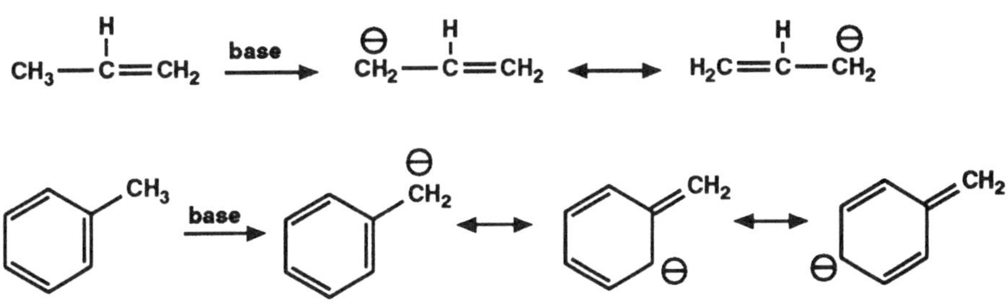

Figure 2.16. Resonance delocalization of allylic and benzylic carbanions

Certain allylic- and benzylic-containing molecules also have been doubly deprotonated leading to C,C-dianions that share the resultant electron-rich π-delocalized systems. These delocalized dianions show marked and distinct reactivity relative to simple allylic/benzylic monoanions. Moreover, studies of these π-containing hydrocarbon dianions provide essential foundation information regarding the character of allylic dianions that may act in an ambident fashion. Such information is critical to the understanding of more complex, heterofunctionally substituted allylic- and benzylic-containing dianion systems.

The coverage of delocalized dianions will be separated into two categories that cover delocalized dianions: allylic- and benzylic-type. Focused attention shall be given to the preparation and synthetic utility of theses intermediates, since many of the reports of delocalized dianions were not aimed toward synthetic application of the reported yields are not optimized.

2.8.1 Allylic-based delocalized dianions

Several alkenes and dienes bearing allylic positions have been converted to delocalized dianions. Select examples are shown in Figure 2.17 including dianions from *cis*- and *trans*-2-butene, 2,4-hexadiene, isobutylene (2-methyleneallyl), and 2,3-dimethyl-1,3-butadiene among others.

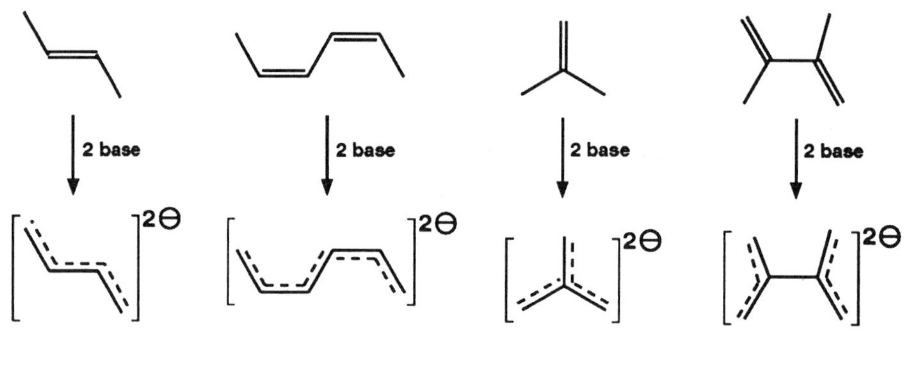

Figure 2.17. Examples of allylic-based delocalized dianions

Bates and coworkers were among the first to report the reactions of delocalized allylic dianions.[247,249-254] The 2-methyleneallyl (isobutylene) and 2-methylpentadiene dianion were shown to react with alkyl halides[249] and dihalides[250] to give the corresponding dialkylated and cyclic products in modest yield (Eqs. 2.76 and 2.77). The dianions were prepared as crystalline dilithium salts (solvated with TMEDA) in many of their early studies. Although the crystallization compromised the overall yields, this method avoided contamination by the monoanion, which stays in solution. The specific reaction between the dianion and electrophile could therefore be better controlled. Lochman's base (KOtBu/BuLi) also has been found to be a superior dimetalating reagent for 2,3-dimethylbutadiene and pentadiene, and has become the base combination of choice for synthetic applications.[251]

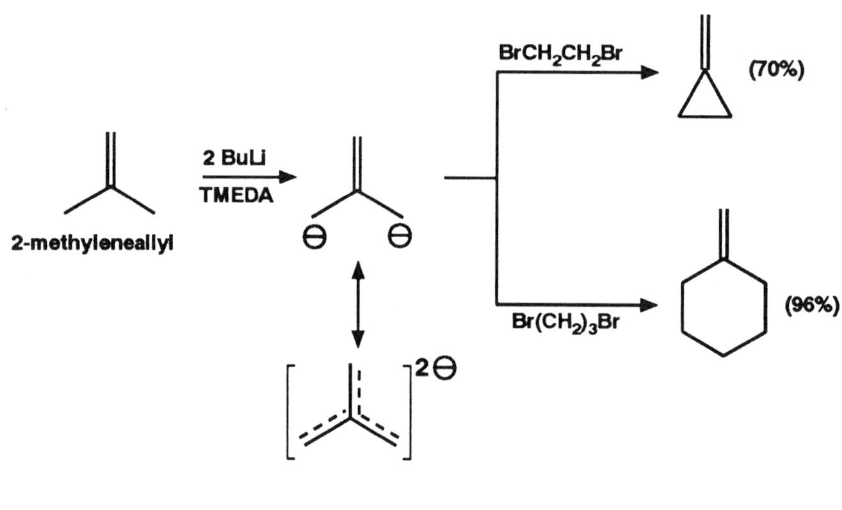

(2.76)

Among the delocalized dianions studied, the methylene allyl dianion has been shown to have good reactivity toward a variety of electrophiles.[252,253] In general, aldehydes, ketones, epoxides and allylic halides react twice with this dianion, but some mono-reaction is always observed. This bilateral reactivity was used to advantage in a novel pyridine synthesis

(Scheme 2.23). Epichlorohydrin does not afford the cyclic product with the dianion of 2-methylene allyl dianion, rather, the dianion acts as a base to give the elimination product, (E)-3-chloro-2-propen-1-ol.

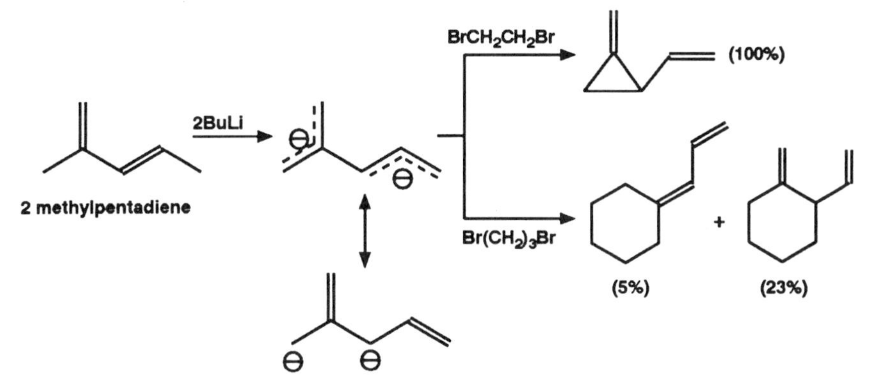

(2.77)

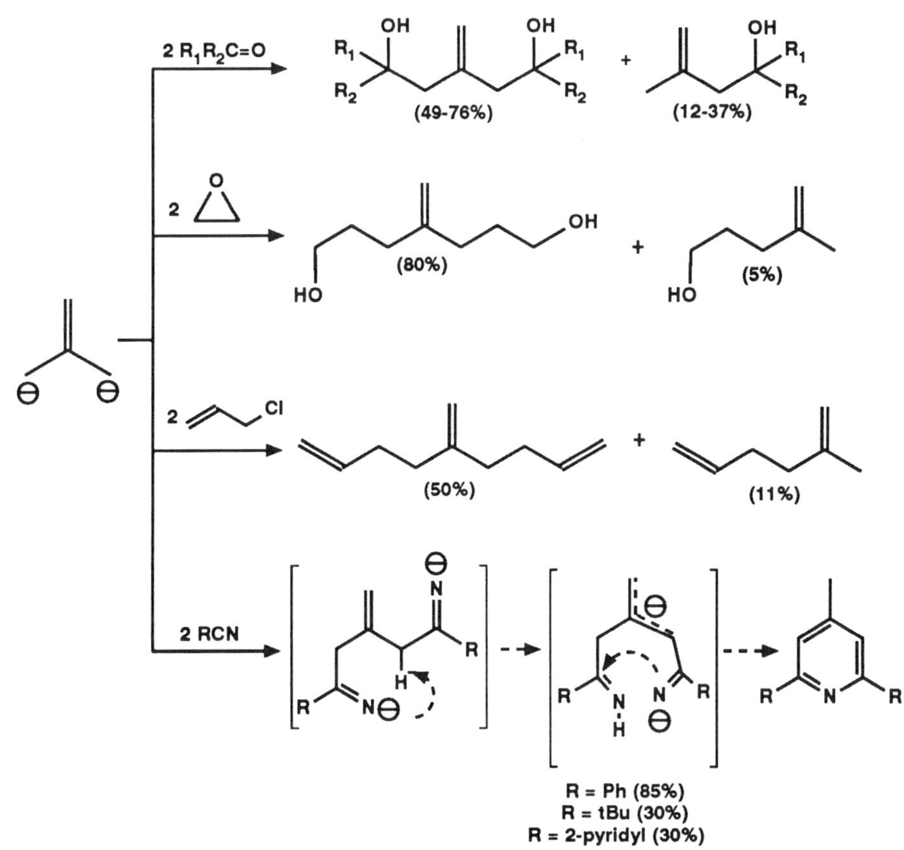

Scheme 2.23. Reactions of the 2-methyleneallyl dianion

Ramon and Yus have published an alternative to the 2-methyleneallyl dianion using naphthalene-catalyzed lithiation of 3-chloro-3-methyl propene.[255]

The dianion of 2,3-dimethylbutadiene also has attracted attention from a synthetic standpoint.[254,256] This dianion was prepared in 70-80% yield and reacts with monofunctional electrophiles to give the 2,3-methylene-substituted bis-adducts (Eq. 2.78). Sulfates (e.g. dimethylsulfate) were found to be far superior to alkyl halides in alkylation reactions when the dianion was crystallized, but the use of Lochman's base to access the dianion in solution led to markedly improved yields of products using alkyl halides.

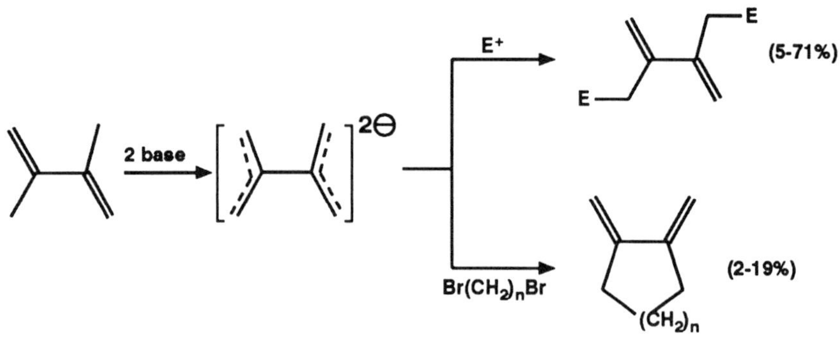

(2.78)

In related work, 3-phenyl-1,5-hexadiene,[257] diisopropene,[258] and 1,4-bis(methylene)cyclohexane[259] have been converted to their delocalized dianions and reacted with electrophiles although little synthetic utility has been explored (Scheme 2).

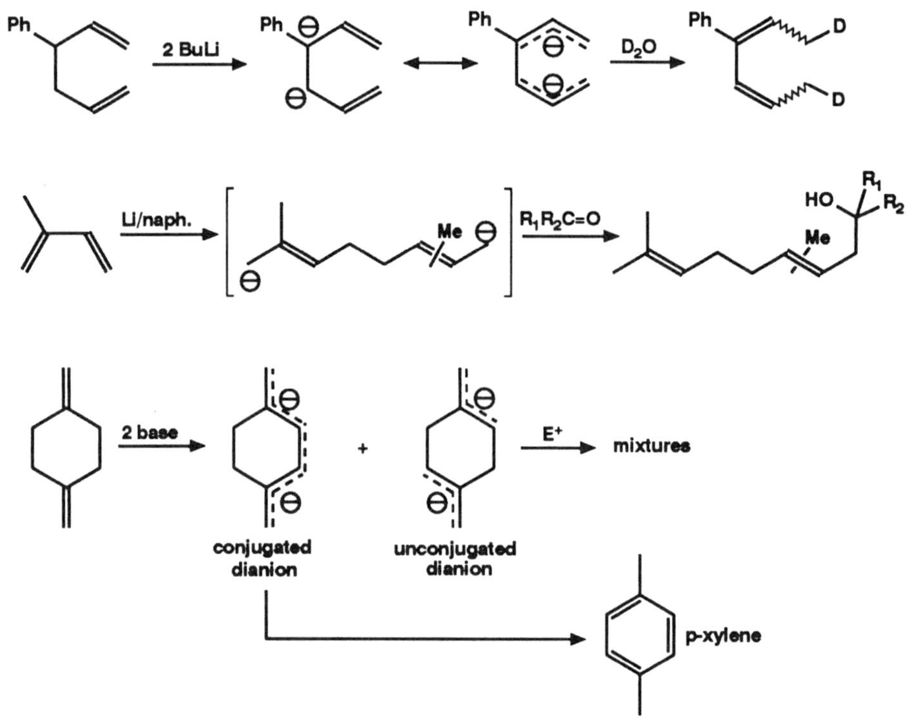

Scheme 2.24. Some reactions of delocalized dianions

2.8.2 Benzylic-based delocalized dianions

As with allylic systems, the reader will find several examples scattered throughout this book of deprotonations occurring at a benzylic site, usually appearing as the second deprotonation in a sequence following OH or NH deprotonation. Examples of this process include lateral and remote deprotonations of toluic acids, phenols, benzamides and their heterocyclic cogeners (Chapters 3 and 4). In general, this type of benzylic carbanion formation has long been used to functionalize aromatic side chains in a single synthetic operation when alternatives to dianion use generally involve protection/deprotection sequences. In this section, benzylic-based dianions that are formed from C,C-bis-deprotonations will be covered.

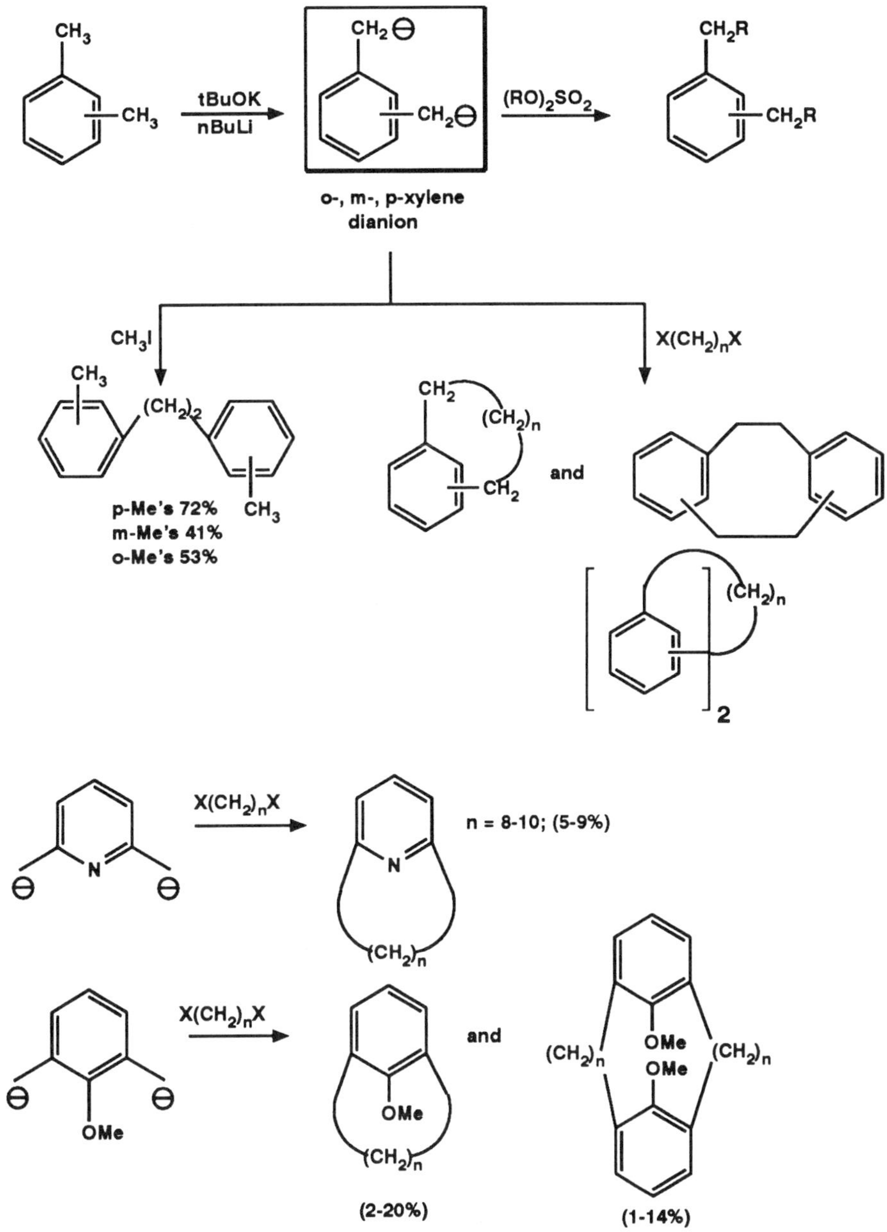

Scheme 2.25. Reactions of dibenzylic dianions

To a lesser extent, benzylic-based dianions that result from initial deprotonation at the benzylic site are less commonplace, although several investigations have showed highly useful synthetic outcomes. In particular, bis-deprotonation of xylenes can be of great synthetic value in the construction of cyclophanes.[247]

Two major types of C,C-benzylic-based dianions have been used in synthesis; dibenzylic dianions and a sequential benzylic-position ring metalation process. The former dianion leads to functionalization at the methyl side chain, whereas the latter can lead to ring substituted products.

Bates and coworkers showed that the dibenzylic dianions of xylene-,[260] lutidine,[260] dimethylbiphenyls,[261] and 2,6-dimethylanisoles[262] can be formed and reacted with electrophiles. The xylene-based dianions reacted with alkylsulfates to give bis-alkylation products, but reaction with methyl iodide led to oxidative coupling (dimeric products) (Scheme 2.25).[260] When the xylene dianions were reacted with certain α,ω-dihalides, cyclophanes were formed in modest yield. The dianion of 2,6-lutidine did not form the corresponding pyridophane with dihalides, but did produce the 2,6-linked chain (Scheme 2.25).[260] The dianion derived from 2,6-dimethylanisole reacts with α,ω-dihalides to give m-cyclophanes (Scheme 2.25).[262]

In a highly interesting variation, substituted dimethylenebiphenyls were converted to their respective dianions and converted to symmetrically substituted biphenyls and biphenyl cyclophanes (Scheme 2.26).[261]

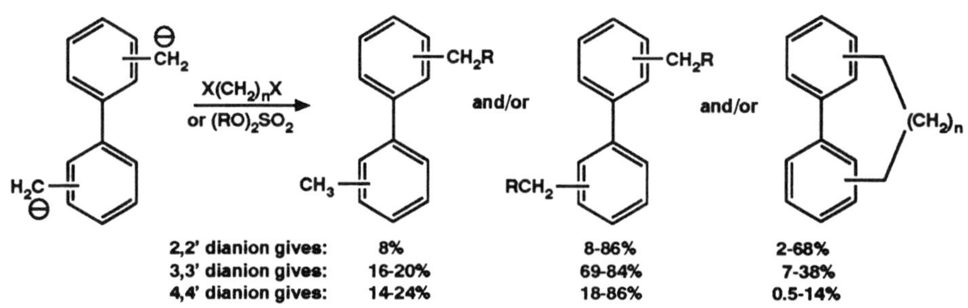

Scheme 2.26. Conversion of dimethylene biphenyls into cyclophanes via benzylic dianions

Benzylic deprotonation may be followed by ring metalation to give a C,C-dianion that may be functionalized by the addition of electrophiles.[263,264] West and Jones showed that toluene gives the benzylic-ring metalated polyanion upon reaction with nBuLi/TMEDA. However, reaction with excess TMSCl led to α,α,*p*-tris(trimethylsilyl)toluene. The α,*p*- and α,*o*-toluene bis-TMS derivatives were the other dominant products formed.

Phenylcyclopropane has been reacted with tBuOK/BuLi to give the benzylic, ring-metalated dianion, but the cyclopropyl anion rapidly rearranges to an allyl anion (Eq. 2.79).[264] Similarly, the tolylcyclopropane dianion leads to ring-opened 1-arylpropenes after the addition of water.

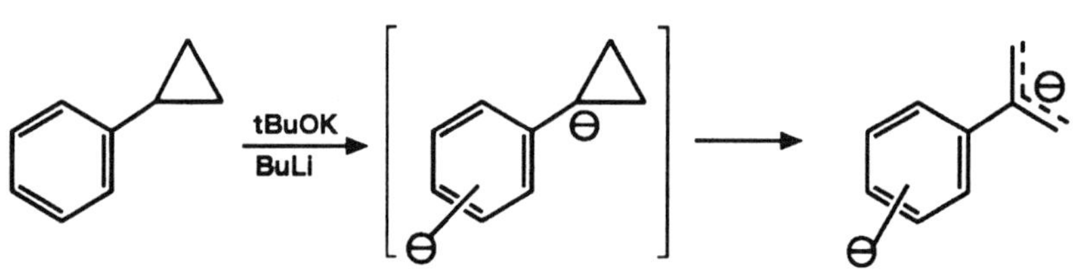

(2.79)

2.9 EXPERIMENTAL PROCEDURES

(4S)-4-Methyl-2-nitro-1,3-hexanediol (Fig. 2.4) - reaction of a nitro-stabilized, geminal C,C-dianion bearing a β-alkoxy substituent with an aldehyde (from Eyer and Seebach).[26]

To a cooled (-90°C) stirred solution of 50 mL THF, 10 mL HMPA and tetrahydropyranyl nitroethanol was added n-BuLi (22 mmol). The resulting solution was warmed to -40°C during three hours, recooled to -90°C and (S)-2-methylbutanal (10 mmol) was slowly added. After the reaction had warmed to -60°C within 90 min, the mixture was cooled to -90°C and quenched with 3 mL (approx. 50 mmol) of acetic acid. The clear, cold solution was combined with 100 mL of ether and washed successively with two 25-mL portions of cold saturated $NaHCO_3$. Each aqueous phase was extracted with one 100-mL portion of ether, and the combined organic phases were washed four times with water, once with brine, dried over $MgSO_4$, filtered and concentrated. Chromatography of the resulting residue on silica gel provided the pure THP-protected product. The deprotection was carried out with 20 mL of methanol and Amberlyst at 45°C for 2 h. Purification was conducted by recrystallization from ether-hexane to afford 64% of the diol.

2-Ethyl-1-phenylvinyl phenyl sulfone (Eq. 2.9) - preparation of α,β-unsaturated phenyl sulfones via a sulfone stabilized geminal dianion (from Umani-Ronchi et al.).[29]

Benzyl phenylsulfone (7 mmol) was dissolved in dry THF (20 mL) and reacted with n-BuLi (15 mmol) in heptane. The mixture was stirred for 1 h at room temperature whereupon a solution of magnesium iodide (15 mmol) in ether-benzene (40:20 mL) was added over 15 min period. After an additional 10 min of stirring, propionaldehyde (7 mmol) was added, and the reaction was brought to reflux for 2 h. The reaction was cooled, poured into water (100 mL), extracted with ether, dried over sodium sulfate, and purified using silica chromatography to give 78% as a mixture of isomers. mp = 87-90°C.

Preparation of ring-fused 2-cyclopentenones by reaction of the vicinal dianion of dimethyl 4-cyclohexene-1,2-dicarboxylate with propynoates (Scheme 2.9) (from Corey and Su).[62]

A solution of n-BuLi (2.2. mmol) was added to a solution of diisopropylamine (2.2 mmol) in 3 mL of THF at 0°C. After stirring for 20 min at 0°C, the resulting LDA solution was cooled to -78°C and a solution of HMPA (2.8 mmol) in 0.5 mL of dry THF was added followed by a solution of dimethyl 4-cyclohexene-1,2-dicarboxylate (1.0 mmol) in 1.0 mL of THF. The solution was stirred at -78°C for 40 min, then allowed to come to room temperature (about 15 min), and stirred an additional 30 min. The deep red dianion solution was then cooled to -45°C and a solution of phenyl 3-t-butylpropiolate (1.0 mmol) in 2 mL THF was added. After stirring for 30 min at -45°C, the reaction mixture was warmed to 0°C and stirred an additional 2 h. The reaction was quenched by the addition of acetic acid. Dilution with 5% HCl, extraction with ether (3 x 10 mL), washing with brine, drying ($MgSO_4$), and evaporation furnished the crude product, which was purified by silica gel chromatography using 5% EtOAC in hexane to give 70%. mp = 64-68°C.

2-Nonanedione (Eq. 2.19) - alkylation of the acetoacetone dianion derived from $NaNH_2$ with n-butyl bromide (from Harris and Harris).[2]

A suspension of $NaNH_2$ (1.1 moles) was first prepared as follows: Commercial anhydrous ammonia (800 mL) is introduced from a cylinder through an inlet tube to a 1 L three-necked flask. To the stirred ammonia solution under a nitrogen atmosphere is added a small piece of sodium. After a blue color appears, 0.25 g of ferric nitrate hexahydrate is added, followed by small pieces of freshly cut sodium until a total of 25.3 g has been added. After the formation of the sodium amide is complete, as indicated by the disappearance of the blue color, the flask is chilled (to 3 in.) to -78°C with a dry ice-acetone bath, and acetylacetone (0.6 mole) in ether (40 mL) contained in a pressure-equilibrating addition funnel is introduced slowly under N_2. The acetoacetone should be added in spurts which fall on the surface of the reaction mixture rather than on the wall of the flask. After the addition of acetoacetone is complete, the cooling bath is removed, the nitrogen purge is stopped, and n-butyl bromide ((0.50 mole) in 40 mL ether is introduced dropwise over 10-20 minutes. The addition funnel is rinsed with a small volume of ether, which is added to the reaction mixture.

After 30 min, 400 mL of ether is added, and the ammonia is removed by cautious heating on a steam bath. Addition of 200 g of crushed ice results in a thick slurry, which is treated with 60 mL of concentrated HCl and an additional 10 g of ice. The mixture is stirred until all solids dissolve, transferred to a separatory funnel, and the aqueous layer made acidic with dilute HCl. The ether layer is separated, and the aqueous layer extracted ether (3 x 100 mL) and the organic phases combined and dried over $MgSO_4$. After filtration and removal of the solvent, the residue is distilled through a 12" Vigreux column to give 2,4-nonanedione (81-82%) as a colorless liquid following a forerun of acetylacetone. bp = 100-103°C @ 19 mm.

Methyl 3-oxohept-6-enoate (Eq. 2.19) - alkylation of the acetoacetone dianion derived from NaH and BuLi with allyl bromide (from Huckin and Weiler).[87]

Approximately 25 mL of THF was added to a 50 mL flask containing sodium hydride (50% dispersion; 11 mmol). The flask was stoppered with a septum, flushed with nitrogen, and cooled to 0°C. Methyl acetoacetate (10 mmol) was added dropwise and the colorless solution was stirred at 0°C for 10 min whereupon n-BuLi (10.5 mmol in hexane) was added resulting in a yellow to orange solution of the dianion. This solution was stirred an additional 10 min before introduction of the electrophile. Allyl bromide (11 mmol) in 2 mL THF was added and the reaction mixture was allowed to warm to room temperature with stirring. The color of the dianion solution faded immediately upon addition of the alkyl halide. Approximately 15 min after the addition of the alkylating agent, the reaction was quenched with 2 mL of concentrated HCl in 5 mL of water and 15 mL of ether. The aqueous layer was further extracted with ether (2 x 10 mL). The organic extracts were combined, washed with water until the aqueous solution gave a neutral pH, and dried over magnesium sulfate. The solution was filtered and the solvents removed under reduced pressure. The residual oil was distilled to the give the desired product in 83% yield. bp = 99-100 @ 15 mm.

S-tert-butyl 3-oxohept-6-enethioate (Eq 2.19) - Alkylation of the dianion derived from the tert-butyl thioester of acetoacetate with allyl bromide (from Booth, Fox and Ley).[108]

S-t-butyl 3-oxobutanethioate (10 mmol) in dimethoxyethane (DME; 5 mL) was added to a suspension of sodium hydride (11 mmol) in DME (40 mL) at 0°C under argon. After 5 min, the mixture was cooled to -30°C, and BuLi (11 mmol in hexane) was added giving an orange solution. After 10 min, allyl bromide (11 mmol) was added to the mixture and and the reaction warmed to room temperature. After 1 h at room temperature, the reaction was poured into saturated NH_4Cl (20 mL) and extracted with ether (2 x 50 mL). The ether extracts were washed with brine, dried, and the solvent was removed under reduced pressure to give the product as a pink oil following chromatography (70%). The product was isolated in a 2:1 keto:enol ratio as determined by nmr.

S-tert-butyl 5-hydroxy-3-oxohept-6-enethioate (Eq 2.19) - Aldol condensation of the dianion derived from the tert-butyl thioester of acetoaceate with acrolylaldehyde (from Clarke and Ley).[137]

The dianion of S-t-butyl 3-oxobutanethioate (3 mmol) in THF was generated as noted above for the alkylation. The deep yellow solution was cooled to -78°C, and freshly distilled acrolylaldehyde (4 mmol) was added dropwise. The mixture was allowed to warm to 0°C over 30 min, poured into saturated NH_4Cl (5 mL) and extracted with ether (3 x 10 mL). The ether extracts were washed with brine, dried, and the solvent was removed under reduced pressure to give the product as a colorless oil following chromatography (83%).

(+/-)-2-[S]-[1-(Benzyloxy)methyl]-5-(E)-[1-(tert-butyloxycarbonyl)ethylidene]-tetrahydrofuran (Scheme 2.12) - reaction of the dianion of β-ketoesters with epoxides and subsequent acid-mediated cyclization to form tetrahydrofurans (from Lygo, O'Connor and Wilson).[119]

tert-Butyl(2-methyl-3-oxo)butanoate (23 mmol) was added dropwise to a stirred suspension of sodium hydride (25 mmol) in THF (250 mL) under argon at 0°C. The mixture was then stirred at room temperature for 30 min, usually giving a thick white precipitate. After cooling to -10°C, n-BuLi (25 mmol) was added, and the mixture stirred an additional 15 min giving a clear solution. 1-(Benzyloxy)prop-2-ene oxide (25 mmol) in THF (7.5 mL) was added and the mixture allowed to warm to room temperature over 3 h. After overnight stirring at room temperature, 1 M HCl (125 mL) was added cautiously, and the resulting mixture extracted with ether (3 x 190 mL). The organic extracts were dried and the solvents removed under reduced pressure to give the crude adduct as a pale yellow oil. The material was immediately dissolved in dichloromethane (100 mL) and solid oxalic acid (7.5 g) added. The mixture was brought to reflux under argon for 2-3 h, cooled to room temperature, and filtered through a pad of silica. Evaporation of the solvent gave the crude tetrahydrofuran, which was purified by chromatography on silica gel (10% ethyl acetate - 90% petroleum ether) to give the desired product in 32% yield as a colorless oil.

Methyl 5-hydroxy-3-oxo-5-phenylhexanoate (Eq. 2.26) - Aldol condensation between the dianion of methyl acetoacetate and acetophenone (from Huckin and Weiler).[121]

Approximately 25 mL of THF was added to a 50 mL flask containing sodium hydride (50% dispersion; 11 mmol). The flask was stoppered with a septum, flushed with nitrogen and cooled to 0°C. Methyl acetoacetate (10 mmol) was added dropwise and the colorless solution was stirred at 0°C for 10 min whereupon n-BuLi (10.5 mmol in hexane) was added resulting in a yellow to orange solution of the dianion. This solution was stirred an additional 10 min before introduction of the electrophile. Acetophenone (11.0 mmol) was added in one portion and the reaction allowed to stir for 10 min before being quenched with concentrated HCl (approx. 2 mL). The reaction was worked up by the addition of water (10 mL) and ether (35 mL). The aqueous layer was separated and further extracted with ether (2 x 35 mL). The organic layers were combined, washed with brine, dried over sodium sulfate and filtered. The solvent was removed by evaporation under reduced pressure to give a yellow oil. The product was purified with preparative thin layer chromatography using chloroform-ethyl acetate (9:1) as eluent to give 77% yield of the desired product.

2-Amino-5-(p-anisyl)-4(1H)pyridone (Eq. 2.30/2.31) - preparation of an β-ketonitrile dianion from 5-methylisoxazole and reaction with a nitrile (from Vinick and Gschwend).[160]

To a stirred solution of diisopropylamine (150 mmol) in 150 mL of THF at -10°C was added n-BuLi (150 mmol). After 15 min, 5-methylisoxazole (75 mmol) was added dropwise. The yellow-orange suspension was stirred at -10°C for 30 min. p-Anisonitrile (68 mmol) in 15 mL THF was added dropwise. The reaction mixture was allowed to warm to room temperature, and stirred an additional 18 hours. Water was added and the pH adjusted to 7 with glacial acetic acid, whereupon the product precipitated from solution. The product was collected by filtration, washed with water and decolorized by digestion in hot methanol to afford 62%.

7-Hydroxy-6-methoxycarbonyl-5-methoxycarbonylmethyl-1-indanone (Scheme 2.14) - acylation of the methyl acetoacetate dianion with dimethylamides followed by intramolecular cyclization to indanones (from Yamaguchi et al.).[143]

Under a nitrogen atmosphere, methyl acetoacetate (2.5 mmol) in THF (2 mL) was added to a suspension of sodium hydride in (2.75 mmol) in THF (1 mL) at 0°C. After 10 min, n-BuLi (2.5 mmol in hexane) was added, and the mixture stirred for 10 min at 0°C. On cooling to -78°C, a THF (2 mL) solution of N,N,N′,N′-tetramethylsuccinamide (0.3 mmol) and BF_3-Et_2O (2.9 mmol) were added successively. The mixture was slowly warmed to 0°C for 1 h, and 1 M HCl was added. After stirring for 3 h at room temperature, the organic materials were extracted with ethyl acetate, washed with aqueous sodium bicarbonate and brine, dried over sodium sulfate, and concentrated. The residue was chromatographed using silica gel (ethyl acetate-benzene) to give the solid product in 67% yield after recrystallization from ether-hexane. mp = 129-131°C.

6-Benzoyl-5-phenylthian-3-one 1-oxide (Scheme 2.19) - formation of a β-ketosulfoxide dianion and reaction with an α,β-unsaturated ester (from Tamura et al.).[194]

ω-(Methylsulfinyl)-acetophenone (2.2 mmol) was dissolved in THF (70 mL) and chilled to -78°C under argon. n-BuLi (4.6 mmol in hexanes) was added dropwise with stirring, and after stirring an additional 20 min, ethyl cinnamate (1.1 mmol) was added. The reaction was allowed to come to room temperature during 1 h with stirring. The reaction was quenched with by the addition of 10 g of wet silica gel, which was removed by filtration. The solvent was removed, and the residue was chromatographed on silica gel using ethyl acetate as eluent to give 68% of the cyclized product. mp = 156-157°C.

4-(tert-Butoxycarbonyl)-3-methoxy-1-(4-methoxyphenyl)-2-azetidinones (Eq. 2.64) - Intramolecular oxidative cyclization of a remote 1,4-(C,C)-dianion (from Hiyama et al.).[221]

To a THF (4 mL) solution of N-[(tert-butoxycarbonyl)methyl]-N-(4-methoxyphenyl)-methoxyacetamide (0.22 mmol) was added t-BuLi (0.48 mmol in pentane) at -78°C. The mixture was stirred at -78°C for 1 h and then poured into a THF suspension of tetraphenylphosphonium bromide (0.48 mmol) which had been cooled to -78°C. After stirring for 30 min, the mixture was cooled to -95°C and N-iodosuccinimide (0.29 mmol) dissolved in THF (1 mL) was added all at once. Thee mixture was warmed to -70°C over a period of 1 h, then poured into saturated NH_4Cl and extracted with ethyl acetate. The organic layer was washed successively with saturated $Na_2S_2O_3$, saturated sodium bicarbonate, and brine, dried over magnesium sulfate, and concentrated in vacuo. The residue was purified by preparative thin layer chromatography (ether:hexane, 99:1) to give 78% of a 12:1, *cis*:*trans* mixture. Cis isomer mp = 101-102°C.

p-Diethylbenzene (Scheme 2.25) - alkylation of the p-xylene dianion with alkylsulfonates (from Bates and Ogle).[260]

In a 250 mL flask fitted with a serum capped side arm, potassium tert-butoxide (22 mmol) and o-xylene (10 mmol) were added to hexane (50 mL). The system was equipped with a condenser. On addition of n-BuLi (22 mmol), a red-orange precipitate formed immediately. The suspension was allowed to reflux under argon for 1 h and cooled to room temperature before the dark-red intermediate (70%) was filtered and rinsed with pentane. To a suspension of the p-xylyl dianion (10 mmol) in hexane (50 mL) was added excess dimethyl sulfate (4 mL). The suspension was then refluxed under argon until the anion color disappeared. Excess dimethylsulfate was destroyed by adding 50 mL of 20% NaOH/EtOH solution and refluxing for 3 h. The solution was washed with water, dried over magnesium sulfate, and the solvent removed in vacuo. The product was isolated by bulb-to-bulb distillation in 63% yield.

References

1. **Hauser, C.R. and Harris, T.M.** Condensation at the methyl group rather than the methylene group of benzoyl- and acetylacetone through intermediate dipotassio salts, *J. Am. Chem. Soc.* **80**, 6360 (1958).
2. **Harris, T.M. and Harris, C.M.** The γ-alkylation and γ-arylation of dianions of β-dicarbonyl compounds, *Org. React.* **17**, 155 (1969).
3. **Thompson, C.M. and Green, D.L.C.** Recent advances in dianion chemistry, *Tetrahedron* **47**, 4223 (1991).
4. **Rabinovitz, M. and Cohen, Y.** Probing the nature of polycyclic conjugated dianions; from carbocyclic to heterocyclic dianions: NMR Studies, π-delocalization and electronic structure, *Tetrahedron* **44**, 6957 (1988).
5. **Kaiser, E.M., Petty, J.D. and Knutson, P.L.A.** Di- and polyalkali metal derivatives of heterofunctionally substituted molecules, *Synthesis* 509 (1979).
6. **Tanaka, K. and Kaji, A.** The synthetic utility of sulfur-containing dianions, *Sulfur Rep.* **1**, 97 (1980).
7. **von Rague Schleyer, P.** Lithium synthetic reagents: dimerization and intramolecular association. Double bridging in "dianions," *Pure Appl. Chem.* **55**, 355 (1983).
8. **Grovenstein, E. Jr.** Structures of organodialkali metal compounds ("dianions"), *Stud. Org. Chem.* **5**, 175 (1987).
9. **Stowell, J.C.** *Carbanions in Organic Synthesis*; J. Wiley & Sons, New York, 1979.
10. **Crowley, P.J., Leach, M.R., Meth-Cohn, O. and Wakefield, B.J.** Do the compounds $PhCH_2Y$ give geminal dianions on addition of two equivalents of strong base, *Tetrahedron Lett.* **27**, 2909 (1986).
11. **Kaiser, E.M., Solter, L.E., Schwarz, R.A., Beard, R.D. and Hauser, C.R.** gem-Dialkali derivatives of certain functionally substituted organic compounds, *J. Am. Chem. Soc.* **93**, 4237 (1971).
12. **Kowalski, C., O'Dowd, M.L., Burke, M.C. and Fields, K.W.** α-Ketone dianions. New reactive intermediates, *J. Am. Chem. Soc.* **102**, 5411 (1980).
13. **Kowalski, C., Weber, A.E. and Fields, K.W.** α-Keto dianion precursors via conjugate addition to cyclic α-bromo enones, *J. Org. Chem.* **47**, 5088 (1982).
14. **Kowalski, C. and Fields, K.W.** Additions of α-keto dianions to sterically congested carbonyls, *J. Am. Chem. Soc.* **104**, 1777 (1982).
15. **Kowalski, C. and Fields, K.W.** Alkynoate anions via a new rearrangement: the carbon analogue of the Hoffman reaction, *J. Am. Chem. Soc.* **104**, 321 (1982).
16. **Kowalski, C., Haque, M.S. and Fields, K.W.** Ester homologation via α-bromo α-keto dianion rearrangement, *J. Am. Chem. Soc.* **107**, 1429 (1985).
17. **Kowalski, C. and Haque, M.S.** Bromomethyl ketones and enolates: alternative products from ester homologation reactions, *J. Org. Chem.* **50**, 5140 (1985).
18. **Sampson, P., Hammond, G.B. and Wiemer, D.F.** A new synthesis of β-keto phosphonates and α-keto silanes, *J. Org. Chem.* **51**, 4342 (1986).
19. **Gornowicz, G.A. and West, R.** Polylithium compounds IV. Polylithiation of nitriles and the preparation of trisilylynamines, *J. Am. Chem. Soc.* **93**, 1714 (1971).
20. **Kaiser, E.M. and Hauser, C.R.** Primary and secondary ionizations of α-hydrogens of phenylacetonitrile by n-butyllithium, *J. Am. Chem. Soc.* **88**, 2348 (1966).
21. **Stevens, R.V. and Wentland, M.P.** Thermal rearrangement of cyclopropylimines. IV. Total synthesis of dl-mesembrine, *J. Am. Chem. Soc.* **90**, 5580 (1968).
22. **Seebach, D. and Lehr, F.** α,α-Doubly deprotonated nitroalkanes. Enhancement of the C-nucleophilicity of nitronates, *Angew. Chem. Int. Ed. Engl.* **88**, 540 (1976).
23. **Seebach, D., Henning, R., Lehr, F. and Gonnermann, J.** Carbon alkylation of α,α- and α,β-doubly deprotonated nitroalkanes, *Tetrahedron Lett.* 1161 (1977).
24. **Seebach, D., Henning, R. and Lehr, F.** Double deprotonation of 3-nitropropene and 4-nitro-1-butene: derivatives of their dianions as novel reagents for C-C linkage, *Angew. Chem.* **90**, 479 (1978).
25. **Yamada, K., Tanaka, S., Kohmoto, S. and Yamamoto, M.** Novel regioselective generation of nitroalkane dianions, *J. Chem. Soc. Chem. Commun.* 110 (1989).

26. **Eyer, M. and Seebach, D.** *l*-2-Nitro-1,3-alkanediols by stereoselective addition of nitroethanol to aldehydes. On the asymmetric electrophilic addition to double bonds, *J. Am. Chem. Soc.* **107**, 3601 (1985).
27. **Williams, T.M. and Mosher, H.S.** A chiral nitronate dianion from d-glyceraldehyde. Enantiospecific syntheses of 2,3-dideoxy-3-nitro furanosides and pyranosides, *Tetrahedron Lett.* **26**, 6269 (1985).
28. **Truce, W.E. and Christensen, L.W.** Preparation and characterization of an α-chloro dicarbanion from 4-[(chloromethyl)sulfonyl]morpholine, *J. Chem. Soc. Chem. Commun.* 588 (1971).
29. **Pascali, V., Tangari, V.P. and Umani-Ronchi, A.** The dianion of benzyl phenyl sulphone. A general method for obtaining α,β-unsaturated phenyl sulphones from aldehydes and ketones, *J. Chem. Soc. Perkin Trans. I.* 1166 (1973).
30. **Bosworth, N. and Magnus, P.** An α-sulphonyl dianion and its E1cB elimination, *J. Chem. Soc. Perkin Trans. I.* 2319 (1973).
31. **Trost, B.M. and Merlic, C.A.** Diastereoselectivity control elements. Acyclic diastereocontrol in formation and reaction of γ-hydroxysulfones, *J. Am. Chem. Soc.* **110**, 5216 (1988).
32. **Giblin, G.M.P., Ramcharitar, S.H. and Simpkins, N.S.** Synthesis and stereoselective chemistry of a novel cyclopentadienyl sulphone, *Tetrahedron Lett.* **29**, 4197 (1988).
33. **Eisch, J.J., Dua, S.K. and Behrooz, M.** [(Phenylsulfonyl)methylene]dilithium as a novel cyclizing and homologizing reagent for bifunctional organic substrates, *J. Org. Chem.* **50**, 3576 (1985).
34. **Mussatto, M.C., Savoia, D., Trombini, C. and Umani-Ronchi, A.** Reactions between 1,1-dilithioalkyl phenyl sulfones and omega-bromo esters. Synthesis of cyclic vinyl ethers, *J. Chem. Soc. Perkin Trans. I.* 260 (1980).
35. **Najdi, S. and Kurth, M.J.** Synthesis of (2R,6R)-(-)-lupetidine; 2,6-disubstituted piperidines as potentially useful C2-symmetric chiral reagents, *Tetrahedron Lett.* **31**, 3279 (1990).
36. **Simpkins, N.S.** *Sulphones in Organic Synthesis*, Pergamon Press, Oxford, U.K., 1993.
37. **McCombie, S.W., Shankar, B.B. and Ganguly, A.K.** Cyclofunctionalization of epoxy alcohol derivatives. 4. Cyclization of sulfonylacetate dianions: a synthesis of "MeBHT," *Tetrahedron Lett.* **30**, 7029 (1989).
38. **Vollhardt, J., Gais, H-J. and Lukas, K.L.** Dilithio(phenylsulfonyl)trimethylsilylmethane: Synthesis, $^{13}C/^{1}H$-NMR characterization, and lithium-titanium exchange, *Angew. Chem. Int. Ed. Engl.* **24**, 696 (1985).
39. **Tanaka, K., Matsui, S. and Kaji, A.** Lithiation of alkoxyalkyl phenyl sulfones. New approach to acyl anion synthesis, *Bull. Chem. Soc. Japan* **53**, 3619 (1980).
40. **Lee, J.W. and Oh, D.Y.** A convenient one-pot synthesis of α-functionalized α,β-unsaturated sulfones, *Synth. Commun.* **20**, 273 (1990).
41. **Yokoyama, Y., Koizumi, T. and Kikuchi, O.** A direct NMR observation of stilbene dianion in solution, *Chem. Lett.* 2205 (1991).
42. **Yokoyama, Y. and Kikuchi, O.** Behavior of triphenylethylene dianion in diethyl ether solution, *Chem. Lett.* 1149 (1992).
43. **Fujiwara, H., Yoshino, A., Yokoyama, Y. and Takahashi, K.** An NMR study of the carbanions produced from triphenylethylene and its methyl-substituted derivatives, *Bull. Chem. Soc. Japan* **65**, 2162 (1992).
44. **Hauser, C.R. and Harris, T.M.** Dicarbanions of dibenzylketone, dibenzylsulfone, and α,β,β-triphenylpropionitrile, *J. Am. Chem. Soc.* **79**, 6342 (1957).
45. **W.G. Kofron, Dunnavant, W.R. and Hauser, C.R.** Alkylations of mono- and dipotassio-2,3,3-triphenylpropionitrile. Equilibrium factor in benzhydrylation of 1,2-dianion. Dehydrocyanation, *J. Org. Chem.* **27**, 2737 (1962).
46. **Kaiser, E.M. and Hauser, C.R.** Influence of metallic cation in carbonyl addition reactions of certain nitriles and sulfones with benzophenone, *J. Am. Chem. Soc.* **89**, 4566 (1967).
47. **Garratt, P.J. and Zahler, R.** The dianion of dimethyl cyclobut-3-ene-1,2-dicarboxylate. Aromaticity vs. coulombic repulsion, *J. Am. Chem. Soc.* **100**, 7753 (1978).
48. **Sekiguchi, A., Nakanishi, T., Kabuto, C. and Sakurai, H.** Molecular structure of phenyl[tris(trimethylsilyl)]ethylene dilithium complex. An orthogonal dianion, *Chem. Lett.* 867 (1992).

49. **Long, N.R. and Rathke, M.W.** Isolation and reactions of the lithium di-enolate of diethyl succinate, *Synth. Commun.* **11**, 687 (1981).
50. **Pohmakotr, M., Reutrakul, V., Phongpradit, T. and Chansri, A.** Dianion of diethyl succinate. Reactions with alkylating agents and carbonyl compounds, *Chem. Lett.* 687 (1982).
51. **Furuta, K., Misumi, A., Mori, A., Ikeda, N. and Yamamoto, H.** A new cyclopentanone annulation of α-bromomethylacrylates, *Tetrahedron Lett.* **25**, 669 (1984).
52. **Furuta, K., Ikeda, N. and Yamamoto, H.** Dianion of 1,2-bis(4′,4′-dimethyl-2′-oxazolin-2′-yl)ethane. A versatile synthetic reagent for annulation, *Tetrahedron Lett.* **25**, 675 (1984).
53. **Koft, E.R. and Williams, M.D.** The formation and alkylation of α-ketoamide dianions, *Tetrahedron Lett.* **27**, 2227 (1986).
54. **Garratt, P.J. and Zahler, R.** Reactions of the dimethyl cyclobutane-1,2-dicarboxylate dianion. A simple synthesis of 1,2-dihydrocyclobuta[b]naphthalene-3,8-dione, *Tetrahedron Lett.* 73 (1979).
55. **Bilyard, K.G., Garratt, P.J., Underwood, A.J. and Zahler, R.** Vicinal diether dianions. Annelation with allylic and benzylic α,ω-dihalides. Synthesis of tricyclobuta[1,2:3,4:6,7]naphthalene, *Tetrahedron Lett.* 1815 (1979).
56. **Noire, P.D. and Franck, R.W.** Regiochemically controlled syntheses of 3-hydroxynaphtho[b]cyclobutene, *Tetrahedron Lett.* **23**, 1031 (1982).
57. **Garratt, P.J. and Porter, J.R.** Reaction of dilithiated 1,2-diesters with β-bromoethyloxirane. Synthesis of annelated γ-lactones, *Tetrahedron Lett.* **28**, 351 (1987).
58. **Garratt, P.J., Pielke, M. and Porter, J.R.** A short, efficient synthesis of (+/-)valerane, *Tetrahedron Lett.* **28**, 589 (1987).
59. **Doecke, C.W. and Garratt, P.J.** Dicyclobuta[a,c]anthracene, *J. Chem. Soc. Chem. Commun.* 873 (1981).
60. **Bilyard, K.G. and Garratt, P.J.** Vicinal diester dianions. Annelation with ω-bromoesters. A simple preparation of bicyclo[4.4.0]decan-2-one derivative valuable in sesquiterpene synthesis, *Tetrahedron Lett.* **22**, 1755 (1981).
61. **Jun, J-G. and Mundy, B.P.** A facile synthesis of propellanes via dianion chemistry, *Bull. Korean Chem. Soc.* **8**, 310 (1987).
62. **Corey, E.J. and Su, W-G.** A novel, one-step annulation process for the synthesis of ring-fused 2-cyclopentenones, *Tetrahedron Lett.* **28**, 5241 (1987).
63. **Ide, J., Inoue, K. and Sakai, K.** Alkylation of the dianion of 2,4-dimethoxycarbonyl-3-methyl-2-cyclopentanone and its application to the synthesis of the Corey's type PG intermediates, *Chem. Lett.* 747 (1978).
64. **Koreeda, M., Liang, Y. and Akagi, H.** Easy generation of the dianions of 3-isobutoxycyclopent-2-en-1-ones and their reactions, *J. Chem. Soc. Chem. Commun.* 449 (1979).
65. **Koreeda, M. and Mislankar, S.G.** Chemistry of the dianions of 3-heteroatom substituted cyclopent-2-en-1-ones: An expedient route to *dl*-coriolin, *J. Am. Chem. Soc.* **103**, 7203 (1983).
66. **Trost, B.M. and Latimer, L.H.** Generation and alkylation of the dianion (homoenolate) of a 1-indanone, *J. Org. Chem.* **42**, 3212 (1977).
67. **Garratt, P.J. and Hollowood, F.** Stereoselective control in the alkylation and annelation of anions and dianions derived from 5-norbornene-2,3-dicarboximides, *J. Org. Chem.* **47**, 68 (1982).
68. **Beslin, P. and Dlubala, A.** Reactivity of a novel ambident dianion formed by double deprotonation of β-thiosubstituted dithiopropanoates: a lithio acrylate equivalent, *Tetrahedron Lett.* **27**, 1687 (1986).
69. **Hampton, K.G., Harris, T.M. and Hauser, C.R.** Mono- versus dialkylation of acetylacetone with alkyl halides through dialkali salts. Metallic cation effect, *J. Org. Chem.* **30**, 61 (1965).
70. **Meyer, R.B and Hauser, C.R.** Alkylations of acetoacetone and certain other β-diketones at the terminal methyl group through dicarbanions, *J. Org. Chem.* **25**, 158 (1960).
71. **Hampton, K.G., Harris, T.M. and Hauser, C.R.** 2-Nonanedione, *Org. Syn.* **47**, 92 (1967).
72. **Hampton, K.G., Harris, T.M. and Hauser, C.R.** Relative reactivities of certain disodio-β-diketones toward alkyl halides in liquid ammonia, *J. Org. Chem.* **31**, 1035 (1966).
73. **Harris, T.M. and Carney, R.L.** Synthesis of 3,5,7-triketoacids and esters and their cyclizations to resorcinol and phloroglucinol derivatives. Models of biosynthesis of phenolic compounds, *J. Am. Chem. Soc.* **89**, 6734 (1967).

74. **Pendarvis, R.O. and Hampton, K.G.** Synthesis of dioxocarboxylic acids, *J. Org. Chem.* **39**, 2289 (1974).
75. **Hampton, K.G. and Flannery, R.E.** Alkylation of disodiopentane-2,4-pentanedione with haloacetals, *J. Chem. Soc. Perkin Trans. I.* 2308 (1973).
76. **Sugiyama, N., Takano, T. and Kashima, C.** Synthesis of omega-phenyl-3-acetoxy-2,4-dioxoalkanes and their enol contents, *Bull. Chem. Soc. Japan* **40**, 2698 (1967).
77. **Hampton, K.G., Harris, T.M. and Hauser, C.R.** Alkylation of unsymmetrical β-diketones through isomeric disodio slats in liquid ammonia. Structural effects on ease of dianion formation, *J. Org. Chem.* **31**, 663 (1966).
78. **Harris, T.M. and Hauser, C.R.** Benzylation at the terminal methyl group of certain unsymmetrical β-diketones through one of two intermediate dicarbanions, *J. Am. Chem. Soc.* **81**, 1160 (1959).
79. **Hampton, K.G., Harris, T.M. and Hauser, C.R.** Arylation of β-diketone dicarbanions with diaryliodonium chlorides to form γ-derivatives, *J. Org. Chem.* **29**, 3511 (1964).
80. **von Schriltz, D.M., Hampton, K.G. and Hauser, C.R.** Stereospecificity in the α-phenylethylation of the dicarbanions of certain β-diketones in liquid ammonia, *J. Org. Chem.* **34**, 2509 (1969).
81. **Stork, G., Kraus, G.A. and Garcia, G.A.** Regiospecific aldol condensations of the kinetic lithium enolates of methyl ketones, *J. Org. Chem.* **39**, 3459 (1974).
82. **Greene, J.C., Wolfe, J.F. and Hudlicky, T.** Reaction of dialkali salts of benzoylacetone with 2-chloroquinoline. Evidence for a SRN1 mechanism in heteroaromatic nucleophilic substitution, *J. Org. Chem.* **37**, 3199 (1972).
83. **Hay, J.V., Hudlicky, T. and Wolfe, J.F.** SRN1 mechanism in heteroaromatic nucleophilic substitution. Photostimulation and entrainment of the reaction of lithoacetone with 2-chloroquinoline, *J. Am. Chem. Soc.* **97**, 374 (1975).
84. **O'Sullivan, W.I. and Hauser, C.R.** 2-Acetyl-6-methyoxycoumaran-3-one. Benzylation at the terminal methyl group, *J. Org. Chem.* **25**, 839 (1960).
85. **Wolfe, J.F., Harris, T.M. and Hauser, C.R.** Condensations at the methyl group of ethyl acetoacetate by means of potassium amide or sodium hydride, *J. Org. Chem.* **29**, 3249 (1964).
86. **Davies, M.J., Heslin, J.C. and Moody, C.J.** Rhodium carbenoid mediated cyclisations. Part 4. Synthetic approaches to oxepanes related to zoapatanol, *J. Chem. Soc. Perkin Trans. I.* 2473 (1989).
87. **Huckin, S.N. and Weiler, L.** Alkylation of dianions of β-keto esters, *J. Am. Chem. Soc.* **96**, 1082 (1974).
88. **Weiler, L.** Alkylation of the dianion of β-keto esters, *J. Am. Chem. Soc.* **92**, 6702 (1970).
89. **Winkler, J.D. and Hey, J.P.** Inside-outside stereoisomerism III. The synthesis of *trans*-bicyclo-[4.3.1]decan-10-one, *Tetrahedron Lett.* **29**, 4691 (1988).
90. **Schlessinger, R.H., Wood, J.L., Poss, A.J., Nugent, R.A. and Parsons, W.H.** An intramolecular Diels-Alder route to racemic quadrone, *J. Org. Chem.* **48**, 1147 (1983).
91. **Sum, P-H. and Weiler, L.** Ring formation via β-keto ester dianions, *Can. J. Chem.* **55**, 996 (1977).
92. **Sum, P-H. and Weiler, L.** Alkylation of β-keto ester dianions with α-chloroethers, *J. Chem. Soc. Chem. Commun.* 91 (1977).
93. **Casey, C.P. and Marten, D.F.** A new general synthesis of isoprenoid chains employing the reaction of dialkylcuprates with β-acetoxy α,β-unsaturated esters, *Synth. Commun.* **3**, 321 (1973).
94. **Sum, P-H. and Weiler, L.** Stereoselective synthesis of latia luciferin, *Tetrahedron Lett.* 707 (1979).
95. **Mori, K. and Fujiwhara, M.** Synthesis of enantiomerically pure (10R,11S)-(+)-juvenile hormone 0, *Leibigs Ann. Chem.* 41 (1989).
96. **Clark, R.D. and Heathcock, C.H.** Synthesis of spiro lactones and ketones via conjugate addition to cyclopropylmalonates and β-keto esters, *Tetrahedron Lett.* 529 (1975).
97. **Skeean, R.W., Trammel, G.L. and White, J.D.** Cyclization of olefinic β-ketoesters. A novel synthesis of $\Delta^{8(14)}$-podocarpen-13-one, *Tetrahedron Lett.* 525 (1976).
98. **Winkler, J.D., Hey, J.P. and Darling, S.D.** Studies directed toward the synthesis of taxane diterpenes: a remarkable rearrangement, *Tetrahedron Lett.* **27**, 5959 (1986).

99. **Kang, S.H., Kim, W.J. and Chae, Y.B.** Total synthesis of (+/-)-trinoranastreptene, *Tetrahedron Lett.* **29**, 5169 (1988).
100. **Pinnick, H.W. and Chang, Y-H.** Cyclopropyl imine rearrangement. Total synthesis of (+/-)-isoretronecanol and (+/-)-trachelanthamidine, *Tetrahedron Lett.* 837 (1979).
101. **Piers, E. and Ruediger, E.H.** Thermal rearrangement of 1(carbomethoxy)-6-*exo*-(1-alkenyl)bicyclo[3.1.0]hex-2-ene systems. A convenient synthesis of functionalized bicyclo[3.2.1]octa-2,6-dienes, *J. Org. Chem.* **45**, 1727 (1980).
102. **Corey, E.J. and Hahl, R.W.** Synthesis of a limonoid, azadiradione, *Tetrahedron Lett.* **30**, 3023 (1989).
103. **Asao, K., Iio, H. and Tokoroyama, T.** Total synthesis of (+/-)-ageline, a physiologically active constituent of *Agelas* sponges, *Tetrahedron Lett.* **30**, 6401 (1989).
104. **Sims, R.J., Tischler, S.A. and Weiler, L.** Lactone synthesis via the intramolecular alkylation of β-keto ester dianions, *Tetrahedron Lett.* **24**, 253 (1983).
105. **Nakayama, M., Ohira, S., Okamura, Y. and Soga, S.** Total synthesis of (+/-)-coronafacic acid, *Chem. Lett.* 731 (1981).
106. **Salzmann, T.N., Ratcliffe, R.W. and Christensen, B.G.** Total synthesis of (-)-homothienamycin, *Tetrahedron Lett.* **21**, 1193 (1980).
107. **Hubbard, J.S. and Harris, T.M.** Condensations of β-ketoamides at the γ-position via dianions, *Tetrahedron Lett.* 4601 (1978).
108. **Booth, P.M., Fox, C.M.J. and Ley, S.V.** Preparation of acyltetronic acids using t-butyl acetothioacetate: total synthesis of the fungal metabolites carolic, carlosic, and carlic acids, *J. Chem. Soc. Perkin Trans. I.* 121 (1987).
109. **Berry, N. M., Darey, M.C.P. and Harwood, L.M.** Direct 4-alkylation of 1,3-cyclohexanediones via dianionic species, *Synthesis* 476 (1986).
110. **Berry, N.M., Casy, G. and Harwood, L.M.** Efficient regioselective preparation of 2,6-disubstituted derivatives of 3-acetoxy-5,5-dimethyl-2-cyclohexenone, *Synthesis* **12**, 978 (1988).
111. **Mellor, M. and Pattenden, G.** Direct alkylation of dianions of cyclic β-diketones. A general synthesis of 4-substituted cyclopentane-1,3-diones, *Synth. Commun.* **9**, 1 (1979).
112. **Wachter, M.P. and Harris, T.M.** Condensations at the 6α-position of triacetic lactone via the dianion, *Tetrahedron* **26**, 1685 (1970).
113. **Giles, M., Hadley, M.S. and Gallagher, T.** β-Ketoester dianions as regiospecific enolate equivalents for N-substituted pyrrolidin-3-ones, *J. Chem. Soc. Chem. Commun.* 1047 (1990).
114. **Lermer, L., Neelnad, E.G., Ounsworth, J.P., Sims, R.J., Tischler, S.A. and Weiler, L.** The synthesis of β-keto lactones via cyclization of β-keto ester dianions or the cyclization of Meldrum's acid derivatives, *Can. J. Chem.* **70**, 1427 (1992).
115. **Kieczykowski, G., Roberts, M.R. and Schlessinger, R.H.** Ring-opening reactions of α-oxy epoxides with tert-butyldilithioacetoacetate, *J. Org. Chem.* **43**, 788 (1978).
116. **Mori, K., Sasaki, M., Tamada, S., Suguro, T. and Masuda, S.** Synthesis of optically active 2-ethyl-1,6-dioxaspiro[4.4]nonane (chalcogran), the principal aggregation pheromone of *Pityogenes Chalcographus* (L), *Heterocycles* **10**, 111 (1978).
117. **Yamaguchi, M. and Hirao, I.** Reaction of oxiranes with dianion of methyl acetoacetate in the presence of boron trifluoride etherate, *Chem. Lett.* 337 (1985).
118. **Lygo, B. and O'Connor, N.** Stereoselective synthesis of (+/-)-methyl nonactate and (+/-)-methyl 8-*epi*-nonactate, *Tetrahedron Lett.* **28**, 3597 (1987).
119. **Lygo, B., O'Connor, N. and Wilson, P.R.** Reaction of dianions derived from β-ketoesters with epoxides-utility in the preparation of synthetically useful tetrahydrofurans, *Tetrahedron* **44**, 6881 (1988).
120. **Lygo, B.** Stereoselective synthesis of (+/-)-methyl homononactate and (+/-)-methyl 8-*epi*-homononactate, *Tetrahedron* **44**, 6889 (1988).
121. **Huckin, S.N. and Weiler, L.** Aldol type condensation of β-keto ester dianions, *Can. J. Chem.* **52**, 2157 (1974) and references therein.
122. **Harris, T.M., Boatman, S. and Hauser, C.R.** Condensations of carbonyl compounds at the γ-position of β-ketoaldehydes by means of potassium amide or sodium hydride, *J. Am. Chem. Soc.* **87**, 3186 (1965).
123. **Ansell, J.M., Hassner, A. and Burkholder, W.E.** Synthesis of the drugstore beetle pheromone, *Tetrahedron Lett.* 2497 (1979).

124. **Light, R.J. and Hauser, C.R.** Condensations of dialkali β-diketones with ketones or aldehydes to form hydroxy β-diketones. Dehydration products. Equilibrium factors, *J. Org. Chem.* **26**, 1716 (1961).

125. **Stipanovic, R.D and Calton, G.J.** Synthesis of yangolide, *Tetrahedron Lett.* 357 (1970).

126. **Light, R.J., Harris, T.M. and Hauser, C.R.** Carbonyl versus conjugate addition of dipotassiobenzoylacetone with certain α,β-unsaturated ketones, *J. Org. Chem.* **26**, 1344 (1961).

127. **Beam, C.F., Bissell, R.L. and Hauser, C.R.** Cyclization of the less nucleophilic center of a β-diketone dicarbanion through a dicarbanion-benzyne intermediate, *J. Org. Chem.* **35**, 2083 (1970).

128. **Hampton, K.G., Harris, T.M., Harris, C.M. and Hauser, C.R.** Condensations at the terminal methyl group of 1,3,5-triketones by means of sodamide in liquid ammonia, *J. Org. Chem.* **30**, 4263 (1965).

129. **Peterson, J.R., Winter, T.J. and Miller, C.P.** Efficacious entry into substituted 5,6-dihydro-4-hydroxy-2H-pyran-2-ones and 2,3-dihydro-4H-pyran-4-ones. Utilizing ketonic dianions, *Synth. Commun.* **18**, 949 (1988).

130. **Kresze, G., Morper, M. and Bijev, A.** Synthese von Prodigiosin-analogen, *Tetrahedron Lett.* 2259 (1977).

131. **Kondo, K., Umemoto, T., Yako, K. and Tunemoto, D.** A new stereoselective synthesis of dl-prostaglandin $F_{2\alpha}$, *Tetrahedron Lett.* 3927 (1978).

132. **Suzuki, Y., Knoche, H.W. and Daly, J.M.** Analogs of host-specific phytotoxin produced by *Helminthosporium maydis*, Race T, *Bioorg. Chem.* **11**, 300 (1982).

133. **Batty, D., Crich, D. and Fortt, S.M.** Synthesis of 1α25-dihydroxyvitamin D_3 a ring model by an acyl radical cyclisation, *J. Chem. Soc. Chem. Commun.* 1366 (1989).

134. **Nagasawa, K., Zako, Y., Ishihara, H. and Shimizu, I.** Stereoselective synthesis of 1α-hydroxyvitamin D_3 A-ring synthons by palladium-catalyzed cyclization, *Tetrahedron Lett.* **32**, 4937 (1991).

135. **Willard, P.G., Grab, L.A. and de Laszlo, S.E.** Synthesis of costatolide, a halogenated monoterpenoid from the red alga *Plocamium costatum*, *J. Org. Chem.* **48**, 1123 (1983).

136. **Kashihara, H., Shinoki, H., Suemune, H. and Sakai, K.** Novel synthesis of indan derivatives, *Chem. Pharm. Bull (Japan)* **34**, 4527 (1986).

137. **Clarke, T. and Ley, S.V.** Use of S-t-butyl acetothioacetate in the preparation of milbemycin seco-analogues, *J. Chem. Soc. Perkin Trans. I.* 131 (1987).

138. **Berube, G. and Fallis, A.G.** An intramolecular cycloaddition-sigmatropic rearrangement approach to (+/-) gascardic acid, *Tetrahedron Lett.* **30**, 4045 (1989).

139. **Matsumoto, M. and Watanabe, N.** Synthesis of 4-(4-indolyl)-3-oxobutanoic acid derivatives from 5-halo-4-oxo-4,5,6,7-tetrahydroindoles, *Heterocycles* **24**, 3149 (1986).

140. **Honda, T., Kondoh, H., Okuyama, A., Hayakawa, T., Tsubuki, M. and Nagase, H.** Chelation controlled aldol reaction of tetronic acid dianion with ketones, *Heterocycles* **33**, 67 (1992).

141. **Huckin, S.N. and Weiler, L.** Claisen condensation of the dianion of β-keto ester, *Tetrahedron Lett.* 2405 (1972).

142. **Yamaguchi, M., Hasebe, K. and Minami, T.** A biomimetic synthesis of polycyclic polyoxygenated aromatic compounds via polyketides, *Tetrahedron Lett.* **27**, 2401 (1986).

143. **Yamaguchi, M., Shibato, K., Nakashima, H. and Minami, T.** Synthesis of phenols by the intramolecular condensation of β,β,δ,δ'-tetraoxoalkanedioates. A novel BF_3-promoted Claisen condensation of acetoacetate dianion with esters and amides, *Tetrahedron* **44**, 4767 (1988).

144. **Yamaguchi, M., Horiguchi, A., Ikeura, C. and Minami, T.** A synthesis of aryl C-glycosides via polyketides, *J. Chem. Soc. Chem. Commun.* 434 (1992).

145. **Narasimhan, N.S. and Ammanamanchi, R.** Mechanism of acylation of dilithium salts of β-keto esters: an efficient synthesis of anibine, *J. Org. Chem.* **48**, 3945 (1983).

146. **Hanamoto, T. and Hiyama, T.** A facile entry to β,δ-dihydroxy esters, *Tetrahedron Lett.* **29**, 6467 (1988).

147. **Hoffman, R.W., Ladner, W., Steinbach, K., Massa, W., Schmidt, R. and Snatzke, G.** Absolute Konfiguration von Stigobinon, *Chem. Ber.* **114**, 2786 (1981).

148. **Sandifer, R.A., Bhattacharya, A.K. and Harris, T.M.** Acylation of multiple anions of poly-β-ketones by hydroxy- and alkoxy benzoates. Cyclization of the resultant tetraketones to benzophenones and xanthones, *J. Org. Chem.* **46**, 2260 (1981).
149. **Ehrig, V. and Seebach, D.** Reaktionen polymetallierter Carbonoylverbindungen mit Nitroolefin. Eine neue RingschluB- und Anellierungsmethode, *Chem Ber.* **108**, 1961 (1975).
150. **Pyne, S.G., Spellmeyer, D.C., Chen, S. and Fuchs, P.L.** Conjugate addition of β-keto ester dianions to vinyl sulfones: a new procedure for seven ring annulation. Synthesis of a chiral cytochalisin C intermediate via an intramolecular Diels-Alder reaction of a chiral Z diene, *J. Am. Chem. Soc.* **104**, 5728 (1982).
151. **Solladie, G. and Ghiatou, N.** Synthesis of optically active β,γ-diketo-p-tolylsulfoxides, *Tetrahedron Asymmetry* **3**, 33 (1992).
152. **Bringmann, G.** A short biomimetic synthesis of the isoquinoline and the naphthalene moieties of ancistrocladus alkaloids from common β-carbonyl precursors, *Tetrahedron Lett.* **23**, 2009 (1982).
153. **Bringmann, G.** Aufbau und Cyclisierung zentral modifizierter β-Pentaketone: Synthese monocyclischer Isochinolinaalkaloid-Vorstufen, *Leibigs Ann. Chem.* 2105 (1985).
154. **Hubbard, J.S. and Harris, T.M.** α,α′-Dianions of aliphatic ketones and the 1,3,5-trianion of 2,4-pentanedione: strongly nucleophilic carbonyl synthons, *Tetrahedron Lett.* 4601 (1978).
155. **Trimitsis, G.B., Hinkley, J.M., TenBrink, R., Poli, M., Gustafson, G., Erdman, J. and Rop, D.** Unusual mode of alkylation of certain ketone dianions, *J. Am. Chem. Soc.* **99**, 4838 (1977).
156. **Trimitsis, G.B., Hinkley, J.M., TenBrink, R., Faburada, A.L., Anderson, R., Poli, M., Christian, B., Gustafson, G., Erdman, J. and Rop, D.** Formation and reactions of 1-phenyl-2-propanone dianion and related systems with electrophilic reagents, *J. Org. Chem.* **48**, 2957 (1983).
157. **Bays, J.P.** Phenylacetone dianion: alkylation with iodomethane, *J. Org. Chem.* **43**, 38 (1978).
158. **Hauser, C.R. and Harris, T.M.** Formation and certain condensations of the dicarbanions of dibenzylketone and dibenzylsulfone by means of potassium amide in liquid ammonia, *J. Am. Chem. Soc.* **81**, 1154 (1959).
159. **Jiang, J.B. and Urbanski, M.J.** Synthesis of novel bicyclic 2-amino-4(1H)-pyridones. Reaction of lactim ethers with α-cyanoacetone dianions, *Tetrahedron Lett.* **26**, 259 (1985).
160. **Vinick, F.J. and Gschwend, H.W.** Preparation and reactivity of acetoacetonitrile dianion, *Tetrahedron Lett.* 4221 (1978).
161. **Boatman, S. and Hauser, C.R.** Condensations at the terminal methyl group of anilino derivatives of benzoylacetone and acetoacetone through their dialkali salts in liquid ammonia, *J. Org. Chem.* **31**, 1785 (1966).
162. **Denmark, S.E. and Ares, J.J.** Stereoselective alkylation of chiral α-nitro keto imine dianions. Observations on the role of amide base, *J. Am. Chem. Soc.* **110**, 4432 (1988).
163. **Ballini, R., Bartoli, G., Castagnani, R., Marcantoni, E. and Petrini, M.** Direct α′ alkylation of 2-nitro ketone dianions, *SynLett.* 64 (1992).
164. **Field, L. and McFarland, J.W.** Grignard reagents of sulfones II. Reactions with carbonyl compounds, *J. Am. Chem. Soc.* **75**, 5582 (1953).
165. **Kaiser, E.M. and Hauser, C.R.** 1,3-Dianion of dimethyl sulfone and 1,1-dianion of benzyl phenylsulfone, *Tetrahedron Lett.* 3341 (1967).
166. **Kaiser, E.M., Beard, R.D. and Hauser, C.R.** Preparation and reactions of the mono- and dialkali salts of dimethyl sulfone, dimethyl sulfoxide, and related compounds, *J. Organomet. Chem.* **59**, 53 (1973).
167. **Miles, M.L. and Hauser, C.R.** Terminal benzoylation of certain β-ketosulfones to form diketosulfones by means of sodium hydride. Dibenzoylation of dimethylsulfone, *J. Org. Chem.* **29**, 2329 (1964).
168. **Truce, W.E. and Klinger, T.C.** Synthesis and configurational assignments of diasteromeric β-hydroxy sulfones, *J. Org. Chem.* **35**, 1834 (1970).
169. **Hauser, C.R. and Harris, T.M.** Formation and certain condensations of the dicarbanions of dibenzylketone and dibenzylsulfone by means of potassium amide in liquid ammonia, *J. Am. Chem. Soc.* **81**, 1154 (1959).
170. **Hauser, C.R. and Harris, T.M.** Dicarbanions of dibenzylketone, dibenzylsulfone, and α,β,β-triphenylpropionitrile, *J. Am. Chem. Soc.* **79**, 6342 (1957).

171. **Bordwell, F.G., Philips, D.D. and Williams, J.M. Jr.** The Ramberg-Backlund reaction of benzyl α-halobenzyl and halomethyl sulfones, *J. Am. Chem. Soc.* **90**, 435 (1968).
172. **Paquette, L.A., Meisinger, R.H. and Gleiter, R.** Unsaturated heterocyclic systems. XCII. Bis deprotonation of bridged bicyclic sulfones. Electronic interactions in rigid bridgehead α-sulfonyl dicarbanions as a function of carbon framework unsaturation, *J. Am. Chem. Soc.* **95**, 5414 (1973).
173. **Paquette, L.A., Photis, J.M. and Ewing, G.D.** 1,2,3,4-Tetramethylcyclooctatetraene and 1,2,3,8-tetramethylcyclooctatetraene. Stable [8]annulene bond shift isomers, *J. Am. Chem. Soc.* **97**, 3538 (1975).
174. **Photis, J.M. and Paquette, L.A.** Lithium aluminum hydride promoted ring contraction of cyclic five-membered α-sulfonyl carbanions and its application to the synthesis of 1,2-dialkylcyclobutenes, *J. Am. Chem. Soc.* **96**, 4715 (1974).
175. **Grossert, J.S., Buter, J., Asveld, E.W.H. and Kellogg, R.M.** Sulfonyl 1,3-dicarbanions. Scope and limitation of formation. Oxidation with copper salts to form alkenes, *Tetrahedron Lett.* 2805 (1974).
176. **O'Sullivan, W.I., Travares, D.F. and Hauser, C.R.** Some condensations at the methylene and terminal methyl groups of benzenesulfonyl acetone through its mono- and dipotassio salts, *J. Am. Chem. Soc.* **83**, 3452 (1961).
177. **Cooke, F. and Magnus, P.** Intermolecular and intramolecular alkylation of mono- and dianions derived from β-ketosulfone, *J. Chem. Soc. Chem. Commun.* 519 (1976).
178. **Carroll, N.M. and O'Sullivan, W.I.** Benzylation and benzoylation of methyl phenacylsulfone, *J. Org. Chem.* **30**, 2830 (1965).
179. **Scholz, D.** New synthetic methods. 7. β-Ketosulfones, useful ethylenediide equivalents for the preparation of olefins and the synthesis of 6-nonen-1-ol, the sex attractant of the Mediterranean fruit fly, *Leibigs Ann. Chem.* 98 (1983).
180. **Grossert, J.S., Hoyle, J. and Hooper, D.L.** The alkylation of an α,α′-dianion in a β-oxosulfone leading to the preparation of axial 2-substituted thiane 1,1-dioxides, *Tetrahedron* **40**, 1135 (1984).
181. **Lygo, B. and O'Connor, N.** Epoxide ring opening by dianions derived from β-ketosulfones. Synthetic studies on the 7-hydroxy-4-oxopropenoate system and medium ring lactones, *SynLett.* **1**, 282 (1990).
182. **Baldwin, J.E., Adlington, R.M., Ichikawa, Y. and Kneale, C.J.** A new method for C-C coupling of terminal alkenes via a sulfonylation-alkylation-desulfonylation sequence: synthesis of E- and Z-α-bisabolenes, *J. Chem. Soc. Chem. Commun.* 702 (1988).
183. **Belletire, J.L. and Spletzer, E.G.** Oxidative coupling V. carbon-carbon bond formation from β-ketosulfone dianions, *Synth. Commun.* 17, 1701 (1987).
184. **Kaiser, E.M. and Beard, R.D.** 1,3-Dialkali salts of dimethylsulfoxide, *Tetrahedron Lett.* 2583 (1968).
185. **Grieco, P.A. and Pognowski, C.S.** Alkylation of the dianion of β-keto sulfoxides. A versatile synthesis of phenyl (2-oxoalkyl)sulfoxides. A general route to ketones, 1,4-diketones, and aldols, *J. Org. Chem.* **39**, 732 (1974).
186. **Grieco, P.A., Boxler, D. and Pognowski, C.S.** Dianions of β-ketosulphoxides. A new, general synthesis of vinyl ketones, *J. Chem. Soc. Chem. Commun.* 497 (1974).
187. **Kuwajima, I. and Iwasawa, H.** Generation of the dianion of phenylsulfinyl acetone and its reactions, *Tetrahedron Lett.* 107 (1974).
188. **Forbes, C.P., Michau, J.D., van Ree, T., Wiechers, A. and Woudenberg, M.** The total synthesis of rac-sceletium alkaloid A_4 and its 3′-demethoxy analog, *Tetrahedron Lett.* 935 (1976).
189. **Forbes, C.P., Wenteler, G.L. and Wiechers, A.** Further applications of endocyclic enamine-enone annulations, *Tetrahedron* **34**, 487 (1978).
190. **Stevens, R.V. and Lee, A.W.M.** Studies on the stereochemistry of nucleophilic additions to tetrahydropyridinium salts. A stereospecific total synthesis of one of the stereoisomers of gephyrotoxin 223, *J. Chem. Soc. Chem. Commun.* 103 (1982).
191. **Corey, E.J. and Chaykovsky, M.** *J. Am. Chem. Soc.* Methylsulfinyl carbanion (CH_3SOCH_2-). Formation and applications to organic synthesis, **87**, 1345 (1965).

192. **Grieco, P.A. and Pognowski, C.S.** Synthesis of 5-alkyl-Δ^2-cyclopentenones and 6-alkyl-Δ^2-cyclohexenones, *J. Chem. Soc. Chem. Commun.* 72 (1975).
193. **Tamura, Y., Shindo, H., Uenishi, J-I. and Ishibashi, H.** Reaction of the α,α'-dianion of β-ketosulfoxides, *Chem. Pharm. Bull. (Japan)* **27**, 3186 (1979).
194. **Tamura, Y., Shindo, H. and Ishibashi, H.** Annelation of α,β-unsaturated esters to thian-3-one 1-oxides using dianion of ω-(methylsulfinyl)acetophenone, *J. Heterocyclic Chem.* **17**, 1637 (1980).
195. **Kosugi, H., Hoshinio, K. and Uda, H.** Cyclisation of polyenic β-ketosulfoxides. A new route to optically active intermediates for cyclic terpenoid synthesis, *J. Chem. Soc. Chem. Commun.* 1173 (1991).
196. **Baldwin, J.E. and Tzodikov, N.R.** Substituent rearrangement and elimination during noncatalyzed Fischer indole synthesis, *J. Org. Chem.* **42**, 1878 (1977).
197. **Taylor, J.D. and Wolf, J.F.** Lithiotriphenylphosphinioacetamide, a new type of 1,3-dianion, *J. Chem. Soc. Chem. Commun.* 876 (1972).
198. **Cooke, M.P. and Goswami, R.** Formation of an ylid dianion, and its application to the construction of an eight-membered ring, *J. Am. Chem. Soc.* **95**, 7891 (1973).
199. **Grieco, P.A. and Pognowski, C.S.** Alkylation of the dianion of β-keto phosphonates. A versatile synthesis of dimethyl (2-oxoalkyl)phosphonates, *J. Am. Chem. Soc.* **95**, 3071 (1973).
200. **Grieco, P.A. and Finkelhoor, R.S.** Dianions of β-keto phosphonates. A two-step synthesis of (+/-)-*ar*-tumerone, *J. Org. Chem.* **38**, 2909 (1973).
201. **Fouque, D., About-Jaudet, E. and Collignon, N.** A convenient synthesis of diethyl-2,4-dioxoalkylphosphonates, *Synth. Commun.* **22**, 219 (1992).
202. **Wada, E., Kanemasa, S. and Tsuge, O.** New synthesis of 2-oxo-3-alkenylphosphonates and hetero Diels-Alder reactions with vinyl ethers leading to 5-substituted 2-phosphinyl-2-cyclohexen-1-ones, *Bull. Chem. Soc. Japan* **62**, 860 (1989).
203. **van den Goorbergh, J.J. and van den Gen, A.** Ethyl-4-diphenylphosphinoyl-2-oxobutanoate: a convenient reagent for the synthesis of γ,δ-unsaturated β-ketoesters, *Tetrahedron Lett.* **21**, 3621 (1980).
204. **Bodalski, R., Pietrusiewicz, M., Monkiewicz, J. and Koszuk, J.** A new efficient synthesis of substituted Nazarov reagents. A Wittig-Horner-Emmons approach, *Tetrahedron Lett.* **21**, 2287 (1980).
205. **Taylor, E.C. and Davies, H.M.L.** Rhodium (II) acetate catalyzed reaction of ethyl 2-diazo-3-oxopent-4-enoates: Simple routes to 4-aryl-2-hydroxy-1-naphthoates and β,γ-unsaturated esters. The dianion of ethyl 4-(diethylphosphono)acetoacetate as a proprionate homoenolate equivalent, *Tetrahedron Lett.* **24**, 5453 (1983).
206. **Moorhoff, C.M. and Schneider, D.V.** Comments on the reaction of ethyl 4-(diethoxyphosphinyl)-3-oxobutanoate and related phosphonate esters with enals, *Tetrahedron Lett.* **28**, 559 (1987).
207. **Goswami, R.** A novel β-ketophosphonate 1,4-dianion. Tin-lithium exchange, *J. Am. Chem. Soc.* **102**, 5974 (1980).
208. **Stowell, M.H.B., Witte, J.F. and McClard, R.W.** The phosphonylphosphinyl dianion: A convenient synthon for the preparation of biologically interesting phosphonylphosphinyl (P-C-P-C) compounds, *Tetrahedron Lett.* **30**, 411 (1989).
209. **Brenner, S. and Bovete, H.** Monometallation of 3-butenenitrile, *Tetrahedron Lett.* 1377 (1974).
210. **Padwa, A., Bullock, W.H., Dyszlewski, A.D., McCombie, S.W., Shankar, B.B. and Ganguly, A.** Alkylation of 2-oxy-substituted 1-sulfonylallyl and 1-sulfonylvinyl anions. New routes to functionalized carbocycles and dihydrofurans, *J. Org. Chem.* **56**, 3556 (1991).
211. **Bhanu, S. and Scheinman, F.** 1,3-Disubstitution reactions of 1,3-dilithiopropyne. "One-pot" linear chain extensions at both termini of propyne, *J. Chem. Soc. Chem. Commun.* 817 (1975).
212. **Hirai, K. and Kishida, Y.** Chemistry of 2-substituted thiothiazoline. III. Reactivities of dianion of 2-propargylthiothiazoline and related 2-alkynylthiothiazoline derivatives, *Tetrahedron Lett.* 2117 (1972).
213. **Hirai, K., Matsuda, H. and Kishida, Y.** Chemistry of 2-substituted thiothiazoline. V. Reaction of dianion of 2-propargylthiothiazoline with benzaldehyde, *Chem. Pharm. Bull. (Japan)* **20**, 2067 (1972).

214. **Kreiser, W. and Wurziger, H.** Olefins by 2,3-sigmatropic rearrangements, *Tetrahedron Lett.* 1669 (1975).
215. **Stoit, A.R. and Pandit, U.K.** Models of folate cofactors. An approach to deethyleburnamonine, *Tetrahedron* **45**, 849 (1989).
216. **Wade, P.A., Kondracki, P.A. and Carroll, P.J.** Polynitro-substituted strained ring compounds. 2. 1,2-dinitrospiropentanes, *J. Am. Chem. Soc.* **113**, 8807 (1991).
217. **Takahashi, H., Oshima, H., Yamamoto, H. and Nozaki, H.** *J. Am. Chem. Soc.* **95**, 5803 (1973).
218. **Kende, A.S. and DeVita, R.J.** Synthesis of the fused bicyclic lactam-lactone terminus of neooxazolomycin by a novel dianion cyclocondensation, *Tetrahedron Lett.* **1988**, *29*, 2521.
219. **Utaka, M., Hojo, M. and Takeda, A.** Regio- and diastereoselective aldol condensation of 1,2-cyclohexanedione dianions with aldehydes, *Chem. Lett.* 1471 (1985).
220. **Kawabata, T., Sumi, K. and Hiyama, T.** A new synthesis of β-lactams through stereoselective oxidative coupling of the dianion of acyclic amides, *J. Am. Chem. Soc.* **111**, 6843 (1989).
221. **Kawabata, T., Minami, T. and Hiyama, T.** Stereoselective synthesis of β-lactams by oxidative coupling of dianions of acyclic tertiary amides, *J. Org. Chem.* **57**, 1864 (1992).
222. **Sudweeks, W.B. and Broadbent, H.S.** Generalized synthesis of γ-diketones. I. Addition of dimetalloacetylides to aldehydes. II. Dialkylation of bisdithianes, *J. Org. Chem.* **40**, 1131 (1975).
223. **Lansbury, P.T., Spagnuolo, C.J., Zhi, B. and Grimm, E.L.** Symmetry-enhanced dicarbanion annulation for latent cycloalkenones: applications to alkaloids and terpenes, *Tetrahedron Lett.* **31**, 3965 (1990).
224. **Pohmakotr, M. and Seebach, D.** Doubly deprotonated allylacetophenone, *Angew. Chem. Int. Ed. Engl.* 320 (1979).
225. **Seebach, D., Pohmakotr, M., Schregenberger, C., Weidmann, B., Mali, R.S. and Pohmakotr, S.** d^5-Reactions of doubly deprotonated γ,δ-unsaturated carbonyl derivatives with electrophiles. A novel approach to the synthesis of tetrahydrofuran and tetrahydropyran derivatives, *Helv. Chim. Acta* **65**, 419 (1982).
226. **Cohen, T., Ritter, R.H. and Ouellette, D.** Single and double ring-closure reactions of dianions of bis(diphenyl thioacetals). A new, synthetically useful principle of carbene generation and intramolecular capture, *J. Am. Chem. Soc.* **104**, 7142 (1982).
227. **Mulvaney, J.E., Gardlund, Z.G., Gardlund, S.L. and Newton, D.J.** Addition and metalation reactions of organolithium compounds. I. Products. *J. Am. Chem. Soc.* **88**, 476 (1966).
228. **Hommes, H., Verkruijsse, H.D. and Brandsma, L.** Synthesis of o-halogenophenylacetylenes via the dianion of phenylacetylene, *Tetrahedron Lett.* **22**, 2495 (1981) and references therein.
229. **Crowther, G.P., Sundberg, R.J. and Sarpeshkar, A.M.** Dilithiation of aromatic ethers, *J. Org. Chem.* **49**, 4657 (1984).
230. **Greving, B., Wolterman, A. and Kauffmann, T.** Preparation and properties of cycloocta[1,2-b:4,3-b':5,6-b'':8,7-b''']tetrathiophene, *Angew. Chem. Int. Ed. Engl.* **13**, 467 (1974).
231. **Kauffmann, T., Greving, B., Koenig, J., Mitschker, A. and Wolterman, A.** Synthesis of heterocyclic cyclopolyaromatics containing similar kinds of nuclei, *Angew. Chem. Int. Ed. Engl.* **14**, 713 (1975).
232. **Brown, P.J., Jones, D.N., Khan, M.A., Meanwell, N.A. and Richards, P.J.** Conjugate addition to α,β-unsaturated sulfoxides: syntheses of cyclopentenones and 9-deoxyprostanoids, *Tetrahedron Lett.* **24**, 405 (1983).
233. **Brown, P.J., Jones, D.N., Khan, M.A., Meanwell, N.A. and Richards, P.J.** Alkenyl sulfoxides as precursors to cyclopentenones and prostanoid β-side chain, *J. Chem. Soc. Perkin Trans. I.* 2049 (1984).
234. **Tanaka, K., Terauchi, M. and Kaji, A.** (Allylthio)acetate dianion as a new and convenient reagent for the stereoselective synthesis of (2E,4E)dienoates from alkyl halides, *Chem. Lett.* 315 (1981).
235. **Aoki, S., Fujimora, T. and Nakamura, E.** A protective strategy in carbene complex chemistry. Synthesis of functionalized Fischer carbene complexes via dianion formation, *J. Am. Chem. Soc.* **114**, 2895 (1992).

236. **Hartman, G.D. and Halczenko, W.** Regioselective generation and trapping of mono- and dianions of 3-arylsulfonylfurans. Bidentate carbanion stabilization via sulfones, *Tetrahedron Lett.* **28**, 3241 (1987).
237. **Chakrabartty, S.K. and Levine, R.** Chemistry of pyrazine and its derivatives. XII. Reaction of acetonylpyrazine with phenyllithium in the presence and absence of methyl benzoate, *J. Heterocyclic Chem.* **1**, 196 (1964).
238. **Kamal, M.R. and Levine, R.** The chemistry of pyrazine and its derivatives VIII. Diketones derived from the monoacylation of 2-methyl-6-(acylmethyl)pyrazines and the diacylation of 2,6-dimethylpyrazine, *J. Org. Chem.* **29**, 191 (1964).
239. **Boekelheide, V., Anderson, P.H. and Hylton, T.A.** Transformation of sulfide linkages to carbon-carbon double bonds. Syntheses of [2.2]metaparacyclophane-1,9-dienes, *J. Am. Chem. Soc.* **96**, 1558 (1974).
240. **Davis, S.E., Shealy, N.L., Shealy, K.D., Shaffer, L.M. and Beam, C.F.** Aldol-type condensation of C(α)-dianions of azines, *Can. J. Chem.* **56**, 1236 (1978).
241. **Tanimoto, S., Oida, T., Hatanaka, K. and Sugimoto, T.** Generation of 1,2-ethanebis(trithiocarbonic acid) dianion from 2,2′-[1,2-ethanediylbis(thio)]-bis-1,3-dithiolane and its reaction with alkyl halides, *Tetrahedron Lett.* **22**, 655 (1981).
242. **Takahashi, T., Nagashima, T., Ikeda, H. and Tsuji, J.** An important role played by dianions in the highly efficient macrocycle formation by an intramolecular alkylation method: Application to the synthesis of zearalenone, *Tetrahedron Lett.* **23**, 4361 (1982).
243. **Takahashi, T., Yamada, H., Haino, T., Kido, Y. and Fukuzawa, Y.** Dianion methodology for the construction of large-membered ring systems: application to the synthesis of muscone, *Tetrahedron Lett.* **33**, 7561 (1992).
244. **You, S., Gubler, M. and Neuenschwander, M.** Oxidative coupling of α,ω-di(cyclopentadienyl)alkyl halides, *Chimia* **45**, 387 (1991).
245. **Houminer, Y. and Williams, D.L.** Studies on alkylpyrazine dianions. Preparation of 5,6,11,12-tetrahydrocycloocta[1,2-b:5,6-b′]dipyrazine, *J. Heterocyclic Chem.* **22**, 373 (1985).
246. **Bates, R.B., Camou, F.A., Siahaan, T.J. and White, J.J.** Resonance-stabilized dianions in synthesis, *Symp. Adv. Carbanion Chem.* 570 (1985).
247. **Barry, C.E. III., Bates, R.B., Beavers, W.A., Camou, F.A., Gordon, B. III., Hsu, H.F-J., Mills, N.S., Ogle, C.A., Siahaan, T.J., Suvannachut, K., Taylor, S.R., White, J.J. and Yager, K.M.** Delocalized carbanions in synthesis, *SynLett.* 207 (1991).
248. **March, J.** *Advanced Organic Chemistry; Reactions, Mechanism and Structure*; 3rd Ed., J. Wiley & Sons, New York, 1985.
249. **Bates, R.B., Beavers, W.A., Greene, M.G. and Klein, J.H.** *J. Am. Chem. Soc.* **96**, 5640 (1974).
250. **Bahl, J.J., Bates, R.B., Beavers, W.A. and Mills, N.S.** Reactions of delocalized dicarbanions with dihalides, *J. Org. Chem.* **41**, 1620 (1976).
251. **Bahl, J.J., Bates, R.B. and Gordon, B. III.** 2,3-Dimethylenebutadiene dianion: convenient procedure for allylic metalation of conjugated dienes, *J. Org. Chem.* **44**, 2290 (1979).
252. **Bates, R.B., Beavers, W.A., Gordon, B. III. and Mills, N.S.** Reactions of 2-methyleneallyl dianions with aldehydes, ketones, epoxides, and allyl halides, *J. Org. Chem.* **44**, 3800 (1979).
253. **Bates, R.B., Gordon, B. III., Keller, P.C., Rund, J.V. and Mills, N.S.** Reactions of dianions with nitriles. A new pyridine synthesis, *J. Org. Chem.* **45**, 168 (1980).
254. **Bates, R.B., Gordon, B. III., Highsmith, T.K. and White, J.J.** Reactions of 2,3-dimethylenebutadiene dianion with electrophiles. Synthesis and conformations of 2,3-disubstituted-1,3-butadienes, *J. Org. Chem.* **49**, 2981 (1984).
255. **Ramon, D.J. and Yus, M.** Naphthalene-catalyzed lithiation of 3-chloro-2-chloromethylpropene: a Barbier-type practical alternative to the triphenylenemethane dianion, *Tetrahedron Lett.* **33**, 2217 (1992).
256. **Gordon, B. III., Blumenthal, M., Mera, A.E. and Kumpf, R.J.** Delocalized carbanions: the facile synthesis of disubstituted butadienes and diols, *J. Org. Chem.* **50**, 1540 (1985).
257. **Magid, R.M. and Wilson, S.E.** Thermal reactions of allylic mono- and dianions, *Tetrahedron Lett.* 19 (1971).
258. **Suga, K. and Watanabe, S.** The reaction of diisoprene dianion with ketones, *Synthesis* 91 (1971).

259. **Meyer, S.D., Mills, N.S., Runnels, J.B., de la Torre, B., Channing Rudd, C. and Johnson, D.K.** Dianions of 1,4-Bis(methylene)cyclohexane. Conformational effects in cyclic and acyclic dilithiated hydrocarbons, *J. Org. Chem.* **56**, 947 (1991).
260. **Bates, R.B. and Ogle, C.A.** Arenes disubstituted with primary alkyl groups from xylylene dianions, *J. Org. Chem.* **47**, 3949 (1982).
261. **Bates, R.B., Camou, F.A., Kane, V.V., Mishra, P.K. and White, J.J.** Preparation and reactions of symmetrical dimethylenebiphenyl dianions; conformations of [0.n]-o-cyclophanes. A simple NMR method for determining twist angles in biphenyls, *J. Org. Chem.* **54**, 311 (1989).
262. **Bates, R.B., Gangwar, S., Kane, V.V., Suvannachut, K. and Taylor, S.R.** Ethoxymetacyclophanes from 2,6-dimethylanisole, *J. Org. Chem.* **56**, 1696 (1991).
263. **West, R. and Jones, P.C.** Polylithiation. II. Polylithiation of toluene and the formation of poly(trimethylsilyl)toluenes, *J. Am. Chem. Soc.* **90**, 2656 (1968).
264. **Ogle, C.A., Black, K.C. and Sims, P.F.** Preparation and ring-opening of benzylic and allylic cyclopropyl dianions, *J. Org. Chem.* **57**, 3499 (1992).

Chapter 3

DIANIONS FROM O, C-BIS DEPROTONATIONS

3.1 INTRODUCTION

Although C,C-bis-deprotonations are considered as the most prominent among dianion intermediates, O,C-dianions must be equally recognized as powerful synthetic tools. O,C-Dianions, which are formed by initial deprotonation of an OH-containing functional group, have a long and interesting history, probably sparked by the development of the Ivanov reagents (magnesium-based α-carboxylate dianions) over 60 years ago.[1] To date, four functional group types have served as starting materials for the production of O,C-dianions including carboxylic acids, alcohols, phenols, and oximes (below), and their dianion chemistry assembled in Sections 3.2 though 3.5. Each of these O,C-dianions have been the subject of prior reviews varying in the extent of coverage.[1-5] Sections 3.2 (carboxylic acid

carboxylic acid-based dianion

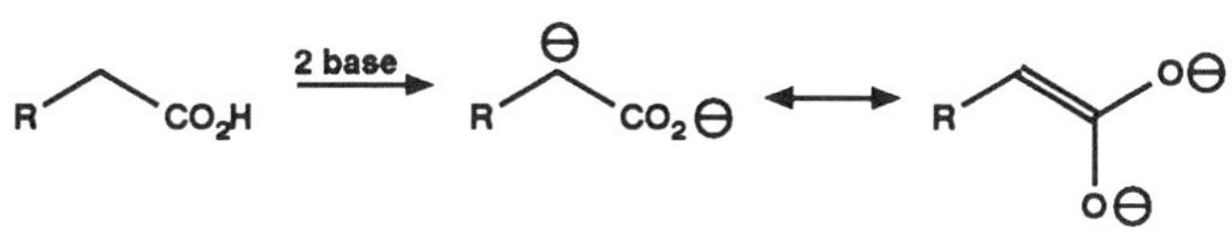

hydroxyl-based dianion

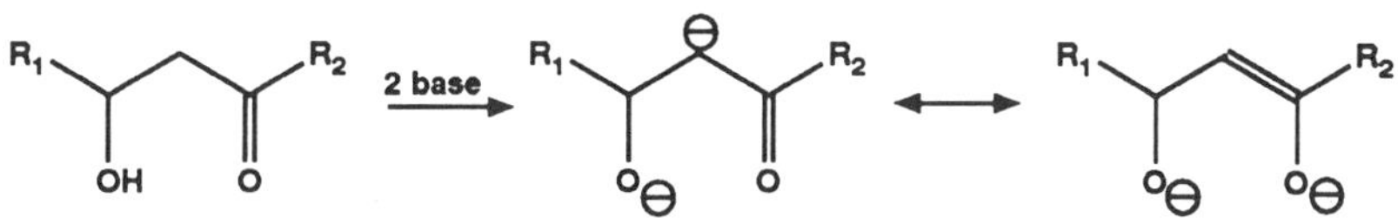

phenol-based dianion

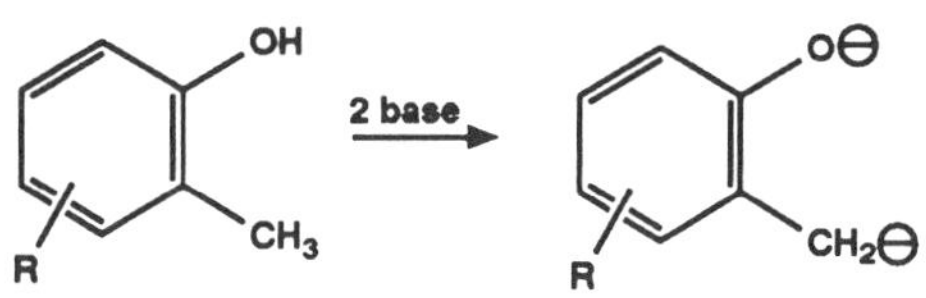

oxime-based dianion

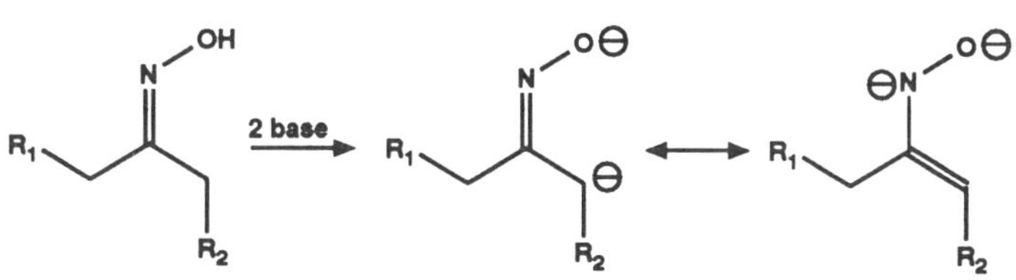

dianions) and 3.3 (hydroxyl-based dianions) are further subdivided reflecting the diversity of dianion types within these categories. Carboxylic acid-based O,C-dianions may be derived from aliphatic or aromatic/heterocyclic carboxylic acids, and the carbanion site may be generated at enolate (α), remote, benzylic, or ring metalated positions giving a variety of possible combinations for synthetic use. Hydroxyl-based O,C-dianions give a similar medley of intermediates but their development has been more evident over the past two decades. The use of hydroxyl-based dianions that possess an asymmetric center at the alcohol carbon has been a particularly interesting and powerful variation offering new pathways in asymmetric control. The phenol/cresol-based dianions and the oxime-based dianions have been explored to a more modest level, but each leads to regiospecifically defined products that are largely unobtainable though other means. The exclusive production of the *syn*-oxime dianion has been a noteworthy synthetic addition to dianion chemistry.

This chapter intends to provide the reader with coverage of the four major types of O,C-dianions and their use in the construction of carbon-carbon bonds. Most of the coverage presented is founded in the methodology rather than application in total synthesis although several examples have been included that show their utility in synthetic planning. As a result of this bias, many examples of O,C-dianions within synthetic sequences have not been included in this chapter. However, it is hoped that the method and/or dianion type excluded from these studies are covered within the following sections.

3.2 CARBOXYLIC ACID-BASED DIANIONS

The most thoroughly studied member of the O,C-based dianions are those prepared via initial deprotonation of a carboxylic acid -OH group. The relatively low pKa (approx. 4) in combination with the reduced reactivity of the resultant carboxylate carbonyl makes carboxylic acids an ideal precursor to dianions.

Carboxylic acid-based O,C-dianions (Fig 3.1) have been covered in several reviews of related reaction intermediates,[1-5] including comprehensive and specific coverage by Petragnani and Yonashiro.[4] In this latter review, carboxylic acid dianions and their ester enolate counterparts were categorized according to their reaction with a particular electrophile type. This chapter aims to complement that approach by classifying carboxylic acid O,C-dianions according to the carbanion stabilizing group followed by various uses in synthesis. This organizational approach was taken to include the many new carboxylic acid types that have gained attention since the prior review was published.

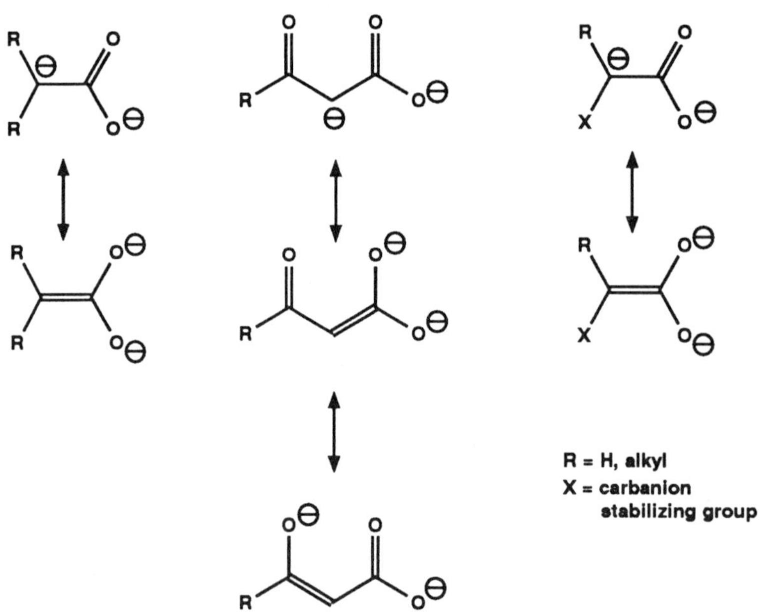

Figure 3.1. Classes of α-carboxylate dianions

There has been a dramatic increase in the use of carboxylic acid dianions over the past two decades. This growth is represented both in the breadth of dianion types and in expanded utility of the most commonly used carboxylic acid dianion intermediate, the α,α′- or 1,3-(O,C)-type. Section 3.2.2 will cover the major uses of α-carboxylate dianions which, in combination with the several excellent prior reviews, should provide the experimenter with a comprehensive account of their applications in organic synthesis. However, given the many new and ingenious variations of carboxylic acid-based dianions, each carboxylic acid-based dianion type will be further subdivided into categories for clarity. For example, α-carboxylate dianions will be further classified according to the absence or presence of specific carbanion-stabilizing groups giving: α-carboxylate dianions with no stabilizing groups (simple aliphatic or hydrocarbon α,α′-carboxylic acid dianions), β-carbonyl-based α-carboxylate dianions (e.g., malonates), and α-carboxylate dianions bearing carbanion-stabilizing groups other than carbonyl (Fig. 3.1).

Following coverage of the α-carboxylic acid dianions, carboxylic acid dianions that are not of the α,α′-type will be reviewed. Three main classes of carboxylic acid dianions will be covered: π-delocalized (or resonance stabilized), benzoic acid-based, and remote carboxylate (Fig 3.2).

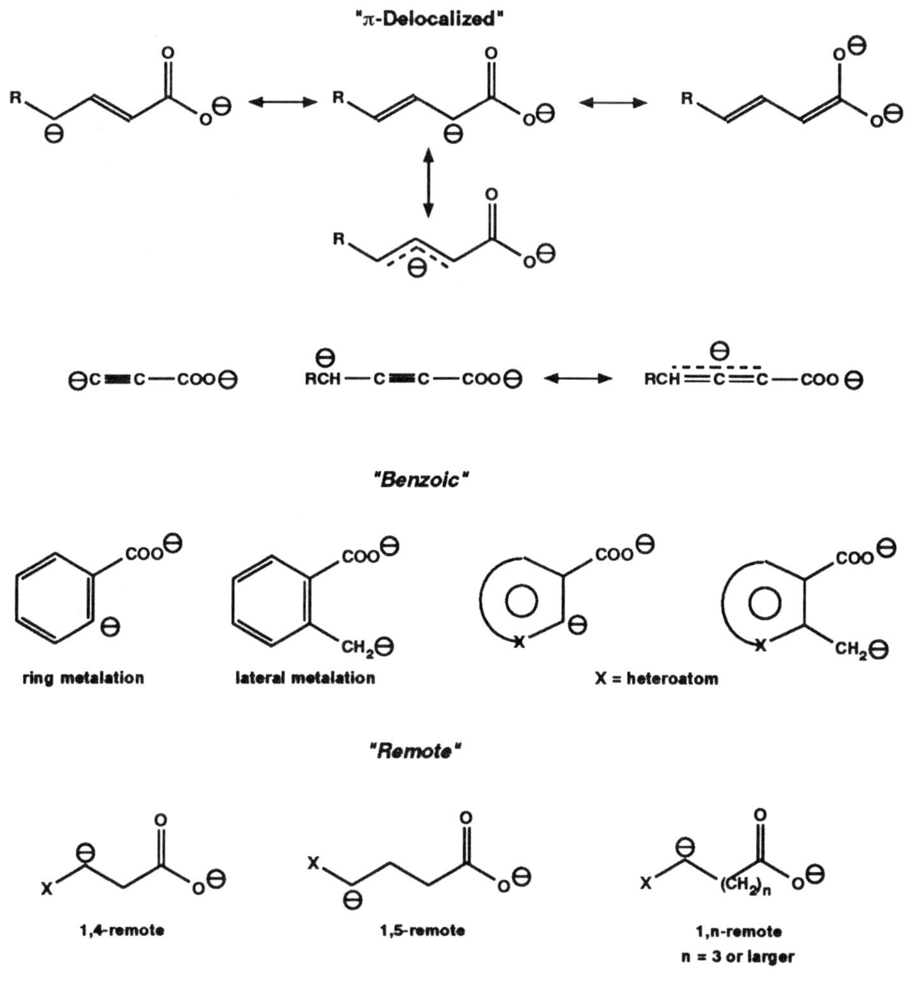

Figure 3.2. Carboxylic acid dianions other than the α-carboxylate type

Resonance-stabilized carboxylic acid dianions result from O,C-deprotonations of α,β-unsaturated- or β,γ-unsaturated carboxylic acids, the second deprotonation occurring at an allylic position. The resultant dianions give carbanions that, through resonance contributors, may undergo reaction at either an enolate (α) or remote (γ) site. Alkynoic acid-based dianions, which also fall into the category of π-delocalized dianions, result from initial COOH deprotonation followed by deprotonation at the terminal alkyne (RC≡CH) position or at an available propargylic (γ) position. Benzoic acid-based carboxylic acid dianions, and their heterocyclic congeners, which also may be considered as remote dianions, clearly differ from prior dianions in that the second deprotonation generally occurs at a site influenced by the aromatic moiety, namely, ring metalation or lateral (benzylic) deprotonation. Last, carboxylic acid dianions that do not fit any of these prior categories are generally classified as remote, since at least one intervening neutral fragment is present between the carboxylate carbonyl and the newly formed carbanion site. These remote carboxylate dianions are divided into 1,4-, 1,5- and 1,Z-types (Z = 6 or greater) indicating the location of the carbanion relative to the oxyanion position.

3.2.1 Generation of carboxylic acid-based dianions

The ease of deprotonation of carboxylic acids permits for a range of bases and countercations to be employed, both factors altering the reactivity of the ensuing dianion. Such reactivities can be further attenuated by additives (HMPT), which have been shown to increase the reactivity of certain carboxylate dianions by dissociating aggregate complexes. Several investigators also have chosen to enter the dianion-forming step by starting with preformed, isolable carboxylate salts. This strategy permits dianion generation with one equivalent of alkyllithium or amide base, which reduces alkyllithium use, and is experimentally easy to conduct.

When generation of a carboxylic acid O,C-dianion is being considered for use in synthesis, several features of these dianions are worth noting. First, the pKa of the enolate C-H of carboxylic acid dianions (also known as 1,3- or enolate-type) is similar to the pKa of simple ester enolates (pKa approx. 24),[6] indicating that bases of similar strength could be used to form α-carboxylate dianions. However, pKa determinations conducted using an equilibration method in 20% HMPA/THF gave values as high as 31.[6] Regardless of pKa, investigators have conducted enolate deprotonation using mostly hindered amide bases (e.g., LDA). Second, the experimenter should be advised that certain carboxylate salts (particularly lithium as cation) may undergo addition reactions with organolithium reagents to form ketones following workup. This addition process may dominate the reaction path under certain conditions (Eq. 3.1), and has been found to be dependent upon the stability of the O,O-dilithio intermediate.[7] Third, prior reviews of carboxylic acid dianions have mostly

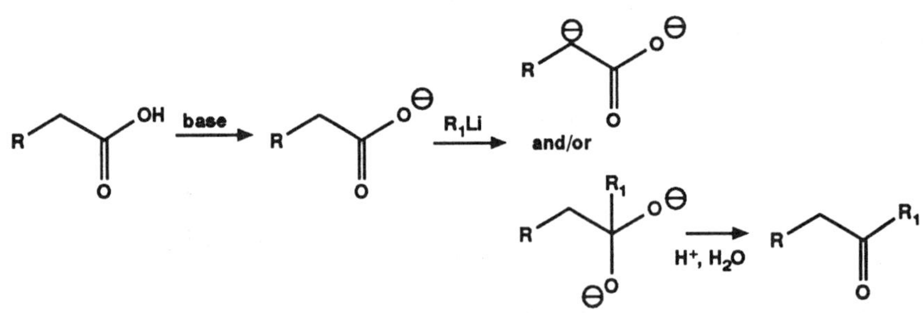

(3.1)

covered the α-carboxylate type (α,α′- or 1,3), where the second deprotonation occurs at the enolate position. The past two decades have revealed that carboxylic acid dianions in which the carbanion occurs at a site other than at the enolate position also have become popular. These newer carboxylic acid based dianions have greatly expanded the versatility of this dianion class, but also indicates that possible competition for the enolate deprotonation site should be considered when a new carboxylic acid dianion type is envisioned. Fourth, owing to the location of the anionic groups, few studies in asymmetric induction have been conducted, particularly for the α,α′-dianion types, although some activity and progress in this area has appeared. Fifth, α-carboxylate dianions can offer distinct advantages over the corresponding ester enolates. For example, aliphatic carboxylic acids may be α-alkylated or condensed with carbonyls following dianion formation circumventing protection-deprotection sequences of the carboxylic acid moiety.

3.2.2 Reactions of α-carboxylate O,C-dianions (α,α′-carboxylate dianions)

No general class of O,C-dianions has experienced more use in organic synthesis than the α-carboxylate dianions. They were first reported over 60 years ago[1,2] and known as Ivanov reagents when the alkali metal was magnesium.[2] Their increased appearance in delicate synthetic operations and undergraduate texts bears testimony to both their global acceptance and confidence of use. Prior reviews of carboxylic acid dianions were dominated by the α-carboxylate intermediates, testimony to their frequency of use. The foundation member of this class, the acetic acid dianion (and analogs), provides a valuable and adaptable two-carbon homologation agent.

This section endeavors to acquaint the reader with the chemistry of α-carboxylate dianion intermediates, particularly how individual types of α-carboxylate dianion type react with different electrophiles, for example, alkylating agents, carbonyls, and acylating agents. In certain cases, sequences representing subsequent manipulation of the product into more complex molecules have been included. This section will differ from prior coverage in that distinct types of α-carboxylate dianions have been identified, and organized in three different section headings; (a) simple alkanoic and (aryl)alkanoic acid dianions, (b) alkanoic acid dianions bearing a β-carbonyl group, and (c) alkanoic acids dianions bearing carbanion-stabilizing groups other than carbonyl. Coverage of these intermediates is restricted to carbanions generated α to a carbonyl where the corresponding enolate resonance structure (dioxomethylene dianion) should be invoked (Fig. 3.1).

α-Carboxylate dianions undergo alkylation, condensation, acylation, and miscellaneous carbon-carbon forming reactions in good yields. Each of these reaction types will be described in further detail with the intention of updating literature coverage from prior reviews, and augmenting any expanded utility of the intermediate where appropriate.

α-Carboxylate dianions with no added stabilizing groups

Simple alkanoic (and phenylacetic) acids may be bis-deprotonated using a variety of bases to form the α-carboxylate dianions. These dianions have been previously denoted as either 1,3-O,C-carboxylic acid dianions, where the numbers represent the atom positions of the anionic sites, or as α,α′-carboxylic acid dianions, which are named similarly to the C,C-dianions of ketones and β-dicarbonyls. For simplification, the general term "α-carboxylate dianion" will be used throughout to describe these intermediates.

Simple straight-chain and branched carboxylic acids bearing alkyl or aryl groups may be bis-deprotonated and reacted with alkyl halides to afford the α-functionalized product (Eq. 3.2) (Table 3.1).[1-5,8] Primary or allylic/benzylic halides have been most used as the

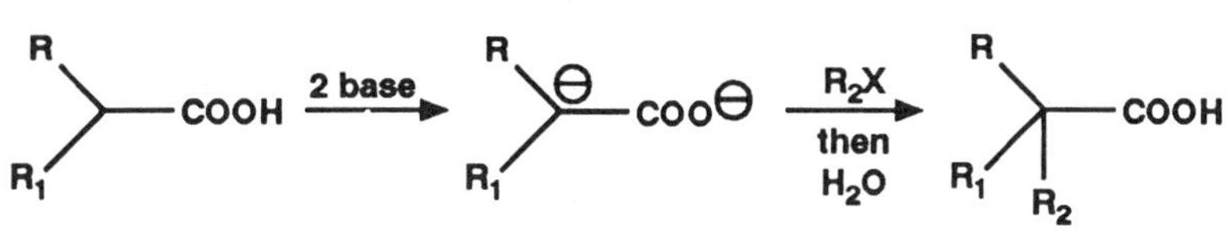

(3.2)

Table 3.1

Reaction of α-Carboxylate O,C-Dianions with Alkyl Halides

R	R_1	Base	R_2-X	% Yield	Ref.
H	H	LDA	BnBr	34-82	9
H	H	LiNaphth	alkyl-X	50-70	1
Me	H	LDA	BuBr	53-78	9,10
Me	H	LDA	allyl-Br		11
Me	H	LDA	CH_2=C(Me)CH_2Br		11
Et	H	LDA	BuBr	55	10
nPr	H	RLi/LDA	MeI	63	12
nPr	H	LDA	BuBr	84	10
c-C_3H_5	H	LDA	MeI	32	13
c-C_3H_5	H	LDA	BnBr	22	13
iPr	H	LDA	BuBr	83	10
nBu	H	LDA	EtBr	98	9
nBu	H	LDA	PrBr	98	9
nBu	H	LDA	BuBr	42-97	9,10
nBu	H	LDA	n-C_6H_{13}Br	86	10
nBu	H	LDA	n-C_8H_{17}Br	88	10
nBu	H	BuLi	BnCl	90	14
nBu	H	LDA	BrCH_2CO_2Li	74	15
nBu	H	LDA	MeCHBrCO_2Li	67	15
nBu	H	LDA	EtCHBrCO_2Li	60	15
nBu	H	LDA	Me_2CBrCO_2H	47	15
tBu	H	LDA	BuBr	81	10
nC_5H_{11}	H	LDA/HMPA	BuBr	43-93	9,16
c-C_6H_{11}	H	LDA	BuBr	84	10
nC_7H_{15}	H	LDA/HMPA	EtBr, PrBr, BuBr	34-91	9,16
nC_7H_{15}	H	LDA	n-C_8H_{17}Br	92	9
nC_8H_{17}	H	LDA	BuBr	86	10
CH_2=CHC_7H_{12}	H	LDA	MeI	87	17
Me	Me	LDA/HMPA	BuX	75-89	8,16
Me	Me	LDA	allyl-Cl	61	8
Me	Me	LDA	BnCl	46	8
Me	Me	LDA	PhOCH_2CH_2Br	41	8
Me	Me	LDA	EtOCH_2CH_2Br	66	8
Me	Me	LDA	c-C_6H_{11}Br	6	8
Me	Me	LDA	ArO$(CH_2)_n$$CH_2$Br	19-80	18
Me	Me	LDA	BrCH_2CO_2Li	63	15
Me	Me	LDA	MeCHBrCO_2Li	73	15
Me	Me	LDA	EtCHBrCO_2Li	56	15
Me	Me	LDA	Me_2CBrCO_2H	78	15
Me	Me	LDA	BrCH_2CO_2tBu	70	15
Me	Me	LDA	BrCH_2CO_2Me	10	15
Me	Me	LiNaphth	X$(CH_2)_n$X	42-75	1
Me	Et	LDA	BuBr	76	8
Me	Bu	LDA	BuBr	85	9
Et	Bu	LDA	BuBr	78	10
Pr	Bu	LDA	BuBr	72-93	9
Ph	H	BuLi	BnCl	90	14
Ph	H	LDA	BuSCH_2Br	80	19
Ph	H	LDA	MeC≡CCH_2Br	90	20
Ph	H	LDA	TMS-C≡CCH_2Br	84	20

electrophile owing to the basicity of the dianion, which causes elimination with secondary and tertiary halides or the analogous sulfonates.

Alkylating agents other than alkyl halides also have been tried. Petragnani and Yonashiro[15] showed that α-halo carboxylates and α-halo esters react with α-carboxylate dianions to give good yields of succinate derivatives. When α-halo esters are to be reacted, the tert-butyl form of the ester is preferred since competitive reaction of the carbanion at the ester site may occur.

Dicarboxylic acids bearing spacer groups between the α-positions can be prepared by reaction of two equivalents of the α-carboxylate dianion with dihaloalkanes.[1] Propargylic halides react with the α-carboxylate dianions to give γ-alkynoic acids which, when treated with mercuric ion, cyclize to give γ-methylene lactones.[20]

α-Carboxylic acid dianions also may be alkylated with epoxides. Creger showed that acetic acid and isobutyric acid dianions add to the sterically less encumbered position of epoxides to furnish γ-lactones following acid workup (Eq. 3.3).[8,21] This reaction was used

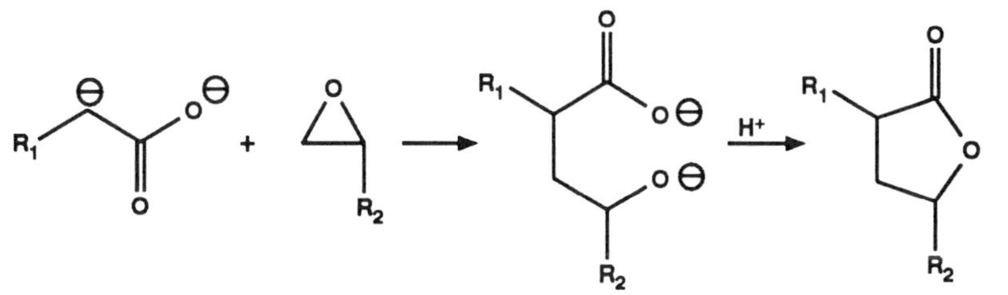

(3.3)

successfully in the preparation of steroidal γ-lactones (Eq. 3.4),[21] and in the total synthesis of carpesiolin (Eq. 3.5).[22] In the later usage, 15 equivalents of the acetic acid dianion in DME at 60°C for 24 h was needed to bring about the reaction.

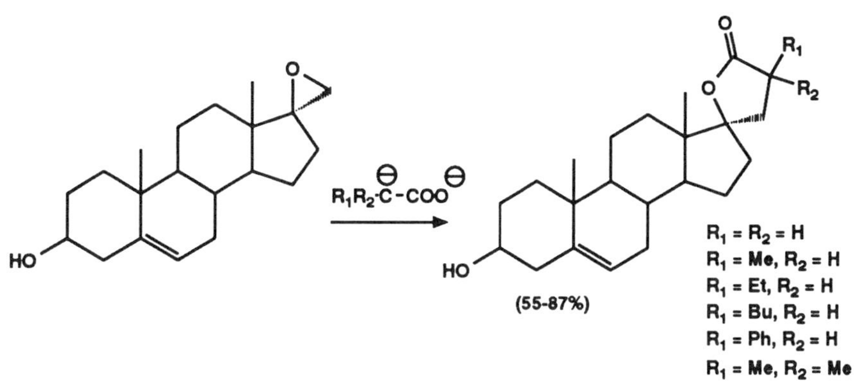

(3.4)

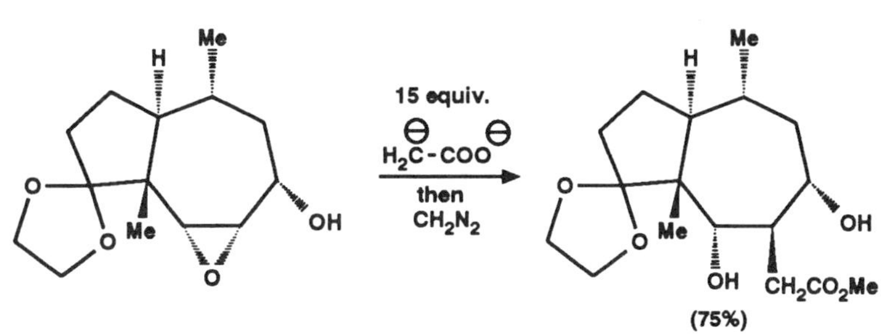

(3.5)

Disodium salts of arylacetic acids, formed with 2 lithium naphthalenide in THF, undergo amidoethylation upon reaction with N-benzoylaziridines (Eq. 3.6).[23]

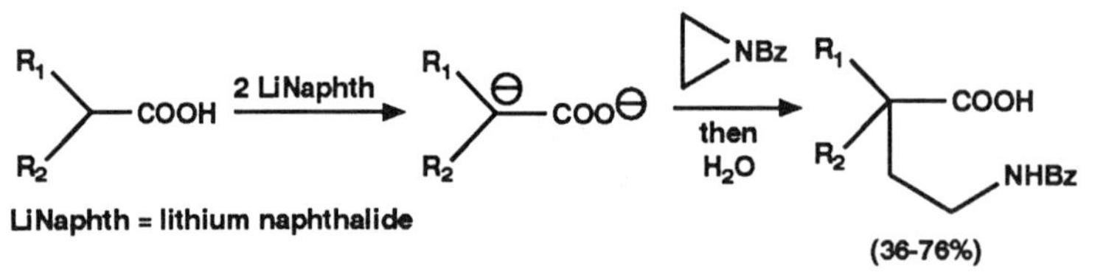

(3.6)

α-Alkanoic acid dianions also may be alkylated by Michael acceptors.[2-4] Ainsworth and coworkers reported a successful conjugate addition of carboxylate dianions to α,β-unsaturated esters, which form glutaric acid monoesters in good yield (Eq. 3.7).[24] Cinnamonitrile undergoes conjugate addition with isobutyric acid dianion, but using acrylonitrile or crotonaldehyde as the electrophile caused polymerization.

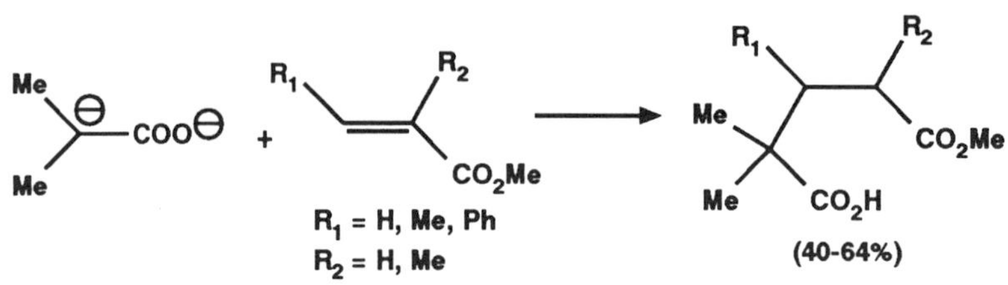

(3.7)

Conjugated ketones undergo preferential 1,4-addition with carboxylate dianions when the condensing partners bear bulky substituents.[25] This reaction will be discussed in more detail (*vide infra*). Nitroolefins alkylate the lithium dianions of alkanoic acids at -100°C (Eq. 3.8).[26] *In situ* reaction of the first formed γ-nitro carboxylic acid with aqueous acid leads to the γ-keto acids in good yield.

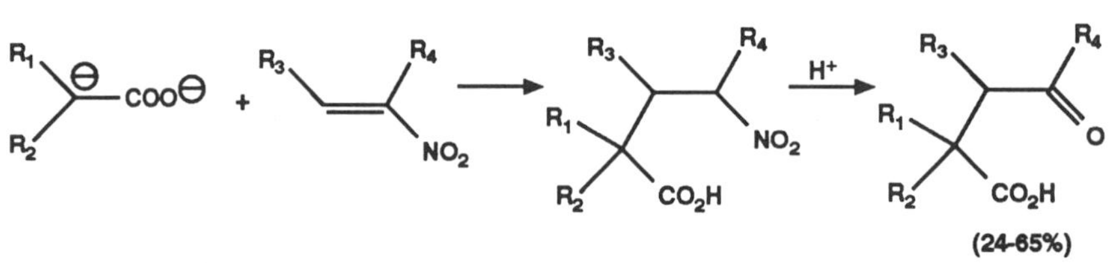

(3.8)

The condensation reaction between simple carboxylic acid dianions and carbonyls gives β-hydroxy carboxylic acids in high yield, and the dianion sequence appears superior to the Reformatsky reaction (Eq. 3.9). An excellent summary of the condensation reaction can be found in the review by Petragnani and Yonashiro.[4] In most cases, lithium amide bases or lithium naphthalenide have been used to generate the dianion species, and yields of the adducts remain good except when steric crowding impairs addition to the carbonyl.

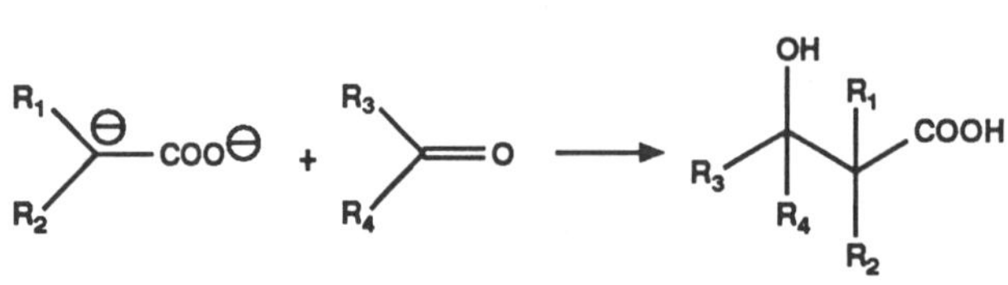

(3.9)

Table 3.2
Reaction of α-Carboxylate O,C-Dianions with Aldehydes and Ketones

R_1	R_2	R_3	R_4	Base	Yield (%)	Ref.
H	H	Ph	H	LDA	70	27
H	H	Ph	Ph	LDA	17	28
Me	H	tBu	H	LDA	70	29
Me	H	Ph	H	LDA	75	29
Et	H	Ph	H	LDA	80	29
iPr	H	Ph	H	LDA	80	29
tBu	H	Ph	H	LDA	73	29
Ph	H	Me	H	LDA	65	29
Ph	H	tBu	H	LDA	95	29
Me	Me	Ph	Ph	LDA	31-74	30
tBu	H	Ph	Ph	LDA	15-92	28,30
Me	Me	Et	Et	LDA	77	30
Me	Me	Ph_2CH	H	LDA	50	30
Et	Et	Ph	H	LDA	98	30
Me	H	Ph	Me	LDA	25	28
Me	Me	Ph	Me	LDA	24	28
$-(CH_2)_3-$		$-(CH_2)_3-$		LDA	86	31
$-(CH_2)_n-$		$-(CH_2)_n-$		LDA	70-82	31
H	H	Ph	Ph	LiNaphth	22-58	9,32
H	H	Me	Me	LiNaphth	60	32
H	H	Et	Et	LiNaphth	65	32
H	H	Me	Ph	LiNaphth	28	32
Me	H	$-(CH_2)_5-$		LiNaphth	40	32
Me	H	Ph	Ph	LiNaphth	60	9
Et	H	Ph	Ph	LiNaphth	60	9

n = 3 - 7

The β-hydroxy acids are important precursors to a variety of molecules: substituted acrylic acids via dehydration,[19] β-lactones via cyclization with $PhSO_2Cl$ and base,[4] and olefins via direct treatment of the hydroxy acid with DEAD/Ph_3P in $CHCl_3$,[33] or by thermolysis of the β-lactones (Scheme 3.1).[28,34,35]

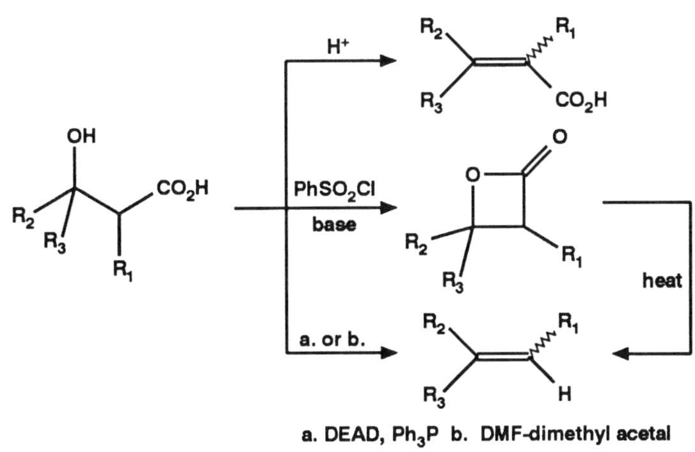

Scheme 3.1. Representative uses of β-(hydroxy)carboxylic acids

Black and coworkers explored the use of β-lactones obtained via the reaction of alkanoic acid dianions with aldehydes and ketones (Eq. 3.10).[36-40] A ring enlargement of the β-lactone via a dyotopic rearrangement was observed to occur in good yield when the alkanoic acid partner bears only one alkyl substituent.[36-38] The rearrangement occurs stereospecifically; the *trans*-substituted β-lactone gives the *trans*-butyrolactone (Scheme 3.2). When the aldehyde component bears a cycloalkyl ring, three contiguous asymmetric centers can be installed.[37] A study of the rearrangement promoters showed that $MgBr_2$ was the most effective in stoichiometric quantity.[38]

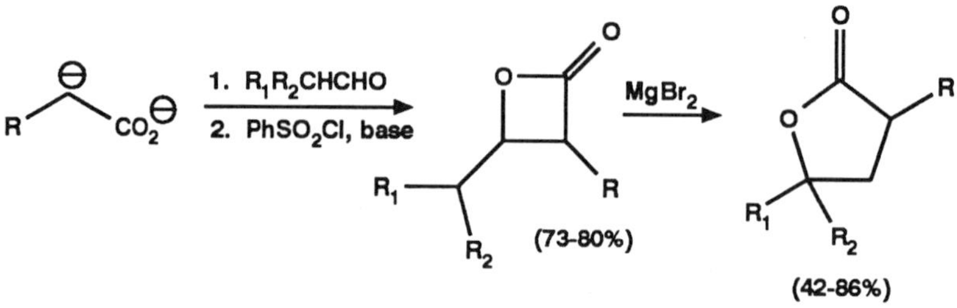

(3.10)

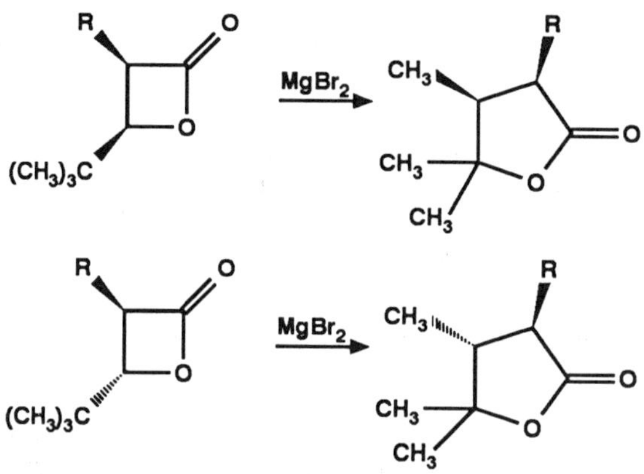

Scheme 3.2 Stereospecificity of the dyotopic rearrangement

When ketones are used as the electrophiles in this sequence, spiro β-lactones result (Eq. 3.11). Unlike the aldehyde derived lactones, which proceed through the dyotopic rearrangement, the spiro β-lactones undergo an ionization-elimination process to give β,γ-unsaturated carboxylic acids.[39]

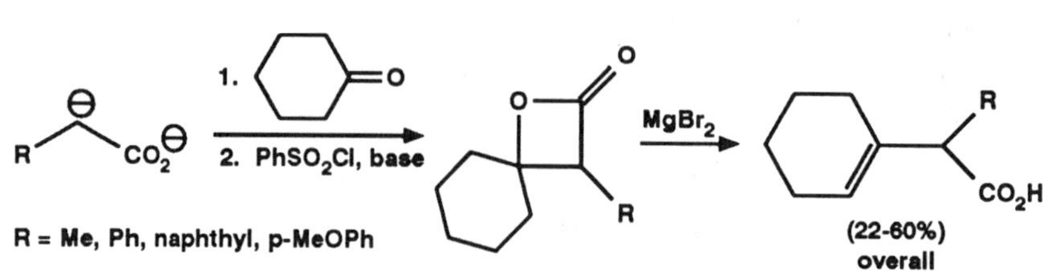

(3.11)

Conjugated aldehydes undergo reaction with α-carboxylate dianions via 1,2-addition to give β-hydroxy-γ-enoic acids, usually in a 1:1 diastereomer ratio (Eq. 3.12).[27,29,30] These poor stereoselectivities could have diminished the use of the condensation reaction, however two developments reversed this situation.

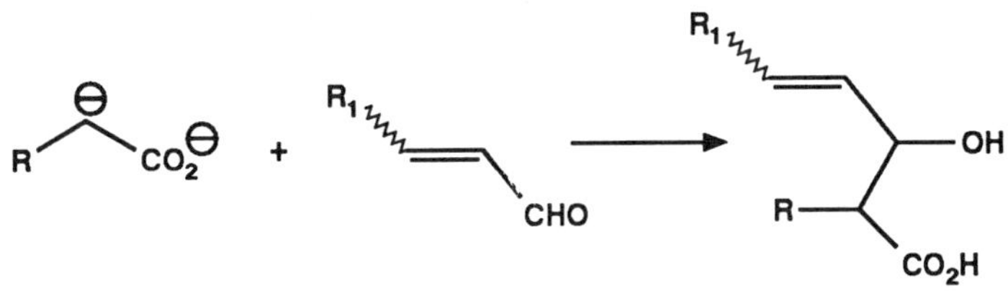

(3.12)

Trost and Fortunak discovered that diastereomeric mixtures of the β-hydroxy acids derived from (E)-enals could be acylated and reacted with 3-5 mol% $(Ph_3)_4Pd$ and 110 mol% TEA to furnish only the E,E-diene, thereby circumventing the stereochemical problem.[41]

The second development that led to increased usage of the aforementioned reaction was a series of highly detailed and thorough investigations by Mulzer and coworkers, who elucidated the stereochemical outcomes of additions of alkanoic acid dianions to a variety of aldehydes and ketones.[25,42,43] The ratio of threo/erythro adducts resulting from addition of carboxylic acid dianions to aldehydes varies with the substituent steric composition of both coupling components and the nature of the countercation (Eq. 3.13).[42] Under

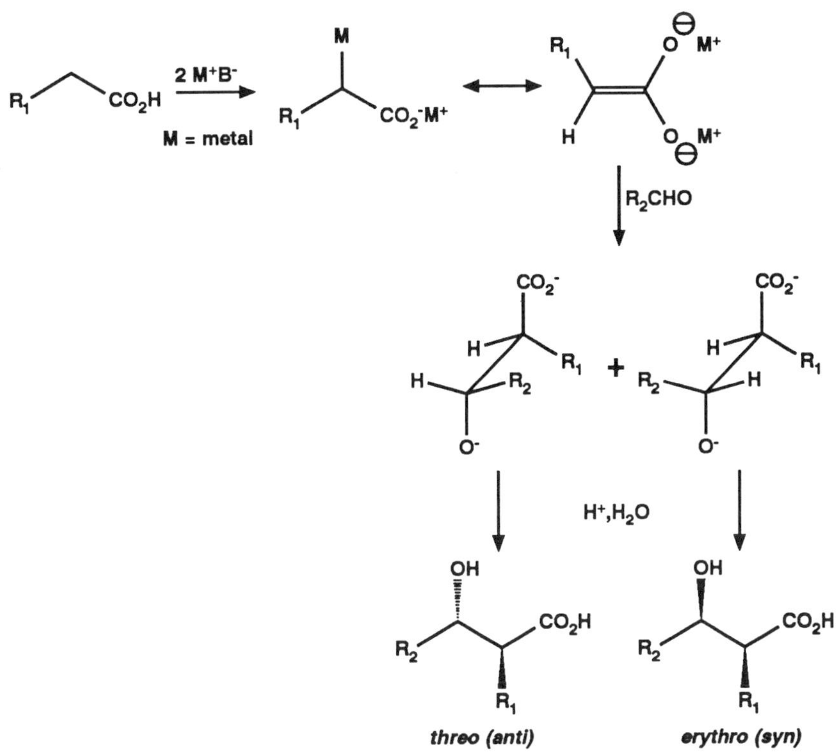

(3.13)

kinetically-controlled conditions (-50°C, 10 min), the threo product is the preferred stereoisomer formed (Table 3.3), and the amount of this isomer increases when sterically demanding substituents (R_1 and R_2) are present.[42] Threo selectivity decreased with the cation charge-to-radius ratio; Mg++, Zn++, and Li+ gave 1:1 ratios of stereoisomers whereas Na+ and K+ gave 4:1 to 32:1 ratios favoring the threo isomer. Under thermodynamic control (rt, 48 h), the threo stereoselectivity was higher than for the kinetic conditions.[42]

Table 3.3
Threo/erythro Stereoisomer Ratios for the Reaction Between α-Carboxylate Dianions and Aldehydes (Eq. 3.13)

R_1	R_2	kinetic threo/erythro	Yield (%)	thermodynamic threo/erythro	Yield (%)
Ph	Me	1.4	65	1.8	60
tBu	Ph	1.5	73	7.3	70
tBu	mesityl	1.8	68	13.4	60
Ph	iPr	1.9	83	14.0	77
Ph	tBu	1.9	95	> 49	90
Ph	Ph	2.1	88	12.0	90
tBu	iPr	3.8	78	9.0	75
tBu	tBu	4.0	85	> 49	83

Dilithio carboxylic acid dianions generally react regiospecifically 1,2 with α,β-unsaturated aldehydes (Eq. 3.14; R_2 = H) to form the unsaturated β-hydroxy carboxylic acids in good yield,[2-4,25] and in certain cases high threo selectivity (> 90:10) can be obtained.

The reaction between alkanoic acid dianions and α,β-unsaturated ketones is far more capricious. Depending upon the substituent patterns almost exclusive 1,2- or 1,4-addition reaction may occur (Eq. 3.14) (Table 3.4).[25]

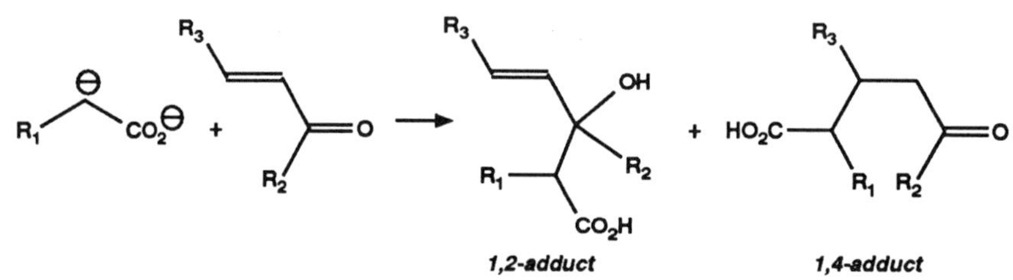

(3.14)

Table 3.4
1,2- vs. 1,4-Addition of Carboxylate Dianions to α,β-Unsaturated Ketones

R_1	R_2	R_3	ratio (1,2 : 1,4)	Yield (%)
H	Me	Ph	> 50	72
H	iPr	Ph	> 50	73
H	tBu	Ph	2.2	45
H	Ph	Me	0.5	65
H	Ph	Bu	1.4	83
H	Ph	Ph	3.5	67
H	Ph	tBu	> 50	71
Me	tBu	Ph	< 0.02	76
Et	tBu	Ph	< 0.02	83
tBu	tBu	Ph	< 0.02	79
Me	Et	Ph	> 50	83
iPr	Et	Ph	> 50	78
Me	Ph	Ph	1.7	85
Et	Ph	Ph	1.7	65
iPr	Ph	Ph	1.0	77
tBu	Ph	Ph	< 0.02	88

The excellent groundwork conducted on the condensation reaction with carboxylic acid dianions by these investigators opened the area for the synthesis of racemic epi-blastmycinone, δ-multistriatine, paraconic esters, and lignan type dilactones.[43] Two additional findings in this latter study are noteworthy. First, if the carboxylic acid dianion was prepared from LDA using α-methylstyrene/lithium rather than BuLi, satisfactory amounts of pure threo stereoisomer can be obtained owing to spontaneous crystallization of the desired material from the reaction mixture. Specifically, the threo β-hydroxy acid resulted as the major product from the condensation between hexanoic acid dianion and β-methylcinnamaldehyde.

Second, the use of optically active, alkoxide-amide bases in place of LDA gave enantioenriched products. In model studies, the phenylacetic acid dianion, generated by reaction with four diferent chiral bases (Eq. 3.15; A-D), was condensed with benzaldehyde, and the ratio and stereochemistry of the product β-hydroxy acids measured. The dianionic bases (-)-A, B, and C all gave the (2S,3R)-threo diastereomer, whereas (+)-A gave the (2R,3S)-antipode. The trianionic base (D) was the most efficient, and afforded the (2R,3S)-product in 85% e.e.

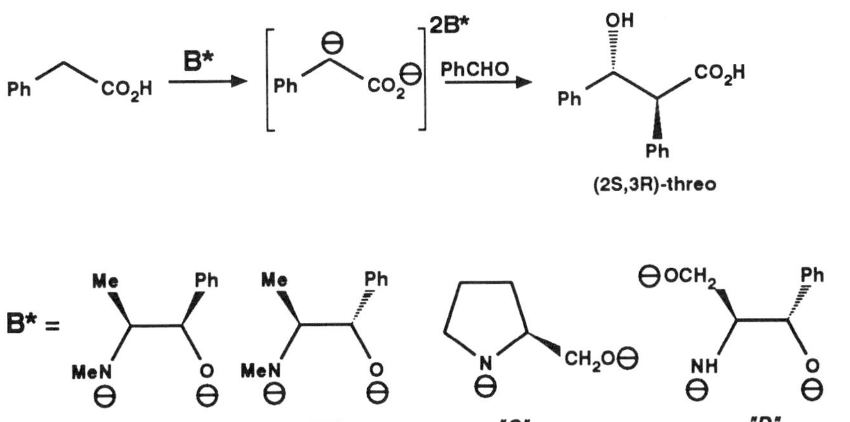

(3.15)

Alkanoic acid dianions can be acylated at the α-position with a variety of reagents to provide β-formyl and β-keto carboxylic acids.[2-4] Carbon dioxide and carbon disulfide give malonic acid derivatives and α-thioester malonates, respectively (Scheme 3.3).[4]

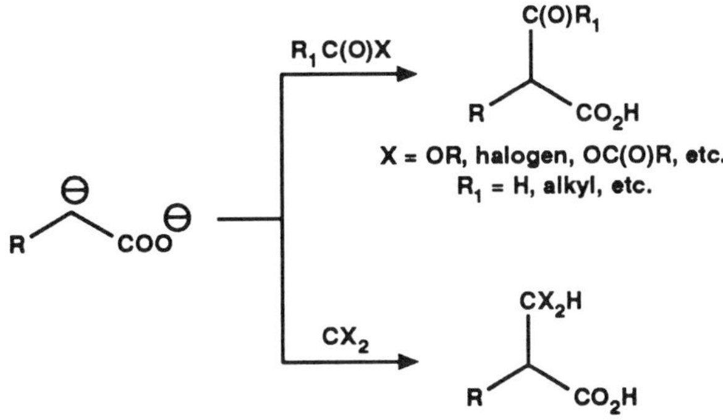

Scheme 3.3. Acylation reactions of alkanoic acid dianions

Esters and disubstituted amides also react with alkanoic acid dianions to give β-keto acids in 10-90% yield.[1,4,24] Since decarboxylation can be easily induced, the procedure can be used to prepare highly substituted ketones (Eq. 3.16).

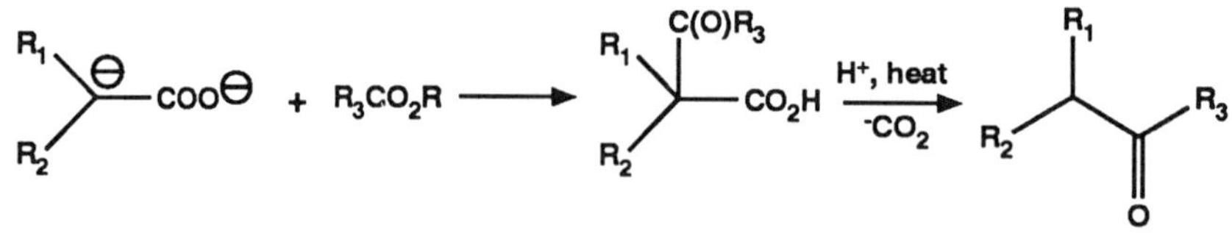

(3.16)

When formate esters or formamides are used as the electrophile, the decarboxylation can be conducted *in situ* to provide a net carboxylic acid to aldehyde conversion (Eq. 3.17).[44,45]

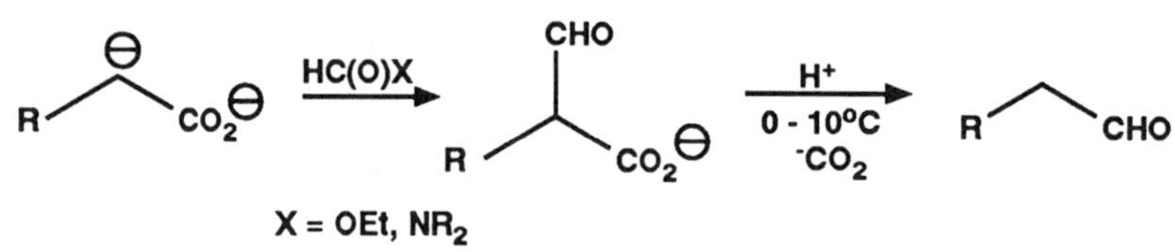

(3.17)

The reactions between an alkanoic acid dianions and a methyl ester leads to β-keto acids that can be trapped as the silyl ester to reduce the opportunity for spontaneous decarboxylation (40-80%).[24]

Acyl halides and anhydrides react satisfactorily with carboxylic acid dianions (Scheme 3.3). This procedure offers a distinct advantage over certain Claisen-type condensations since no cross-over products are obtained. Carbonates and chlorocarbonates react with alkanoic acid dianions to give malonate monoesters (Eq. 3.18).[46-48]

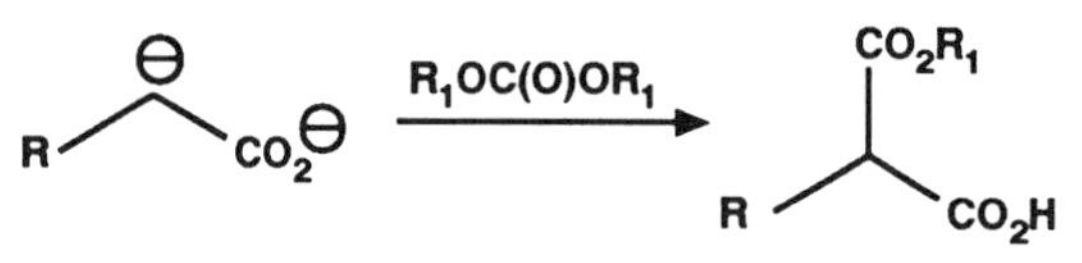

(3.18)

The dianion of alkanoic acids can be quenched with CO_2 or CS_2 to form malonic acids or α-dithioester carboxylic acids, respectively (Scheme 3.3).[4] Following formation of the α-thiocarbonate, this product may be decarboxylated to the dithioester or alkylated at sulfur (following decarboxylation) to afford the ketene dithioacetal.[49]

Functionalization of the α-position of carboxylic acid dianions with heteroatom-based groups also may be accomplished. Of particular interest has been the incorporation of sulfur, oxygen, silicon, halogen, and amines (Scheme 3.4).[1-5] α-Alkanoic acids undergo nucleophilic substitutions at a variety of heteroatoms, but no group holds as much prominence in this transformation as the α-sulfenylation. The α-sulfinyl carboxylic acids have emerged as an important dianion class, and their widespread use as reagents warrants specific coverage in a later section.

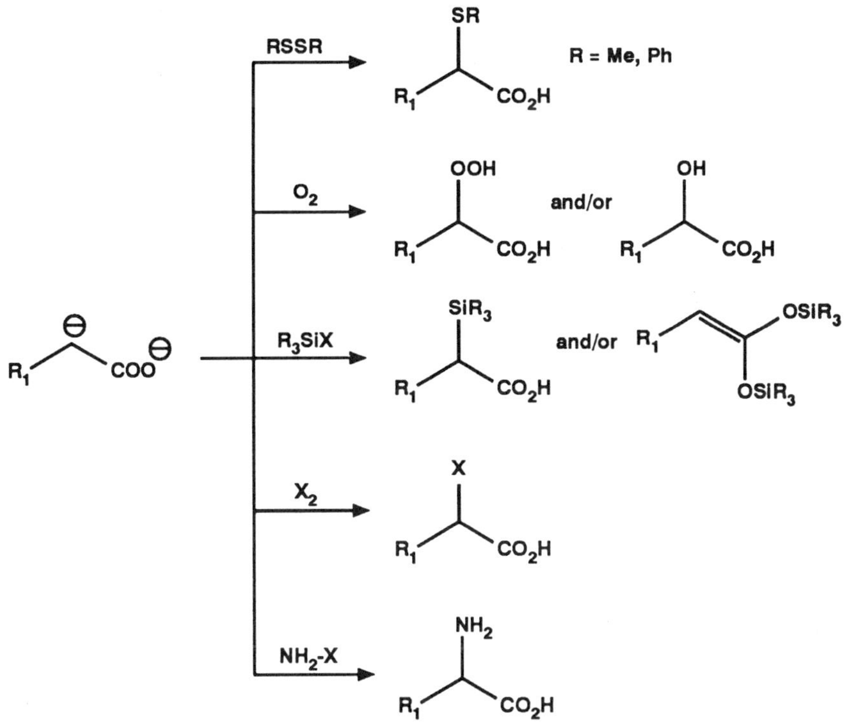

Scheme 3.4. Functionalization of the α-position by heteroatom-based functional groups

Dianions of simple and somewhat complex alkanoic acids may be α-sulfenylated in high yield using dimethyl disulfide or diphenyl disulfide (Eq. 3.19).[50-52] Generally, the alkanoic acid dianion is generated with 2 LDA in a solvent mixture of THF-HMPT (10:1, v:v). As

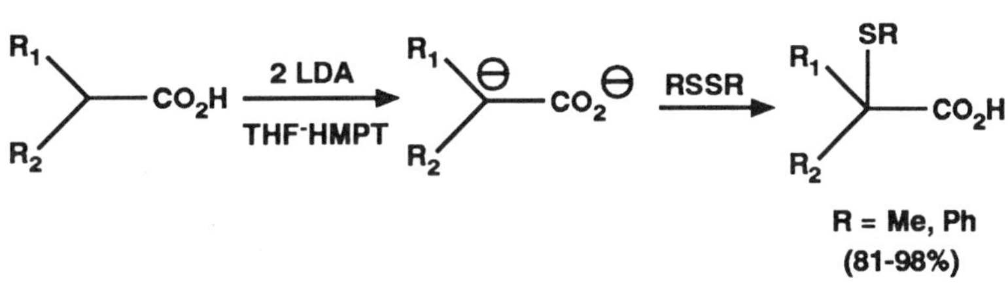

(3.19)

mentioned, a number of important functional group transformations can be conducted with these products, many of which are covered in the review by Trost.[53]

If the α-mercapto acid is desired, the carboxylic acid dianion can be reacted with elemental sulfur (50-93%).[54] α-Selenylation also has been conducted using PhSeCl or PhSeSePh.[4] Like the sulfur compounds, α-phenylselenocarboxylates and -carboxyesters are also useful synthetic intermediates.[55,56]

Most dianions of simple carboxylic acids have been shown to react with TMS-Cl to form the bis(trialkylsilyl)ketene acetals, although the lower molecular weight acids provide mixtures of O- and C-silylation.[57,58] More recently, Larson and coworkers showed that methyl diphenylsilyl chloride gives only C-silylated products of cyclopropanes and higher cyclic homologues (Eq. 3.20).[59]

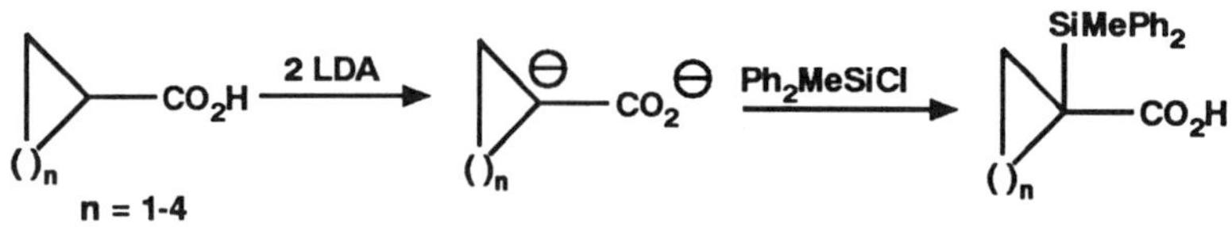

(3.20)

Carboxylic acid dianions can be reacted with halogens to provide the α-halo carboxylic acids.[1,2,4] However, the most widely used application of this reaction, the addition of 0.5 equiv. of iodine (I_2) to the dianion, leads to oxidative coupling forming symmetrical or in certain situations unsymmetrical, succinate derivatives (Eq. 3.21). The mechanism presumably proceeds through reaction of the first formed α-iodocarboxylate with the

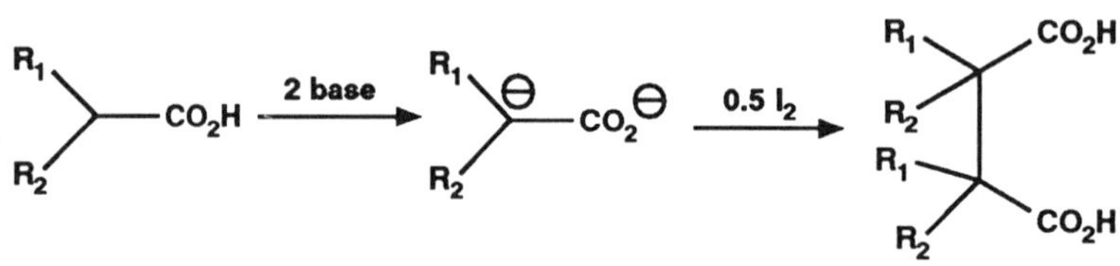

(3.21)

carboxylic acid dianion. Belletire and coworkers fully explored the nature and utility of this oxidative coupling.[14,60-64] The procedure is amenable to a wide array of alkanoic and phenylacetic acid dianions, and the stereoselectivity is illustrated in the coupling of the phenylacetic acid dianion, which forms an 11:1, *dl* to meso isomer mixture.[60] This process served as the defining step for the construction of enterolactone,[60] hinokinin,[62] fomentaric acid,[62] wikstomol,[63] and a central intermediate for the preparation of anti-tumor lignan prototypes (Fig. 3.3).[64]

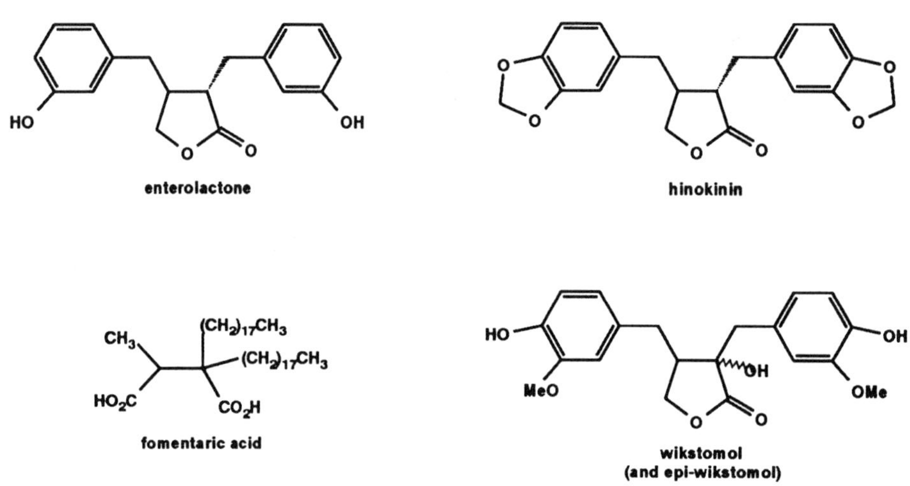

Figure 3.3. Representative molecules prepared by oxidative coupling of carboxylic acid dianions

Oxidative coupling also may be conducted with heterocyclic analogues. Indoleacetic acid was converted to the corresponding dianion with 3 RLi (removal of the indole N-H proton was also needed), and coupled with 0.5 eq. I_2 (Eq. 3.22).[65]

α-Hydroxy- and α-peroxycarboxylic acids are valuable intermediates and products in

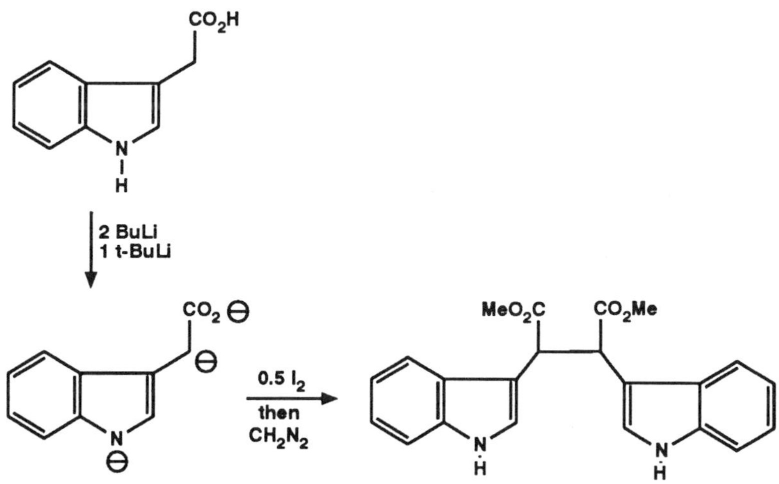

(3.22)

organic chemistry. Several investigators recognized early that the reaction of O_2 with alkanoic acid dianions would provide rapid access to these molecules (Scheme 3.4).[4] Carboxylic acid dianions, generated with 2 LDA can be converted into the α-hydroxy- and/or α-peroxycarboxylic acid depending upon the temperature employed (Eq. 3.23).[66-68] The conditions for the preparation of the peroxy compounds usually give some contamination

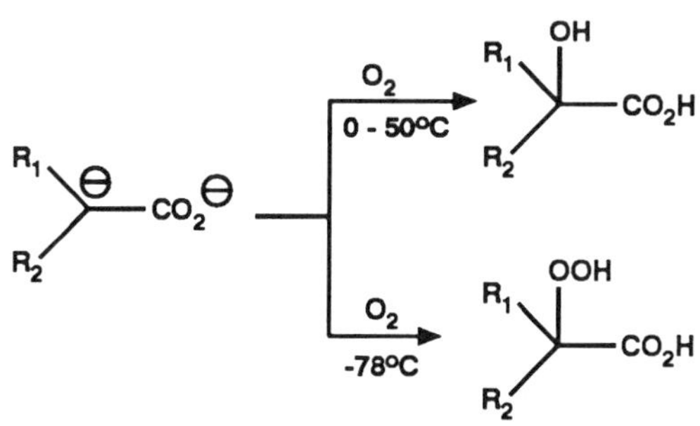

(3.23)

by the hydroxy compounds, although the preparation of the hydroxy acids usually proceeds quantitatively. Specifically, at higher temperatures, the peroxy compounds (intermediate) may be reduced to the hydroxy acid by the starting material. Wasserman and Lipshutz showed that the α-peroxy acids can be eliminatively decarboxylated following treatment with dimethylformamide dimethyl acetal to yield ketones (Eq. 3.24).[67]

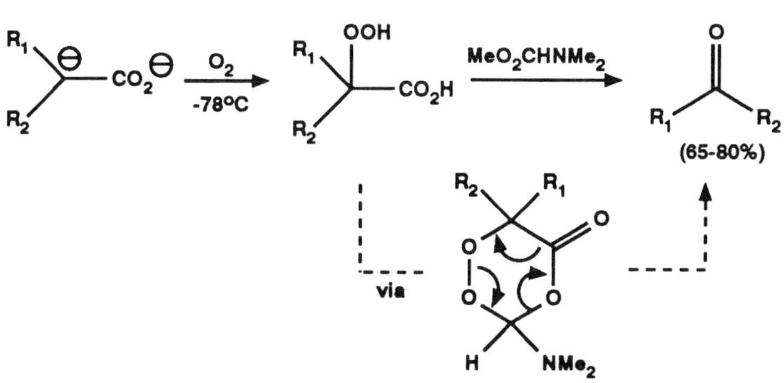

(3.24)

Dianions of carboxylic acids have been shown to react with electrophilic "aminating" reagents to form α-amino acids (Scheme 3.4).[69]

β-Carbonyl-based α-carboxylate dianions

Additional stabilization to the enolate carbanion of carboxylic acid dianions can be provided by a β-carbonyl group (Fig. 3.1). The β-carbonyl group greatly lowers the pKa of the enolate position facilitating access and lowering the resultant basicity of the carbanionic center, which alters the reactivity of this dianion species. Although several types of carbonyl groups can be used, there has not been a significant amount of activity in this area. Reluctance in usage of this dianion class may emanate from facile decarboxylation and/or competitive O-alkylation. Among the work conducted, the malonate monoesters have been the most widely explored, particularly as alternatives to the malonic ester synthesis. The malonic ester synthesis, which reacts an enolate of a malonic diester with an electrophile (e.g., alkyl halide) followed by saponification and decarboxylation, affords two-carbon chain extended esters. A one-pot variation has been introduced that improves yield.[70]

Dianions accessed from monoesters of malonates can be alkylated in the same manner as the malonic esters except that the decarboxylation can now be conducted in single step since no saponification is needed. Experimentally, monoethyl malonate (hydrogen ethyl malonate) was converted to the corresponding dianion with 2 eq. of lithium isopropyl cyclohexyl amide and reacted with a variety of alkyl halides at -78°C. The reaction mixture was warmed to reflux, which induces decarboxylation. Standard workup furnished the esters in yields of 22-98% (Eq. 3.25).[71]

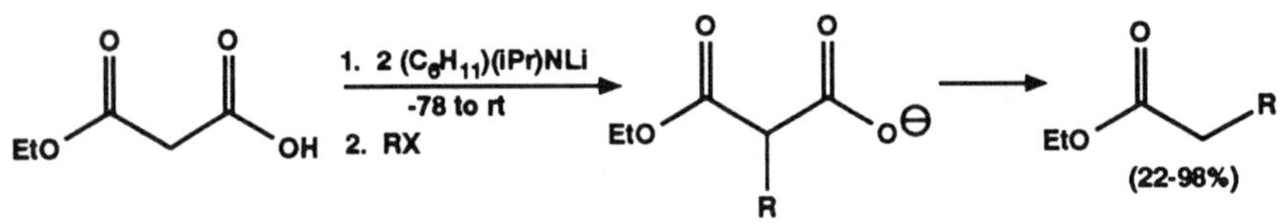

(3.25)

Acylations of the malonate monoester dianions, which lead to β-ketoesters after decarboxylation, also have been conducted with success, but the yields have been low. However, Brown and coworkers formed the dilithio species directly with 2 BuLi, and acylated it with 2-methyl benzoyl chloride in 72% yield.[72]

Most experimenters now generate the hydrogen (m)ethyl malonate dianion by a variation developed by Rapoport and coworkers (Eq. 3.26).[73] First, hydrogen methyl malonate is

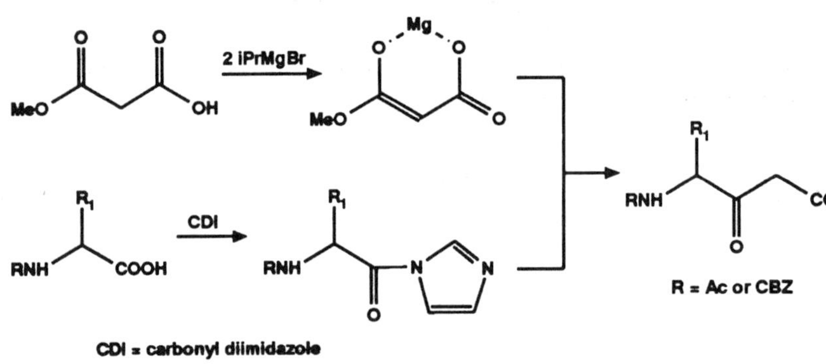

(3.26)

converted into the dianion using isopropyl magnesium bromide as the base. The Grignard base forms a putative chelate structure that dramatically alters the reactivity toward acylation. A further important feature of their method is the use of the imidazolide as the acylating agent rather than halide, ester or anhydride. Excellent yields of β-keto esters (70-95%) were achieved making the procedure amenable to virtually any carboxylic acid. The fact that this acylation procedure works in the presence of either carbobenzyloxy or acetamides is noteworthy as is the preservation of the asymmetric center during reaction with amino acids.[73,74]

The stereoselective alkylation of dianions of malonate monoester has been studied.[75] Alkyl hydrogen malonates bearing different chiral auxiliaries (R*) on the ester moiety were prepared (Eq. 3.27). The enolate alkyl group was varied as methyl ethyl and isopropyl. The asymmetric dianions were accessed by reaction with LDA or LiHMDS and quenched with alkyl halides to afford diastereomer ratios ranging from 1.4:1 to 16:1 (Eq. 3.27). The 8-phenyl menthol ester afforded the highest diastereomer ratio and was used for further study with various alkylating agents (Table 3.5).

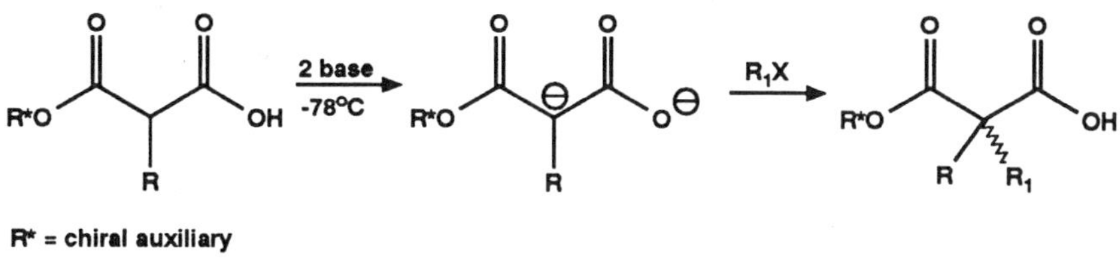

(3.27)

Table 3.5
Alkylation of Asymmetric Alkyl Malonate Monoester Dianions

R*	R	R_1	base	Yield (%)	diastereomer ratio
A	iPr	Me	LDA	87	2.4:1
A	iPr	Me	LHMDS	87	2.4:1
A	iPr	Me	LDA, HMPA	83	1.4:1
A	iPr	Me	LDA, $CuBr{\cdot}SMe_2$	83	1.5:1
B	Et	Me	LDA	78	1.1:1
C	Et	Me	LDA	90	3:1
D	Et	Me	LDA	89	2:1
E	Et	Me	LDA	80	5:1
E	Me	Et	LDA	83	4:1
E	Me	nPr	LDA	72	4:1
E	Me	$CH_2{=}CHCH_2$	LDA	77	7:1
E	Me	$CH_2{=}C(Me)CH_2$	LDA	91	6:1
E	Me	Bn	LDA	72	12:1
E	Me	4-MeOBn	LDA	71	12:1
E	Me	2-MeOBn	LDA	73	16:1
E	Pr	Me	LDA	84	5:1
E	Bn	Me	LDA	61	15:1

A = (1R,3R,4S)-p-menthan-3-yl, B = α-phenethyl, C = 2-(1-naphthyl)-3-borneyl,
D = menthyl, E = 8-phenylmenthyl

α-Carboxylate dianions bearing carbanion stabilizing groups other than carbonyl

Dianions of carboxylic acids that bear α-heteroatom-based substituents have been generated and used successfully in synthetic chemistry. A relatively broad number and variety of dianions bearing heteroatom functional groups have been explored including thioether, silane, selenoether, phosphonate, halogen, ether, and amine moieties (Fig. 3.4). In most instances, the heteroatom-based dianion reagents possess distinct advantages

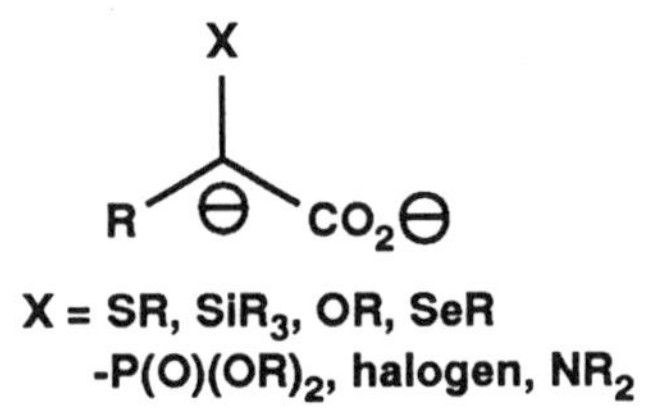

Figure 3.4. Examples of α-heteroatom-based carboxylate dianions

over the hydrocarbon congener, especially in terms of reactivity and potential for subsequent functional group interchanges. For this reason, α-thioether carboxylic acid dianions have attracted the most attention among the heteroatom-based species; the thioether group serving as the pivot for an array of molecule manipulations.[53]

About two decades ago, several groups began to show the tremendous potential of thioacetic acid dianion reagents.[4,76,77] These reagents undergo dianion formation with relative ease using LDA or BuLi, and react with alkylating agents (alkyl halides or epoxides) to give the α-alkylated thioacetic acid (Eq. 3.28). Alternatively, the alkylated carboxylate may be re-deprotonated and realkylated to afford α,α-dialkylated products.[51,52,81] The yields of the monoalkylations are excellent, far superior to those obtained with the corresponding acetic- and phenylacetic acid dianions (Table 3.6).

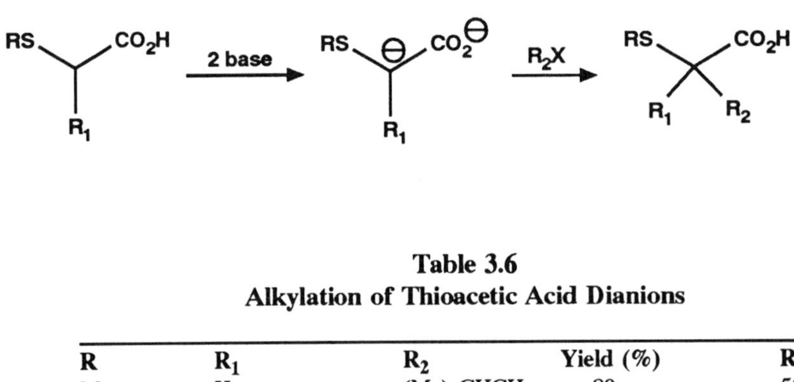

(3.28)

Table 3.6
Alkylation of Thioacetic Acid Dianions

R	R_1	R_2	Yield (%)	Ref.
Me	H	$(Me)_2CHCH_2$	80	51,52
Me	H	iPr	59	51,52
Me	H	$Cl(CH_2)_{11}I$	69	52
Me	Me	$Ar(CH_2)_n$	57-95	78
Me	$-CHCH=CH_2$	Bu	74	79,80
Me	$-CHCH=CH_2$	$n-C_6H_{13}$	76	79,80
Me	$-CHCH=CH_2$	$n-C_{10}H_{21}$	85	79,80
Ph	H	$Me(CH_2)_{11}$	98	81
Ph	H	$Me(CH_2)_3$	96	81
Ph	H	$BrCH_2(CH_2)_3$	75	81
Ph	H	Me_2CH	99	81
Ph	iPr	$Me(CH_2)_{11}$	95	81
Ph	$Me(CH_2)_{11}$	Me	98	81
Ph	H	Me	93	82
Ph	H	Et	100	82

Following isolation of the alkylated thioacetic acid, N-chlorosuccimide (NCS) oxidation gives ketones via the acetals (in alcohol solvent),[51] or the thermodynamically less stable enol thioether.[78] Alternatively, the α-thioacid may be first converted to the primary alcohol, which serves as a precursor to enol thioethers, alkenes, ketones and epoxides.[81] Tanaka and coworkers used the 2-allyl derivatives to form furanones via iodolactonization.[79] Representative reactions are summarized in Scheme 3.5.

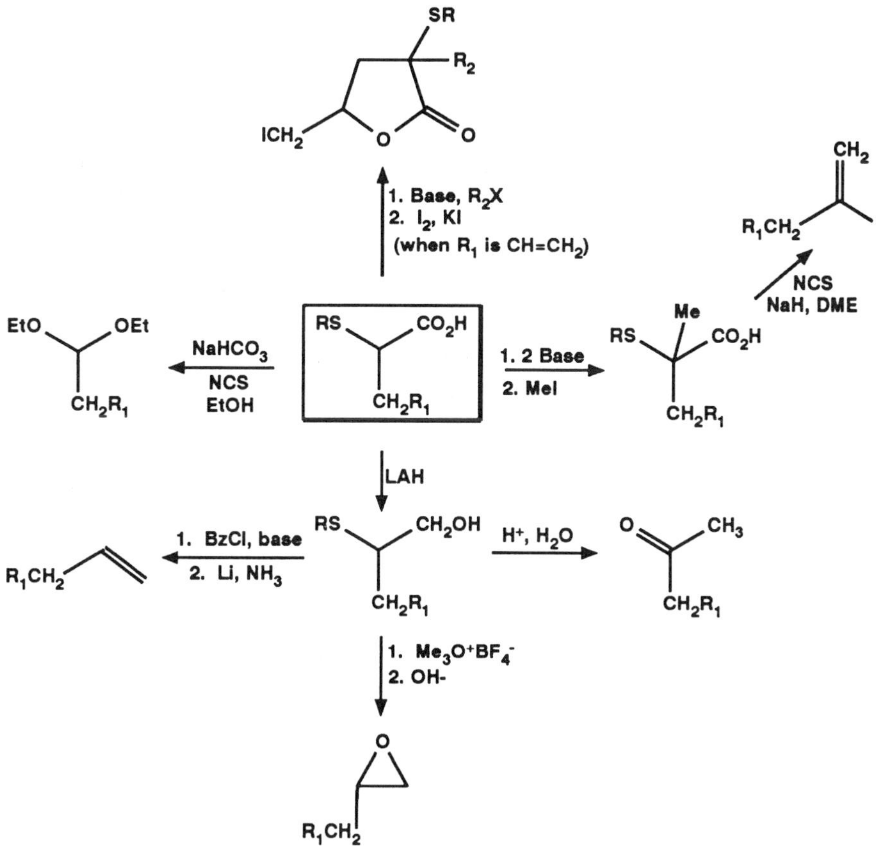

Scheme 3.5. Reactions of alkylated α-thioacetic acids

Some interesting variations of the alkylation theme have been investigated. An intramolecular alkylation with an o-chloro group led to the formation of a dihydroisobenzothiophene derivative.[83] Bates and coworkers showed that glyoxylic acid thioacetals, as either the diethyl or dithianyl derivative, may be alkylated in good yield.[84,85] When the thioacetal is the diethyl, alkylation with BnCl requires HMPA as cosolvent and extended reaction times (18 hours). However, the dithiane analog (Eq. 3.29) may be alkylated in quantitative yield with BnCl in 3 hours. The authors also indicate that the use of KHMDS as base increased the solubility of the resulting dianion leading to better yields.

Nicolaou and coworkers reacted dithiane carboxylic acid dianion with allyl bromides to give the alkylated products in 85-90% yield.[86]

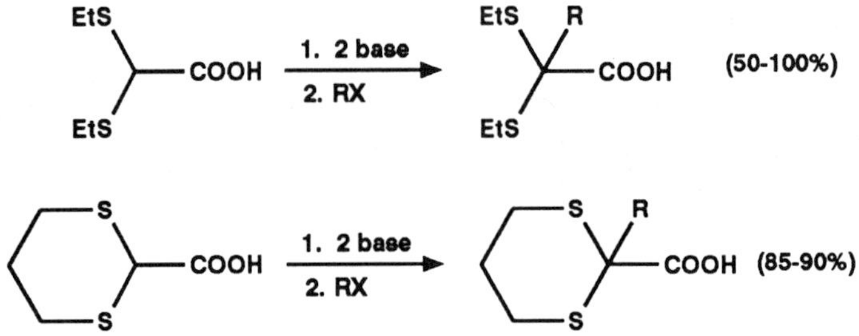

(3.29)

Reaction of α-thioacetic acid dianions with epoxides has been popularized by Iwai and coworkers. The hydroxy-acid products can be cyclized to thioether-substituted butyrolactones (Scheme 3.6).[76,82,87-90] Again, the sulfur-moiety permits chemoselective manipulation of the heterocycle, and butenolides can be formed following oxidation and thermal elimination of phenylsulfinic acid. Pummerer rearrangement gives phenylsulfe(i)nyl butenolides that are good Michael acceptors.[87,88] This strategy has been used in the preparation of lepichlorin. Likewise, the dithioacetal derivatives (Eq. 3.29) also undergo reaction with epoxides and cyclization to lactones.[85]

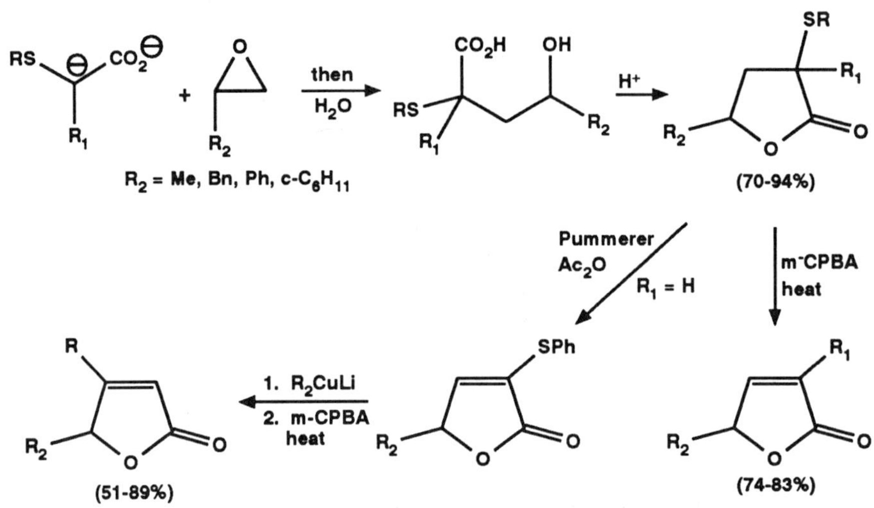

Scheme 3.6. Preparation and reactions of butyrolactones from thioacetic acid dianion addition to epoxides

α-Thio carboxylic acid dianions undergo condensation reactions with aldehydes and ketones.[82,91,92] Saturated carbonyl compounds react with the dianions to give α-thioether-β-hydroxy acids (Eq. 3.30), and α,β-unsaturated ketones give exclusive 1,2-addition.[92] The (phenylthio)acetic acid dianion reacts with saturated acyclic and cyclic esters leading to β-ketoacids that decarboxylate to the α-phenylthio carbonyl compounds in good yield (40-50%).[92] Reaction with α,β-unsaturated esters did not provide any identifiable products. A successful Michael reaction between the (phenylthio)acetic acid or 2-(phenylthio)propionic acid dianion and unsaturated nitro compounds has been reported (Eq. 3.31).[26,93] The intermediate γ-nitro acids are treated with acid to furnish the corresponding keto acids that are transformed into lactones.[26]

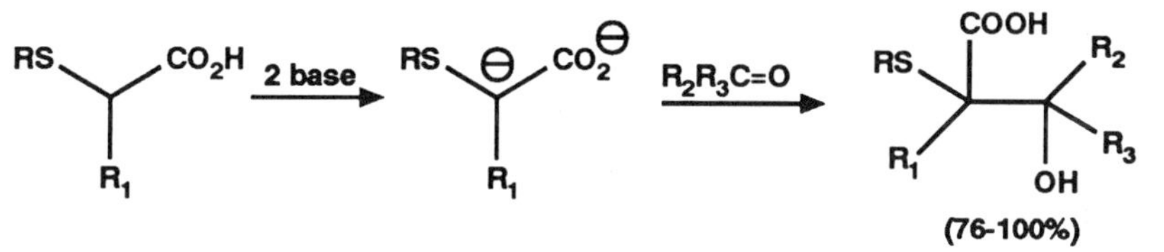

(3.30)

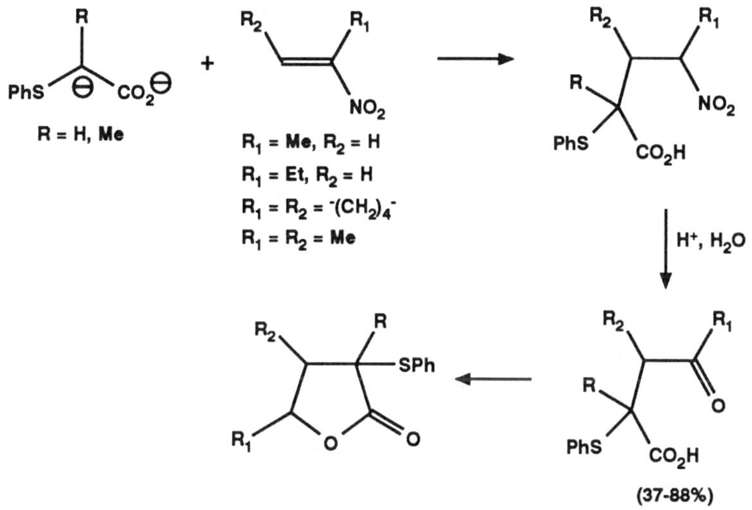

(3.31)

The dianion of (phenylthio)acetic acid has been shown to react with n-propylnitrate to give nitro(phenylthio)methane following expulsion of CO_2.[94] Alkoxy and dialkoxy groups have been used as the stabilizing group at the α-position of carboxylic acid dianions, and have been condensed with aldehydes and ketones.[95-99] Modest to good yields of the alcohol products were obtained (Eq. 3.32). A novel reaction occurred during the [3,3]-rearrangement of a dianion (Eq. 3.33).[100]

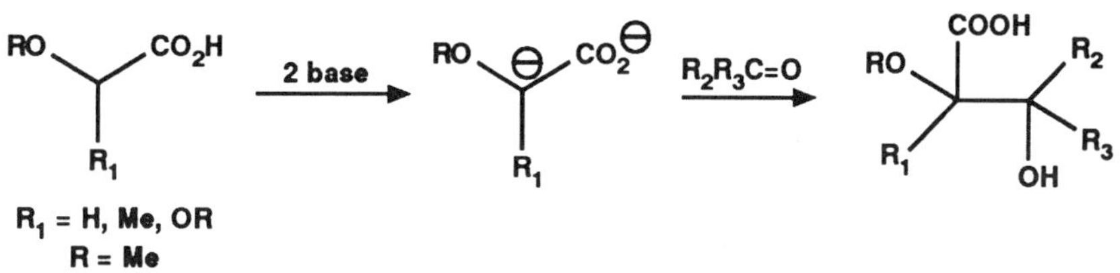

(3.32)

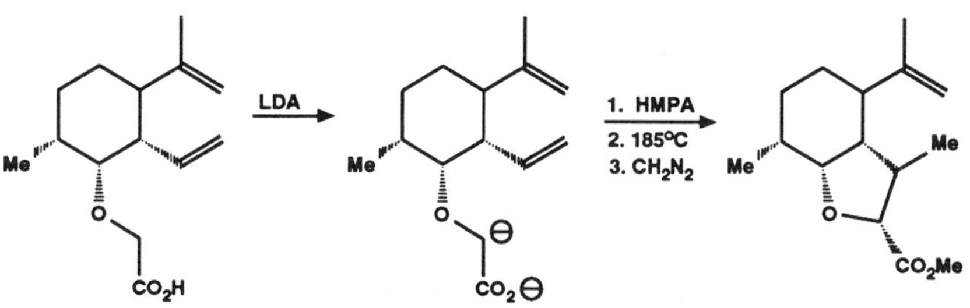

(3.33)

Chloro- and dichloroacetic acid dianions were reacted with alkyl halides and acylating agents, the latter undergoing decarboxylation to form α-chloroketones.[101] Johnson and Bade showed that reaction with ketones results in formation of the intermediate β-chloro alkoxides that close to form epoxides in a reaction reminiscent of the Darzens condensation (Eq. 3.34).[101,102] In a similar manner, conjugate addition takes place when these dianions are added to α,β-unsaturated esters, which undergo cyclization to form cyclopropanes.[101]

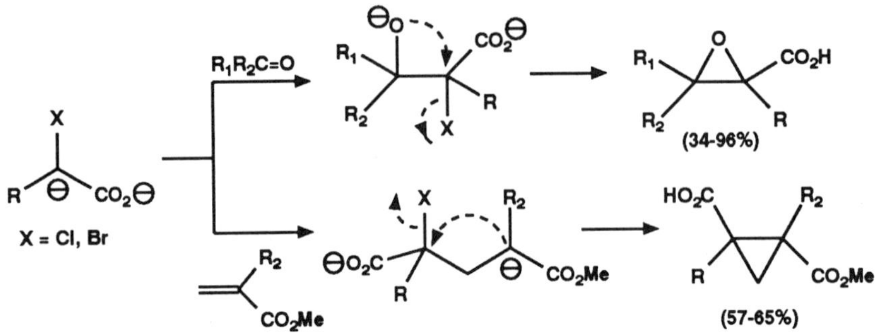

(3.34)

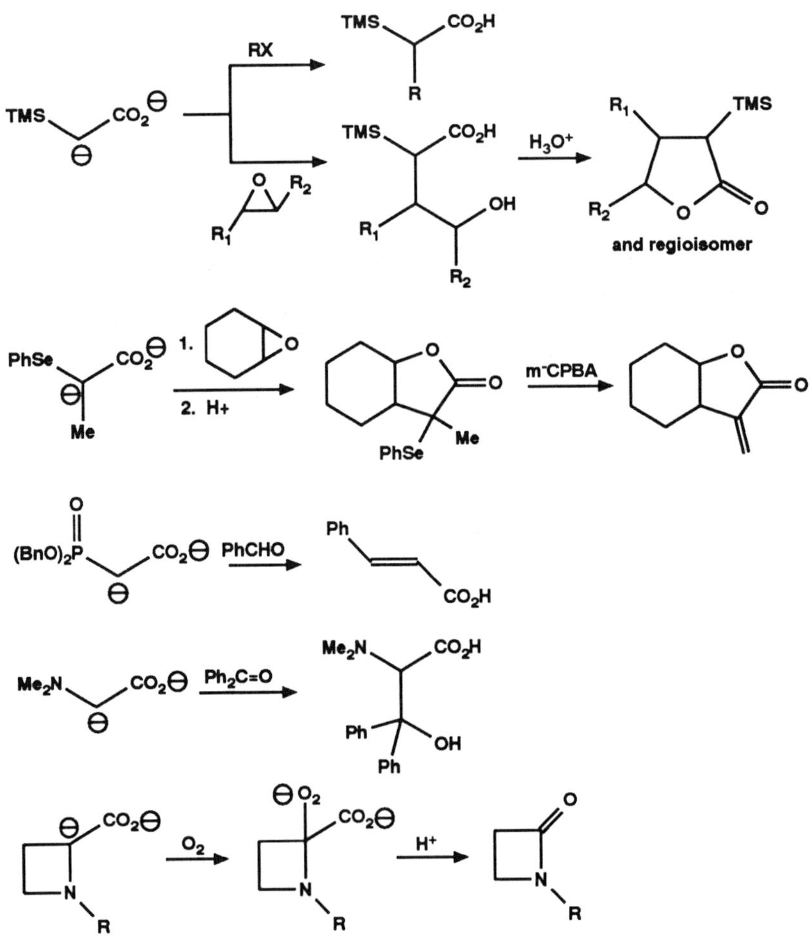

Scheme 3.7. Reactions of miscellaneous heteroatom stabilized α-carboxylate dianions

Careful isolation following addition of chloroacetic acid dianions to ketones can provide the intermediate hydroxy acid, which Black used in a preparation of spirooxetan-2-ones.[103] The analogous dichloroacetic acid dianion can enter into reactions with carbonyls to give epoxides.[104] α-Carboxylate dianions of trimethylsilyl,[105] phenylselenyl,[106] phosphonate,[107] and amine groups[1,108] have been explored (Scheme 3.7).

3.2.3 Reactions of alkenoic and alkynoic acid-based O,C-dianions

The reaction of an [alkyl]-substituted, α,β-unsaturated carboxylic acid with 2 equivalents of base gives an O,C-dianion that can be represented by resonance-delocalized structures (Fig. 3.5). The newly formed allylic carbanion portion of the dianion may prefer the α-(1,3)- or the γ-(1,5)-dianion depending upon several experimental variables including: temperature and/or time of equilibration, substituent effects (of dianion and/or electrophile component), cation, additives, or combinations of these variables. Moreover, eventual capture of the electrophile may depend upon the reversibility of the reaction.

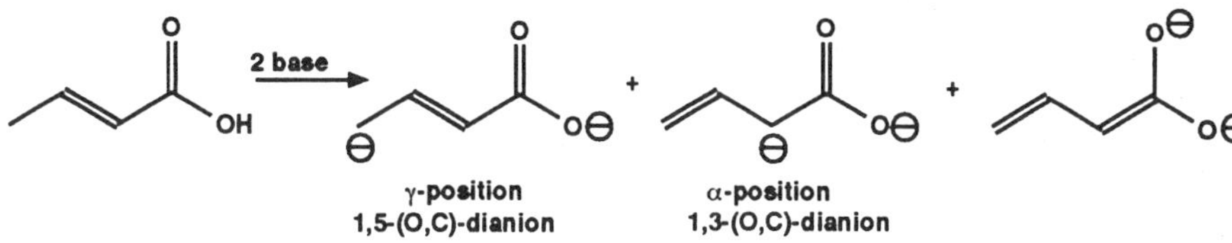

Figure 3.5. Resonance structures for the crotonic acid dianion

Although earlier studies of conjugated acid dianions with alkylating agents indicate a clear preference toward reaction at the α-position (1,3-dianion), more recent investigations suggest that a thorough understanding of both the reacting system and the interplay of variables are essential to establish specific regiochemical outcomes. As an example, the alkylation of the crotonic acid dianion by an alkyl halide (RX) is shown (Scheme 3.8).

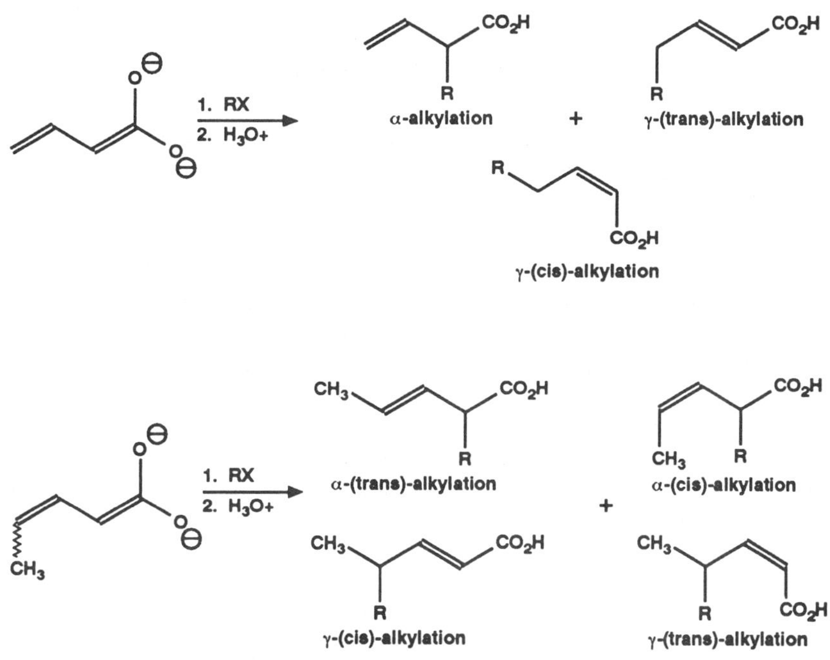

Scheme 3.8. Reaction of the crotonic and pentenoic acid dianion with an electrophile

Following dianion formation, two possible sites of reaction must be considered, α or γ. Further, alkylation at the γ-site can result in the γ-(trans)- and/or the γ-(cis)-product(s). Thus, a minimum of three products are possible for the reaction between an alkenoic acid dianion and a simple electrophile. Reexamination of this reaction as the pentenoic acid dianion leads to an additional stereochemical distinction, since α-alkylation now can afford both *trans* and *cis* geometries (Scheme 3.8).

Other factors must also be considered when using non-symmetric allylic halides, aldehydes, or non-symmetric ketones as electrophiles. In the former case, Sn2 or Sn2′ pathways may occur leading to two possible alkylation products at the α- and/or γ-positions. When certain carbonyl compounds are used as the electrophiles, mixtures of diastereomers also may result.

In summary, early investigations in this area were faced with the regio- and stereochemical control of the dianion reaction. More recent studies, particularly efforts in detailed experimental protocols, have demonstrated that regiochemical carbon-carbon bond construction can be achieved with α,β-unsaturated carboxylic acid dianion chemistry. This section will review the development of these intermediates as a reagent class, provide experimental tactics essential to specific product outcomes, and summarize the important reactions that can be conducted with the various electrophiles.

The first studies with α,β-unsaturated (and β,γ-unsaturated) carboxylic acid dianions probed the positional reactivity and the preference toward isomerization by 2- or 3-hexenoic acid dianions.[109] *Cis*-2-, *trans*-2-, *cis*-3-, and *trans*-3-hexenoic acid were converted to their respective dianions and reprotonated or quenched with MeI. All four dianion species gave 3-hexenoic acid upon reprotonation indicating that the 1,3-O,C-dianion (α-carbanion) is the more stable dianion. The dianion derived from *trans*-2-hexenoic acid gave a 2:1 cis/trans ratio of 3-hexenoic acids upon reprotonation, whereas the dianion generated from cis-2-hexenoic acid gave *trans*-3-hexenoic acid. The dianions derived from *cis*- or *trans*-3-hexenoic acid showed no isomerization upon reprotonation. Exclusive C-2-(α)-alkylation of all four isomers was observed using MeI. Although this study supports α-alkylation, a prior investigation showed exclusive γ-condensation reaction with cyclohexanone with butenoic acid dianions (crotonic or 3-butenoic).[1] Pfeffer and coworkers[110] later found that butenoic acid dianions gave a 4:6 mixture of α:γ products with cyclohexanone, but a 6:4, α:γ alkylation ratio with MeI. In light of the studies with hexenoic acids, the authors suggested that charge density played a major role in the product ratio differences.

In support of these later findings, Umani-Ronchi and coworkers studied the condensation reaction between benzaldehyde and 3-methyl-2(or 3)-butenoic acid dianions and found a 54:46 ratio of α:γ products when the dianion was prepared as the dilithio salt (Scheme 3.9).[111] When the 3-methyl-2-butenoic acid dianion is prepared as the dipotassio salt, exclusive γ-attack occurs (46%). The mixed Li/Sn salt gives exclusive α-condensation-forming diastereomers, whereas the Na/Li or K/Li salts gave α:γ-mixtures varying in composition. However, this cation effect does not alter the alkylation (by MeX) of these dianions, giving products of exclusive α-alkylation.[112] Addition of HMPA could increase the α:γ alkylation ratios as high as 65:35 when the dipotassio salt was employed.

The influence of the cation upon the dianion reactivity was not fully realized until Katzenellenbogen and coworkers studied the alkylation of six different alkenoic acid dianions by allyl halides.[113,114] In general, the α to γ alkylation ratio obtained by reaction with allyl bromide and its congeners ranged from 96:4 to 100:0 when the cations are lithium. In stark contrast, the α to γ ratio ranged from 1:4 to 0:100 when the cation was copper, a complete reversal in regiochemistry. Further analysis of the system showed that the alkylation mechanism proceeded by an Sn2′ mechanism when the copper enolate was unsubstituted at the γ-position but mixtures of Sn2 and Sn2′ products were obtained when one substituent was present at the γ-position. Non-allylic electrophiles reacted slowly with the copper

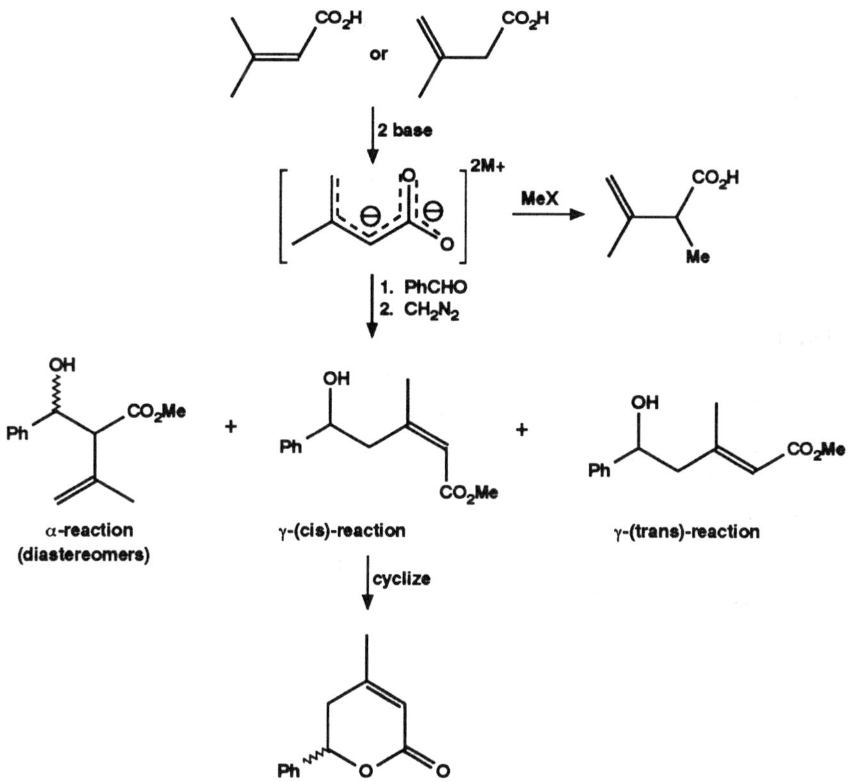

Scheme 3.9. Regiochemical outcomes of the reactions of the 3-(methyl)butenoic acid dianion

dienolates and gave little γ-selectivity. Vinyl epoxides were found to be excellent γ-selective alkylating agents (89-93%) giving only (E)-stereochemistry from tiglic and crotonic acid dianions, while the senecioic dianion gave mostly the (Z)-stereochemistry (Eq. 3.35).[114]

Cardillo and coworkers used a different approach in their efforts to uncover the optimum conditions for selective α- or γ-selective reactivity.[115,116] Using small amounts of HMPT (10% by volume) at low temperatures, kinetically controlled electrophilic reactions can be obtained. At reflux, electrophilic attack at C-2 (α) becomes reversible, whereas C-4 (γ) reaction is not reversible. Further, reaction of the 3-methyl-2-butenoic acid dianion with aldehydes at elevated temperatures gave exclusive Z-stereochemistry of the newly formed olefin (Eq. 3.36).

Exclusive γ-alkylation was also observed upon addition to decanal in THF/HMPA (9:1) at room temperature,[116] the intermediate then used in the synthesis of malyngolide.

The reversibility of the C-2 or α-reaction at elevated temperatures and the effect of the counterion was used to advantage during a synthesis of perillenol.[117] In that study, an exclusive α-Michael reaction occurred between the dilithio tiglic acid dianion and methyl acrylate (-78°C or -30°C). At 0°C, a 4:1, α:γ product distribution was found, which increased to 3:1 using copper as the counterion. Carbon-13 and *ab initio* calculations suggest that the lithium metal of the dienolate lies above the plane of the senecioic acid dianion.[118]

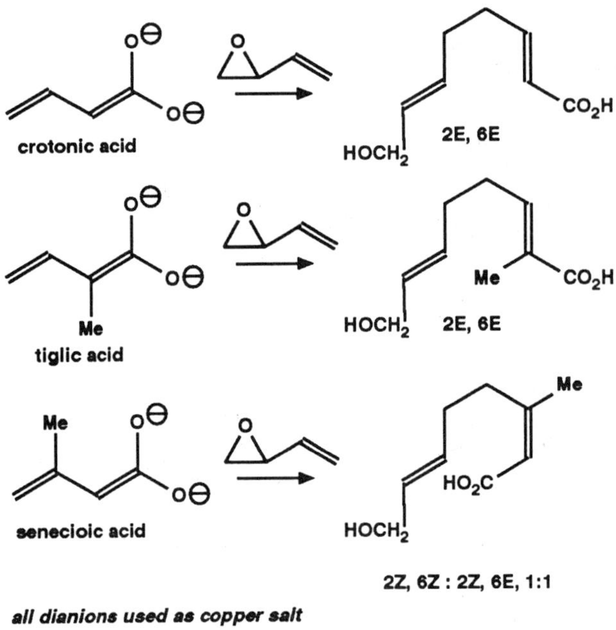

(3.35)

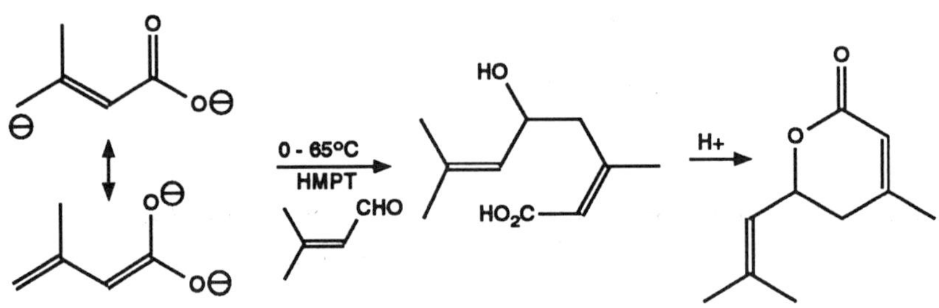

(3.36)

The reaction between α,β-unsaturated carboxylic acid dianions with aldehydes and ketones[119] or Michael acceptors[120] has been studied in detail. Crotonic and tiglic acid dianions react with aldehydes or symmetrical ketones to give α- and γ-product distributions as expected from kinetic or thermodynamic control (Eq. 3.37).[111,112,115,116]

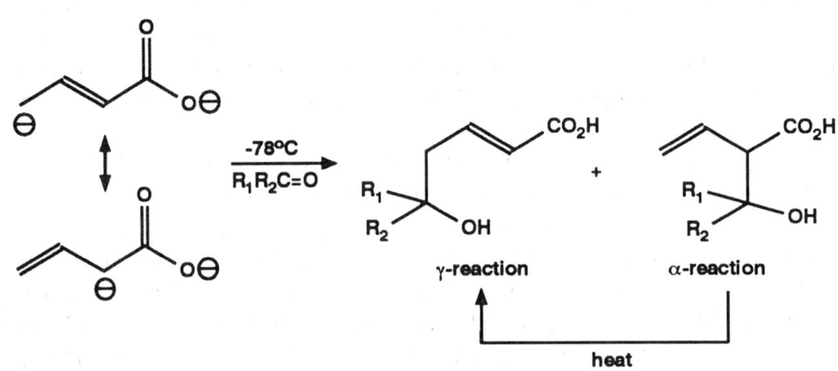

(3.37)

When styryl ketones were reacted with the crotonic acid dianion, four theoretical products can form, α-1,2-addition, γ-1,2-addition, α-1,4-addition, and γ-1,4-addition, although no 1,2-addition was observed (Eq. 3.38).[120] The α-1,4 to γ-1,4 ratio varied greatly with the bulk of the substituent R_1 (Table 3.7).

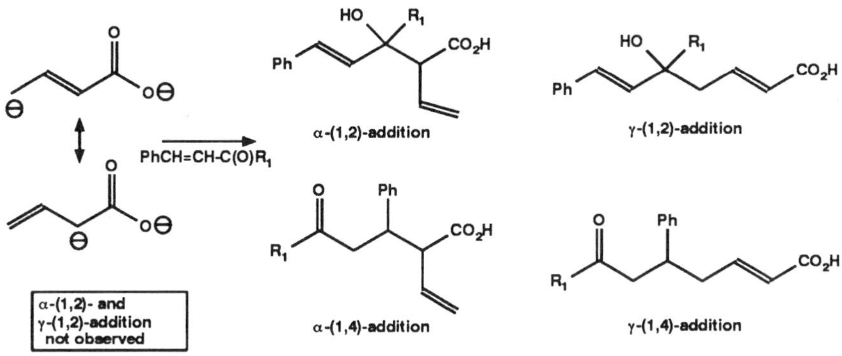

(3.38)

Table 3.7
Effect of Keto-Substituent on γ-Reaction with the Crotonic Acid Dianion (Eq. 3.38)

R_1	α-1,4-product	γ-1,4-product	Yield (%)
Me	20	80	-
Pr	36	64	87
iPr	47	53	70
iBu	55	45	71
tBu	85	15	92
Ph	99	1	81

Close examination of specific Michael addition products showed that only the R,R (or S,S) diastereomer had formed, likely arising from a tandem addition/oxy-Cope rearrangement (Scheme 3.10).[120]

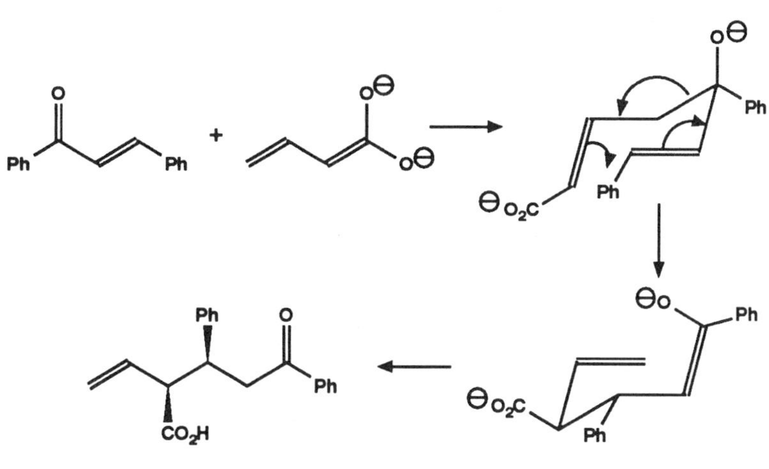

Scheme 3.10. Tandem dianion addition/oxy-Cope rearrangement

Several interesting synthetic uses of the α- and/or γ-reaction products have been conducted. Lactones have been accessed from either itaconic monoester dianions[121] or from α-methyl (or benzyl) cinnamic dianions (Eq.3.39).[122]

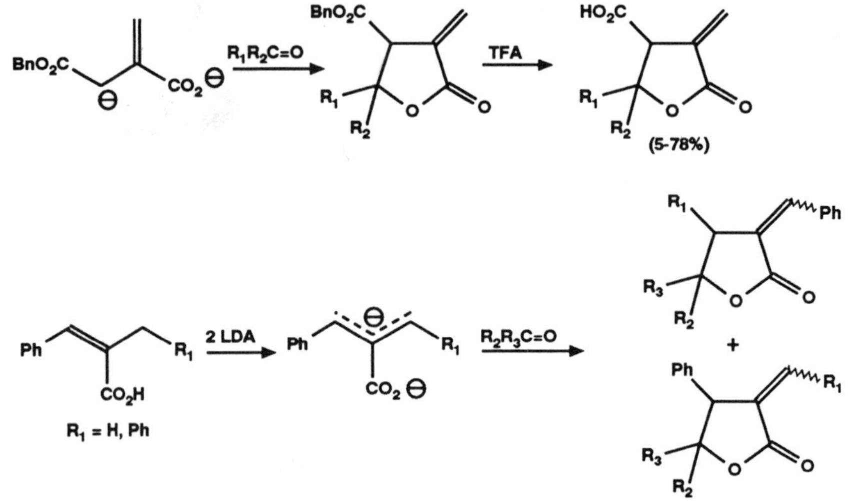

(3.39)

Mestres and coworkers have revealed useful conditions for the oxidative coupling (I_2)[123] and amination using $(PhO)_2P(O)ONH_2$[124] of π-delocalized carboxylic acid dianions, and the crotonic acid dianion has been found to add to haloaromatics via a benzyne intermediate to form 4-(aryl)butenoic acids in modest yield.[125] Heteroatom stabilization as thioethers has provided an interesting entry to butenolide α-formyl acetals.[126] Allan and Fong[127] used a conjugated carboxylic acid dianion in an effort to functionalize and deconjugate nipecotic acid derivatives (Eq. 3.40). The regioselective γ-halogenation with N-chlorosuccinimide is worth noting.

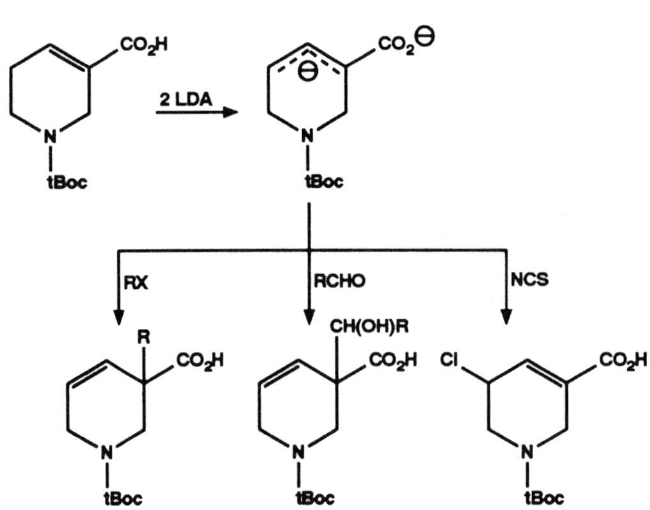

(3.40)

A sub-class of π-delocalized carboxylic acid dianions is that generated by Birch reduction of benzoic acid and its derivatives. Although the Birch reduction does not produce dianions via bis-deprotonation, select examples are chosen to illustrate their utility as O,C-dianions. Despite possible π-delocalization, there has been little evidence found for any γ-reaction of Birch intermediates, and dianions of benzoic acid and its substituted derivatives generated using metal/ammonia conditions undergo predominant α-alkylation in good to excellent yield (Eqs. 3.41 and 3.42).[128,129] The corresponding reaction with epoxides[130] and Michael acceptors[131] has been reported (Eq. 3.43).

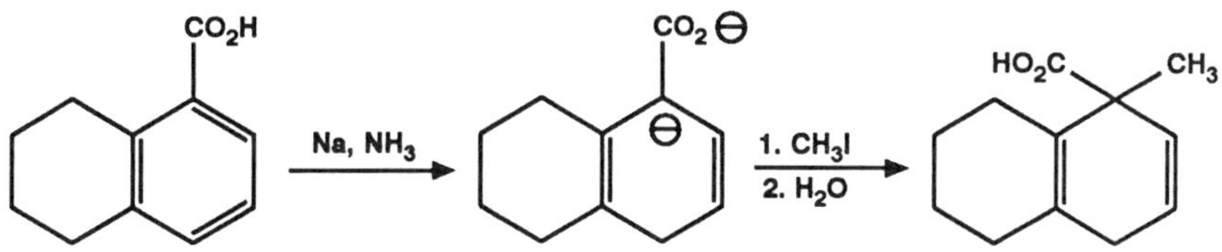

(3.41)

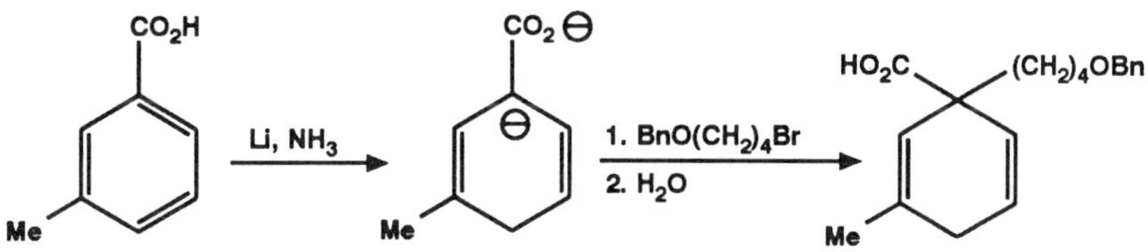

(3.42)

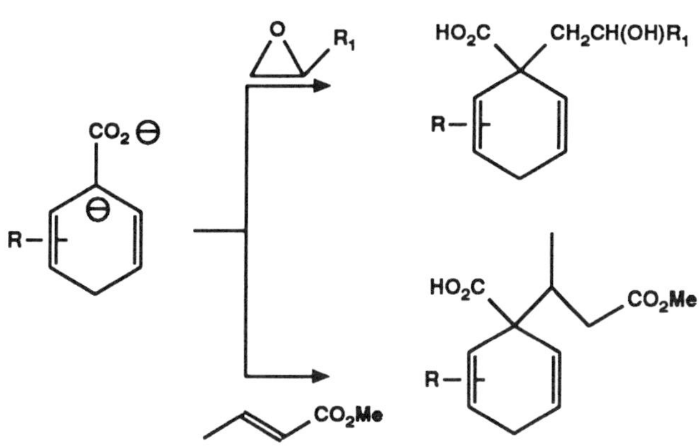

(3.43)

An alternative for γ-functionalization using the aldimine of tiglaldehyde was reported by Vedejs.[132]

Alkynoic acid dianions

A second class of compounds that can be transformed into dianion intermediates of the α,β-unsaturated type are the alkynoic acids. Within this subdivision, two dianion types have been explored; butynoic and propynoic (Fig. 3.6). Butynoic acids are bis-deprotonated successively at the acid-OH and propargylic methyl (or substituted methylene) to form dianions that can be represented as two resonance-delocalized structures, propargylic and vinylic.

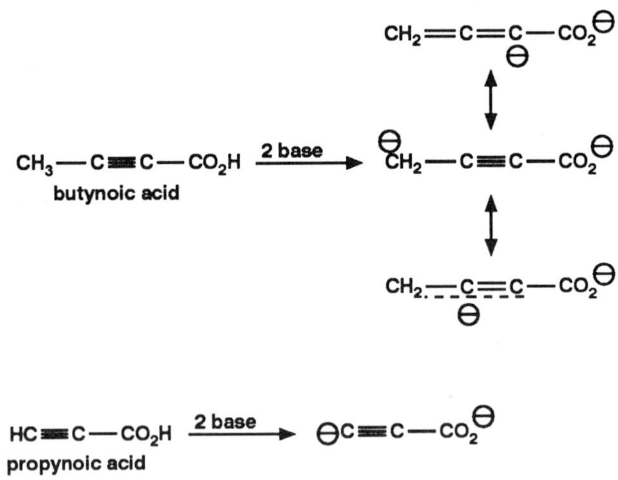

Fig. 3.6. Alkynoic acid dianion types

The butynoic acid dianion reacts with allyl bromide[133,134] and aldehydes (Scheme 3.11)[135,136] to give γ:α product mixtures (Scheme 3.11).

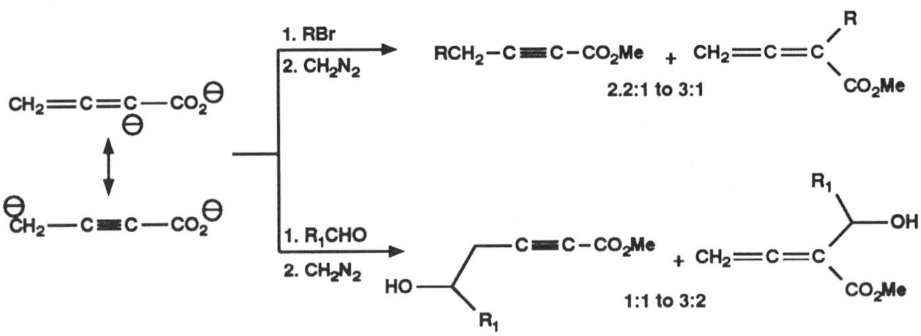

Scheme 3.11. Reaction of butynoic acid dianion with electrophiles

The propynoic acid dianion is a useful 3-carbon chain-extension reagent, and reacts exclusively with electrophiles at the alkyne carbon. Addition to epoxides leads to 5-hydroxy-2--alkynoic acids that have been manipulated to pyranones including (+/-)-pestalotin and optically active hexadecalactone (Scheme 3.12).[137-139]

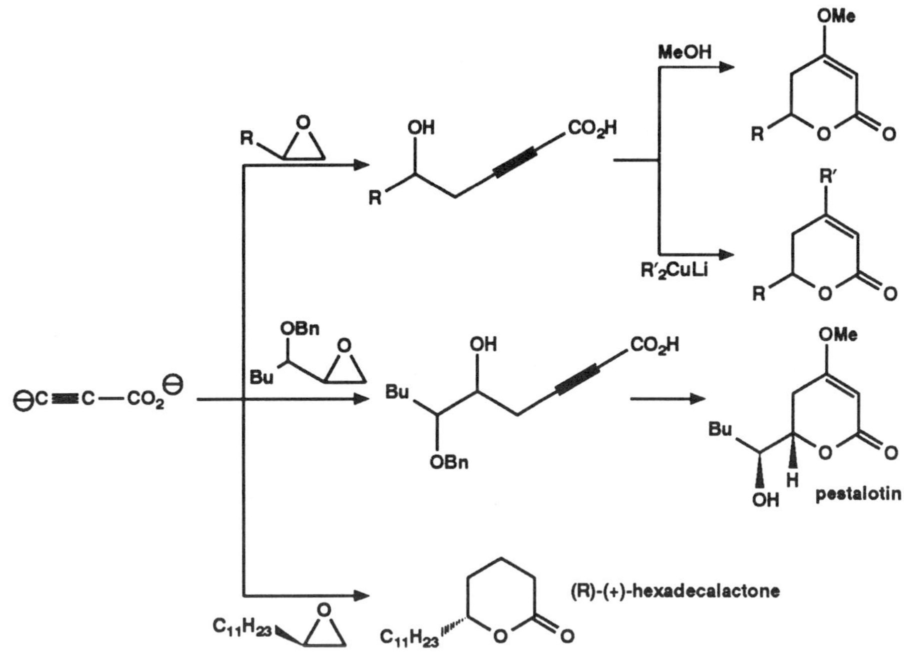

Scheme 3.12. Uses of the propynoic acid dianion in synthesis

3.2.4 Reactions of benzoic acid-based O,C-dianions

Aromatic and heteroaromatic carboxylic acids may be bis-deprotonated to form dianions that vary in the position and reactivity of the carbanion site. Two distinct sites for carbanion formation have been researched, ring metalation and lateral (lateral-type) deprotonation,[140] as illustrated in Figure 3.6. Ring metalation describes carbanion generation at a ring carbon, whereas lateral deprotonation refers to formation of a carbanion at an alkyl (benzylic or benzylic-type) substituent. Frequently, the carbanion "site," whether lateral or metalated, is formed adjacent to the carboxylate anion, although the specific anion positions are governed

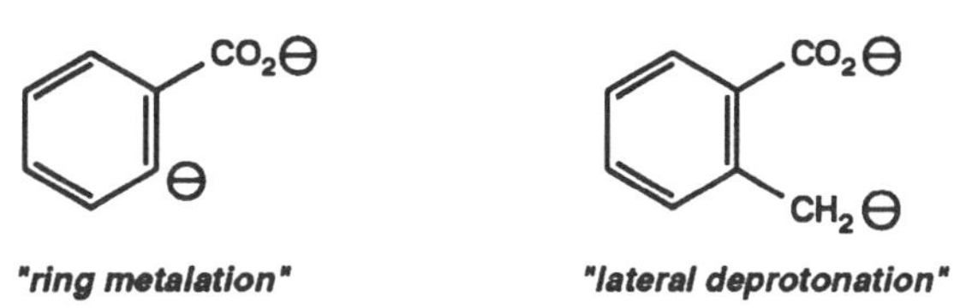

Figure 3.6. Types of aromatic dianions

by the ring substituent pattern. In general, the dianions of benzoic acid and toluic acid have not been as widely explored as certain heteroaromatic species like furan and thiophene. In part, this is due to the increased stabilization imparted to the carbanion center by the heteroatom making dianion access a favored process over carbonyl addition, which occurs with benzoic acid and, to a lesser extent, toluic acid. Further, the widespread use of "ortho" directing groups has obviated the need for simple aromatic dianions.[141,142]

The isomeric toluic acids have been converted to the corresponding dianions and reacted with electrophiles.[10,143] The *o*-toluic acid dianion was generated using 2 BuLi and alkylated in good yield. Further, the *ortho*- and *para*-toluic acid dianions could be coupled with 0.5 I_2. The *meta*-toluic acid dianion, however, was slow to form and gave low yields of product. The *p*-toluic acid dianion was generated with 2 LDA to suppress ketone formation.

The azacyclic analogs of *o*-toluic acid, 2-methyl and 4-methyl nicotinic acid, undergo lateral deprotonation with LDA and condense with aldehydes and ketones to form dehydro azacoumarins following cyclization (Scheme 3.13).[144]

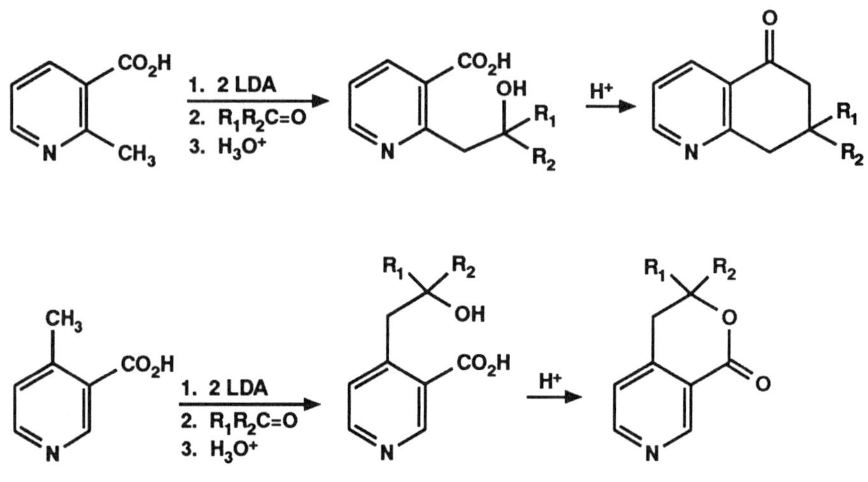

Scheme 3.13. 2-Methyl and 4-methylnicotinic acid dianions

The dianions of furanoic acid and thiophene carboxylic acid, their methyl-substituted derivatives, and their benzo-fused analogs have been the subject of several studies largely by Knight and coworkers. Both the presence and position of a methyl substituent affects the type of dianion that is formed, namely ring metalated or lateral carbanion. Furan-2-carboxylic acid is converted exclusively to the 5-metalated carbanion, whereas furan-3-carboxylic acid forms the 2-metalated carbanion.[145] These dianions do not react well with alkyl halides or epoxides, but condense efficiently with aldehydes and ketones (Scheme 3.14).

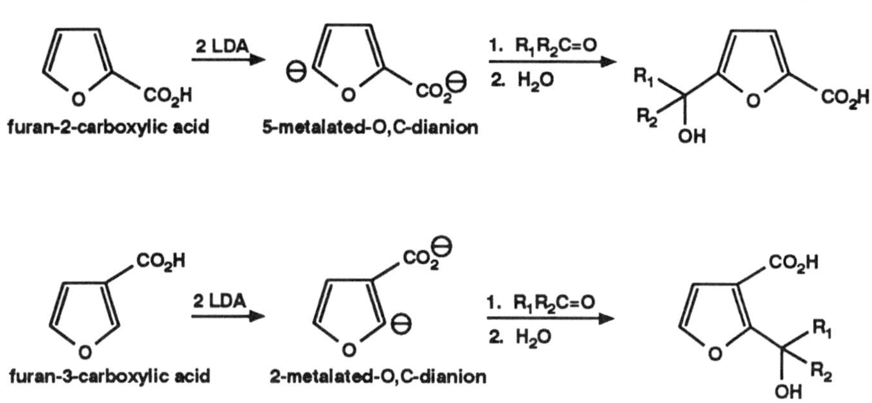

Scheme 3.14. Dianions of 2- and 3-furan-carboxylic acid

When methyl furoic acids are converted to their respective dianions, the substitution pattern regulates whether ring metalation or lateral deprotonation dominates (Scheme 3.15).

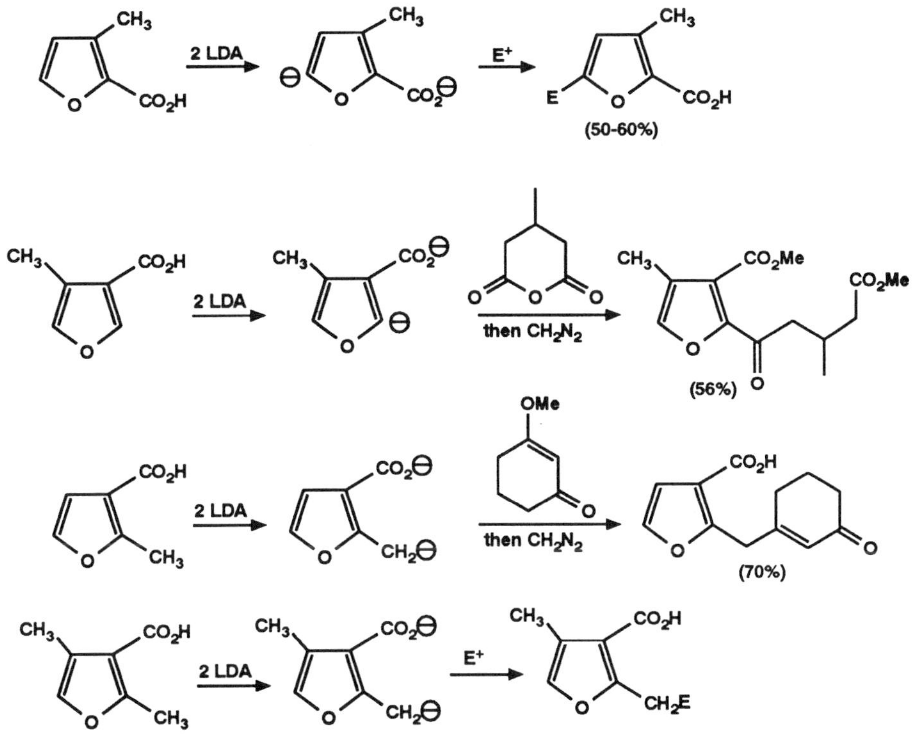

Scheme 3.15. Furanoic acids dianions

3-Methyl-2-furanoic acid[145] and 4-methyl-3-furanoic acid lead to ring metalated 5- and 2-carbanionic species, respectively. 4-Methyl-3-furanoic acid forms the 2-positional ring metalated dianion, and was used in the synthesis of gnididione.[146] Conversely, 2-methyl-3-furanoic acid[147] and 2,4-dimethyl-3-furanoic acid[148] form dianions via lateral deprotonation, which were used in the synthesis of isoalantolactone and furanoeremorphilane, respectively. 2,4-Dimethyl furanoic acid dianion forms the carbanion at the 2-methyl, indicating a lower proton acidity for this site than the 4-methyl substituent. Proximity to the heterocyclic oxygen may be responsible for the difference in acidity.

Benzofuran carboxylic acid dianions also have been generated and reacted with electrophiles, although several additional features play a role in the product outcome.[149-152] Benzofuran-3-carboxylic acid forms the furan ring metalated O,C-dianion, which reacts with TMS-Cl and carbonyl compounds but does not undergo alkylation well (Eq. 3.44).[149,150]

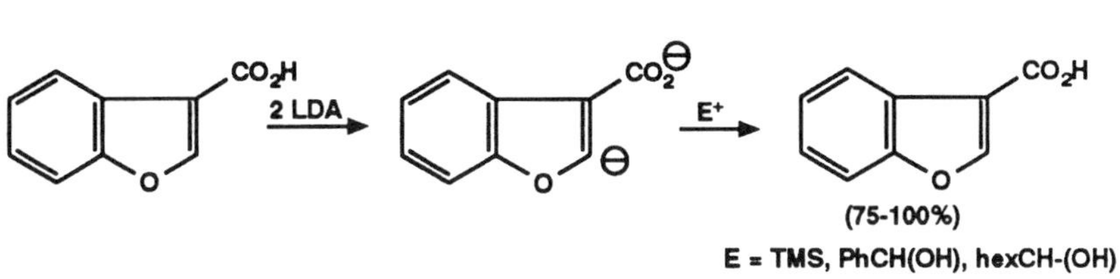

(3.44)

Benzofuran-2-carboxylic acid forms the 3-furan ring metalated dianion but has a transient lifetime and undergoes ring opening to the phenol (Eq. 3.45),[149,150] although trapping with carbon dioxide to form 2,3-benzofuran dicarboxylic acid was possible.[151] The intermediacy

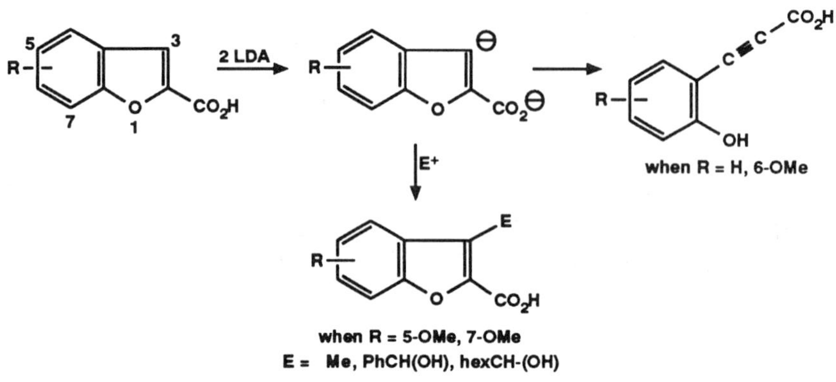

(3.45)

of the benzofuran-2-carboxylic acid dianion could be controlled, in part, by aromatic ring substituents. The 5- and 7-methoxy analogs of the dianion showed no phenol formation, and gave high yields of alkylation and condensation products, whereas the 6-methoxy derivative gave ring opening (Eq. 3.45).[149,150] 3-Methylbenzofuran-2-carboxylic acid forms the lateral O,C-dianion that can be trapped with triisopropylsilyl chloride (TIPS-Cl; R = iPr) and manipulated to benzofuran-2,3-xylylene precursors (Eq. 3.46).[152]

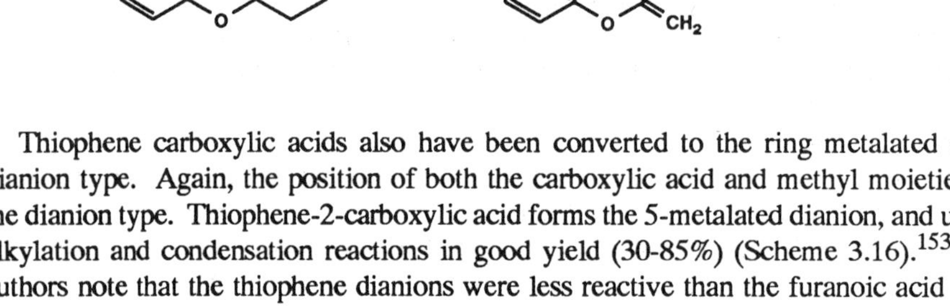

(3.46)

Thiophene carboxylic acids also have been converted to the ring metalated or lateral dianion type. Again, the position of both the carboxylic acid and methyl moieties control the dianion type. Thiophene-2-carboxylic acid forms the 5-metalated dianion, and undergoes alkylation and condensation reactions in good yield (30-85%) (Scheme 3.16).[153,154] The authors note that the thiophene dianions were less reactive than the furanoic acid dianions.

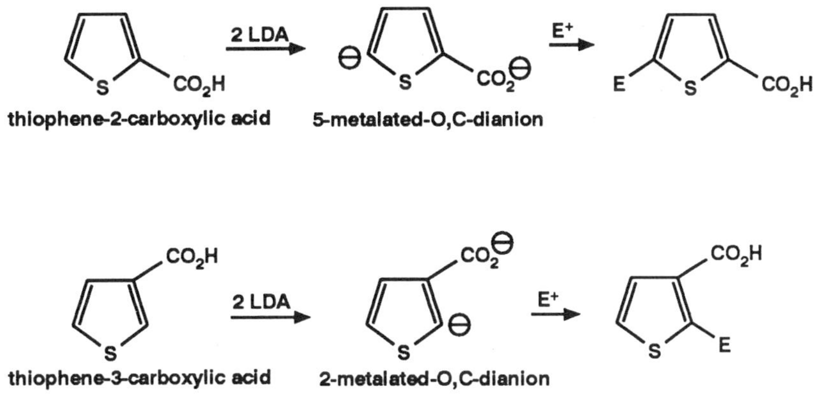

Scheme 3.16. Formation of thiophene carboxylic acid dianions

The isomeric methyl thiophene-2-carboxylic acids were converted into O,C-dianions.[155] 5-Methylthiophene-2-carboxylic acid forms the 5-methyl carbanion, and 4-methylthiophene-2-carboxylic acid forms the 5-ring metalated carbanion. However, 3-methylthiophene-2-

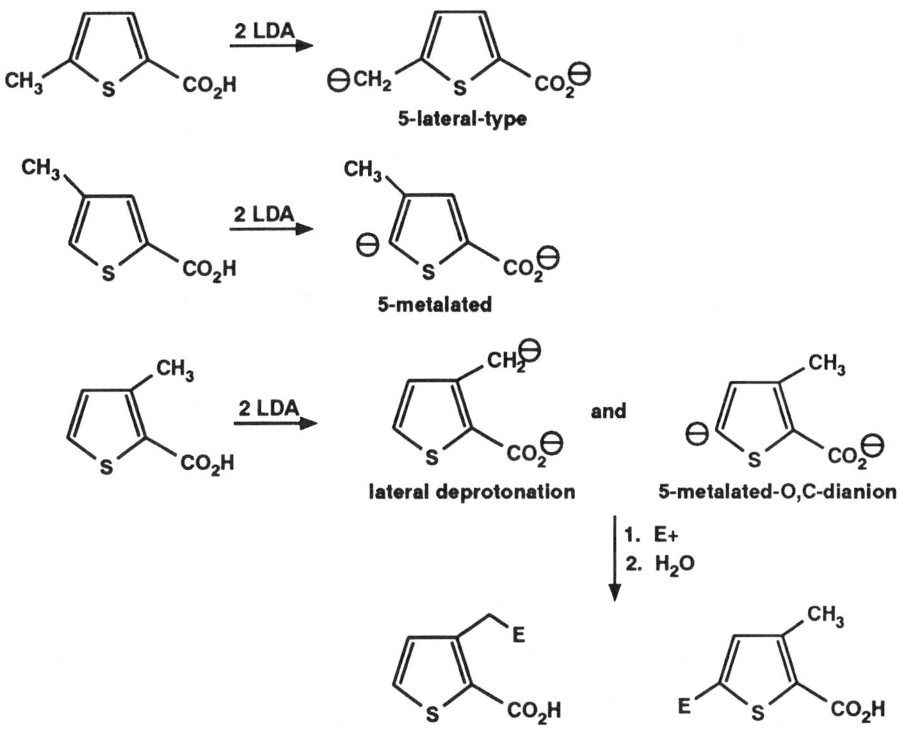

E		
D	74	26
Me	65	35
Me_2CHOH	62	38

Scheme 3.17. Formation of methyl-substituted thiophene carboxylic acid dianions

carboxylic acid forms a mixture of 5-metalated and 3-lateral dianion types (Scheme 3.17) as evidenced by various product mixtures. Last, 3,5-dimethylthiophene-2-carboxylic acid affords the 5-lateral O,C-dianion,[155] and 4-substituted-thiophene-3-carboxylic acid gave the 2-ring metalated dianions.[156] Oxazole-,[157,158] thiazole-,[157,158] and isoxazole-carboxylic acids[159] were converted to dianions and reacted with electrophiles (Scheme 3.18).

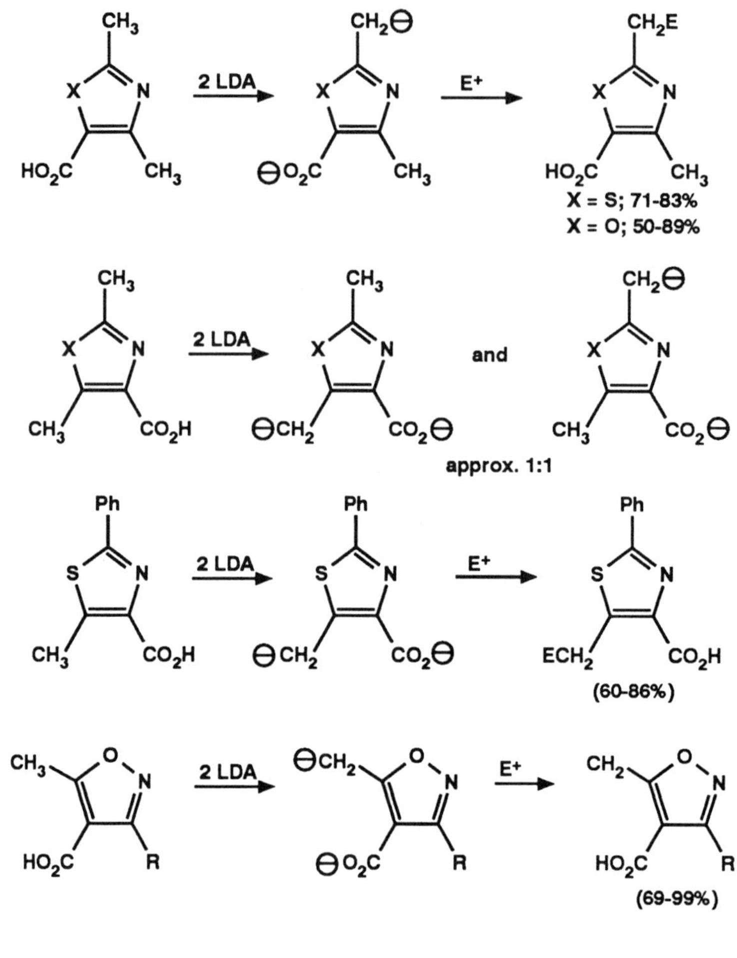

Scheme 3.18. Dianions of various carboxylic acid azoles

3.2.5 Reactions of remote carboxylic acid-based O,C-dianions

As seen in the preceding sections, a wide array of alkanoic, alkenoic, alkynoic, and aromatic acids may be used to form carboxylic acid-based O,C-dianions. A majority of these dianions utilize the carboxylate carbonyl to further promote and stabilize the carbanion at the enolate or α-position giving rise to 1,3-(O,C)- or α,α'-(O,C)-dianions. The remainder of those intermediates previously presented that do not use the carbonyl group to enhance carbanion stability, for example the propynoic (Sect. 3.2.3) and aromatic (Sect. 3.2.4), were placed with similar functional group counterparts or, owing to novel chemistry, given their own classification/section. In the following section, carboxylic acids in which the carbonyl group does not participate in the stabilization of the carbanion center will be covered. In these instances, the carboxylic acid dianion reacts with electrophiles at a carbanion site that appears at least one intervening methylene (or functional group spacer) from the carboxylate carbonyl. These dianion species, known as "remote carboxylate dianions," can be further distinguished by the distance appearing between the oxyanion and carbanion positions, for example 1,4-(O,C)-, 1,5-(O,C)-, or 1,Z-(O,C)-dianions, where Z refers to the numerical carbanion position relative to the oxyanion (Fig. 3.7). In many reports, the remote 1,4-dianion and 1,5-dianion types have been termed homoenolate and bis-homoenolate carboxylate dianions, respectively.[4,5]

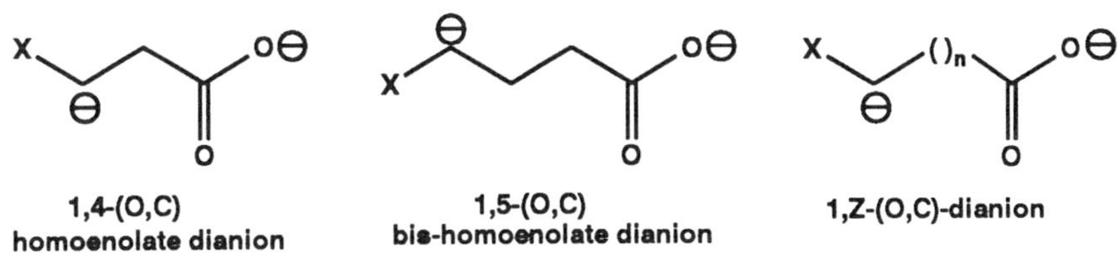

Figure 3.7. Examples of remote carboxylic acid dianions

To access any of the remote carboxylate dianions, the molecule must bear a carbanion-stabilizing group adjacent to the remote carbanion position that is superior to the carboxylate carbonyl, or the carbanion site must be generated through metal-halogen or metal-metal exchange and reacted before thermodynamic equilibrium can occur. In most cases, stabilizing groups are used to isolate the remote position among which the sulfinyl, sulfonyl, carboxamide and carboxyester groups have been the most popular.

1,4-Dianions of carboxylic acids

Kofron and Wideman[160] showed that monoesters of succinic acid are converted to the 1,4-dianion (2 $LiNH_2$) and react with alkyl halides or carbonyls (Eq. 3.47). Importantly, alkylation of the corresponding substituted succinate diester leads to mixtures of alkylation products showing the advantage gained in regiocontrol when using the dianion approach.

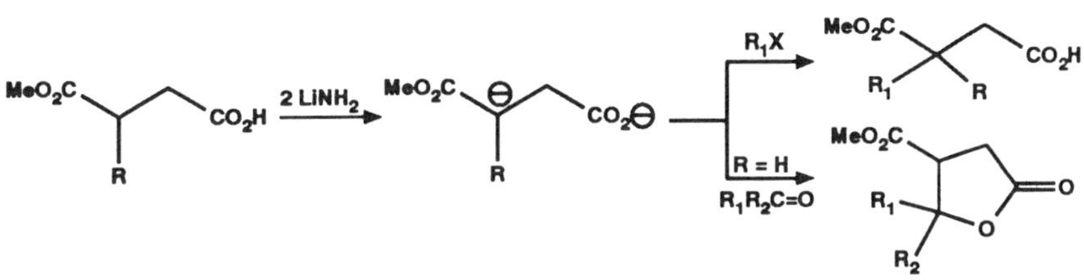

(3.47)

Iwai[161] and Bravo[162] studied 1,4-remote propanoic acid dianions with sulfur-based stabilizing groups at the 3-position. 3-(Phenylsulfinyl)- and 3-(phenylsulfonyl)propanoic acid dianions form the corresponding 1,4-(O,C)-dianions upon treatment with 2 LDA and react with simple carbonyls to give lactones following cyclization (Eq. 3.48).[161] The β-sulfinyl lactone could be converted to the the butenolide after pyrolysis.

Enantioenriched lactones were accessed using an asymmetric sulfinyl group as the remote stabilizing moiety (Scheme 3.19).[162] Cis-pyrolytic elimination led to optically pure butenolides in good yield.

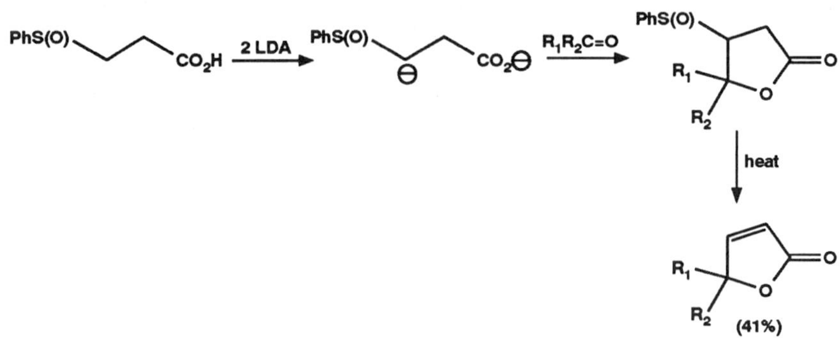

(3.48)

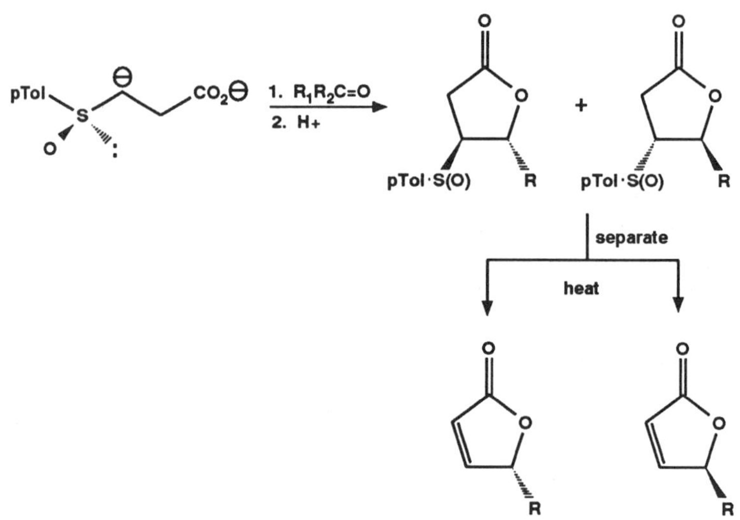

Scheme 3.19. Preparation of optically pure lactones using remote dianions

The lactam carbonyl was a remote stabilizing group in route to thienamycin (Eq. 3.49).[163]

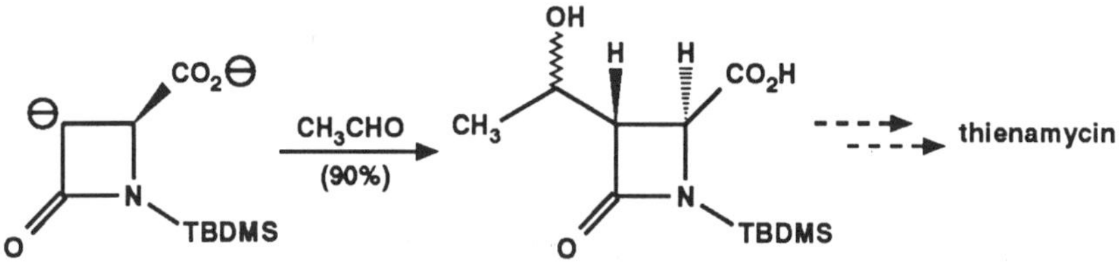

(3.49)

Katritsky and Sengupta reported a 1,4-(O,C)-dianion from trimethylsilylcarbonate.[164] Although this last example does not fit the exact description of a carboxylic acid dianion, it does show how intervening heteroatoms can lead to interesting dianion-based reagents, in this case a methylene transfer reagent.

1,5-Dianions of carboxylic acids

Carboxylic acid dianions with two intervening elements between the oxygen and carbon anion positions (1,5-dianions) have been explored. The β-lactam carbonyl group was used to stabilize the 5-carbanion of 2-azetidinone-4-acetic acid in route to thienamycin,[165] and an N-carboxysuccinimide (generated in situ) was converted to the remote dianion and condensed with the imidazolide of N-CBZ-valine during a synthesis of andrimid (Eq. 3.50).[166]

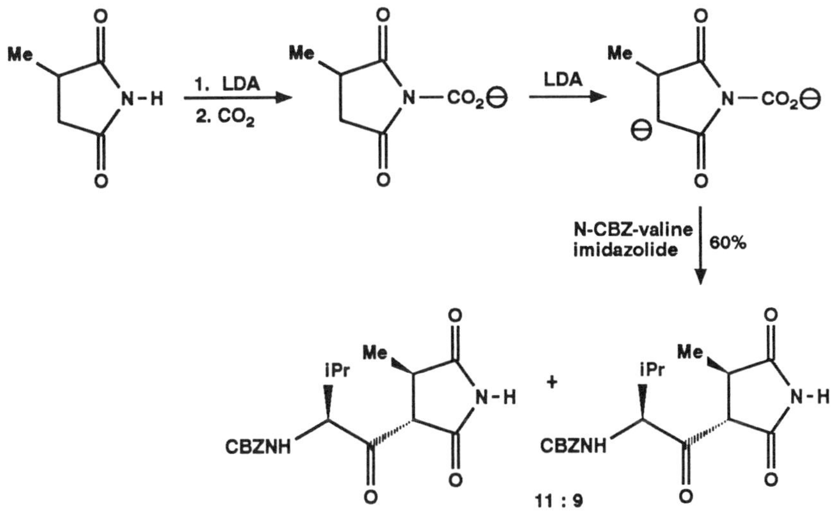

(3.50)

Thompson and coworkers reacted the dianion of 4-(phenylsulfonyl)butanoic acid (4-PSBA; using 2 BuLi) with various electrophiles (Scheme 3.20).[167-175] Condensation with aldehydes and ketones leads to 5-(phenylsulfonyl)-δ-lactones following trifluoroacetic

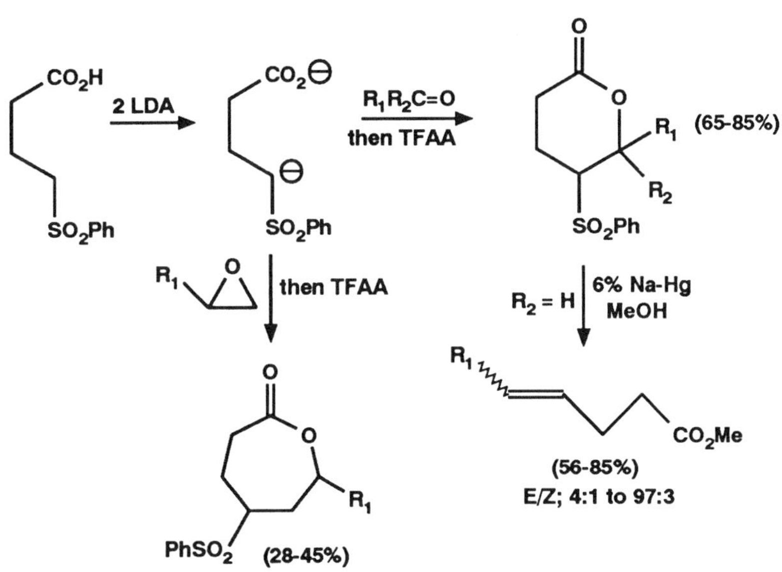

Scheme 3.20. Reaction of the dianion of 4-PSBA with electrophiles

anhydride-assisted cyclization[167,168] in a reaction sequence similar to Iwai[161] who used the dianion of 3-(phenylsulfonyl) propanoic acid dianion. The lactones undergo reductive ring opening with 6% Na-Hg in methanol to afford γ,δ-unsaturated methyl esters,[168] a net 4-carbon chain elongation of the starting carbonyl compound (R_1). The stereochemistry of the alkene follows the steric composition of the carbonyl component. This process was used in the preparation of sulcatole.[169] Epoxides serve as electrophiles for the 4-PSBA dianion giving ε-lactones after cyclization although lower yields were obtained.[170]

The most important carbon-carbon bond forming reaction that can be performed with the dianion of 4-PSBA is the addition to activated imines (Scheme 3.21). Under standard

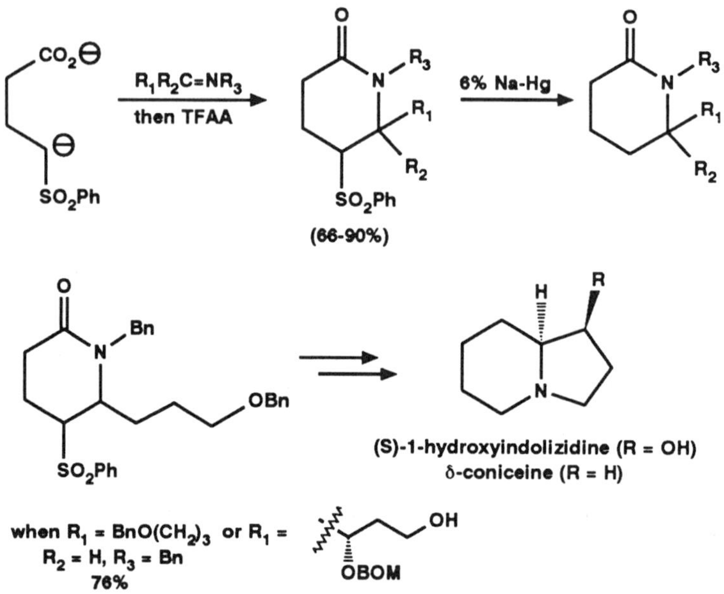

Scheme 3.21. Addition of 4-PSBA dianion to imines

reaction conditions the dianion of 4-PSBA adds to imines in poor yield, however, addition of a molar equivalent of BF_3 etherate to the imine prior to addition of the dianion "activates" the C=N bond toward addition.[171] The resultant amino-carboxylate dianion is cyclized *in situ* with trifluoroacetic anhydride (TFAA) to give lactams. The procedure was amenable to Schiff bases, aryl-alkyl imines and dialkyl imines (Scheme 3.21; $R_1 = R_2 = -(CH_2)_5$), and was used as the pivotal step in the construction of δ-coniceine,[172] a histrionicotoxin analog,[173] (S)-1-hydroxyindolizidine,[174] and several alkyl indolizidines.[175] Unlike the lactones, the sulfone group can be removed with Na-Hg without ring opening.[172,174,175]

Bravo and coworkers used the dianion of α-methylene-γ-(tolylsulfinyl)butyric acid to access α-methylene and α-methyl lactones (Eq. 3.51).[176]

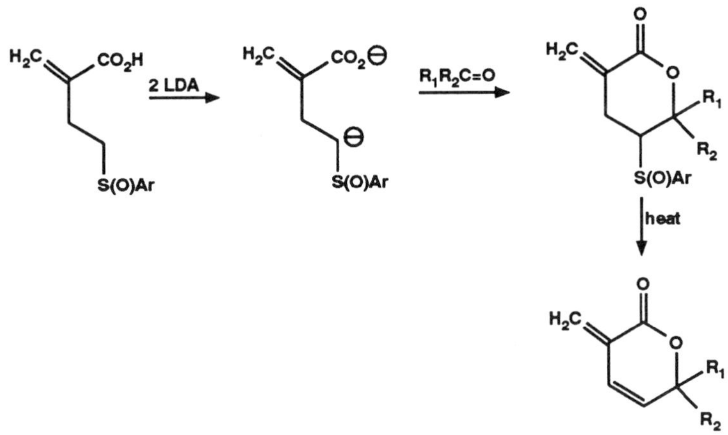

(3.51)

1,Z-Dianions of carboxylic acids

A summary of these dianions are provided in Scheme 3.22. Both 1,6- and 1,7-O,C-dianions of carboxylic acids were prepared from the corresponding 3-keto- and 4-keto acids,

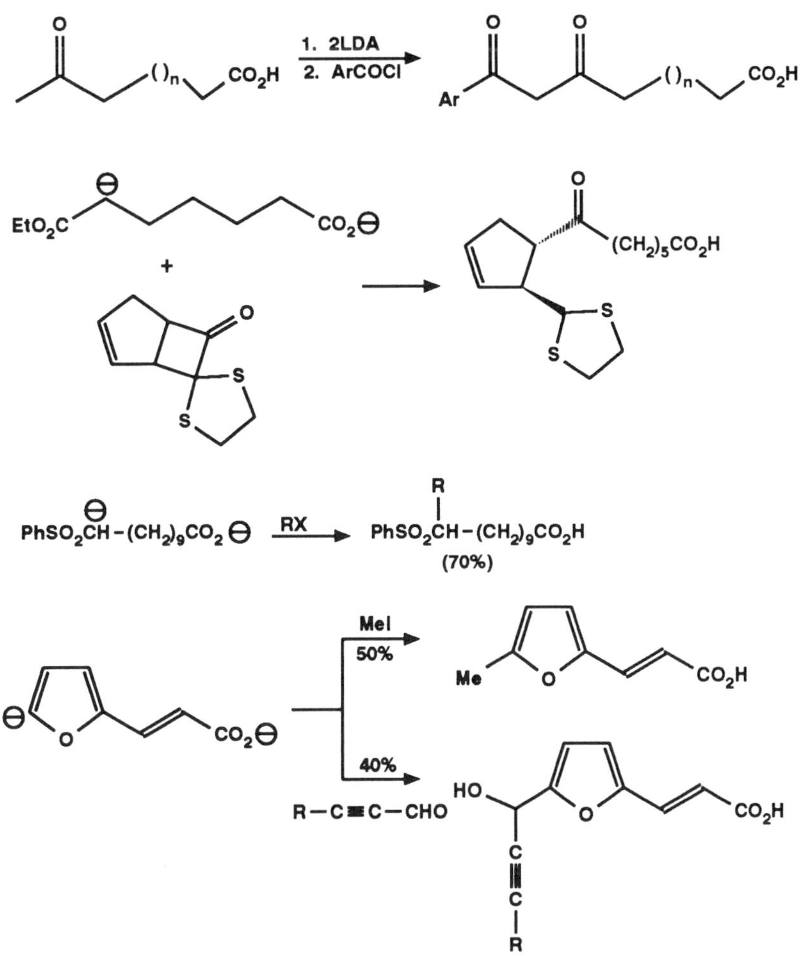

Scheme 3.22. Synthetic uses of 1,Z-remote carboxylate dianions

respectively, and acylated with benzoyl chlorides.[177] Reaction was observed only at the terminal methyl position. Ester[178] and sulfone[179] groups also have been used to induce a remote carbanion position including a 1,12-(O,C)-dianion used by Julia and Badet in route to exaltolide.[179] Knight and Nott used their knowledge of regioselective furan ring metalation to prepare remote carboxylate dianions used in the synthesis of wyerone and dihydrowyerone.[180]

3.3 HYDROXYL-BASED DIANIONS

Dianions that are generated from initial deprotonation of a hydroxyl group have been reviewed.[3,5,181] The hydroxyl group is ubiquitous in organic chemistry, and despite significant progress in the area of hydroxyl-based O,C-dianions, continued opportunities exist for their incorporation in synthetic strategies. The purpose of this section is to review the major classes of hydroxyl-based dianions (Fig. 3.8), and illustrate their major applications.

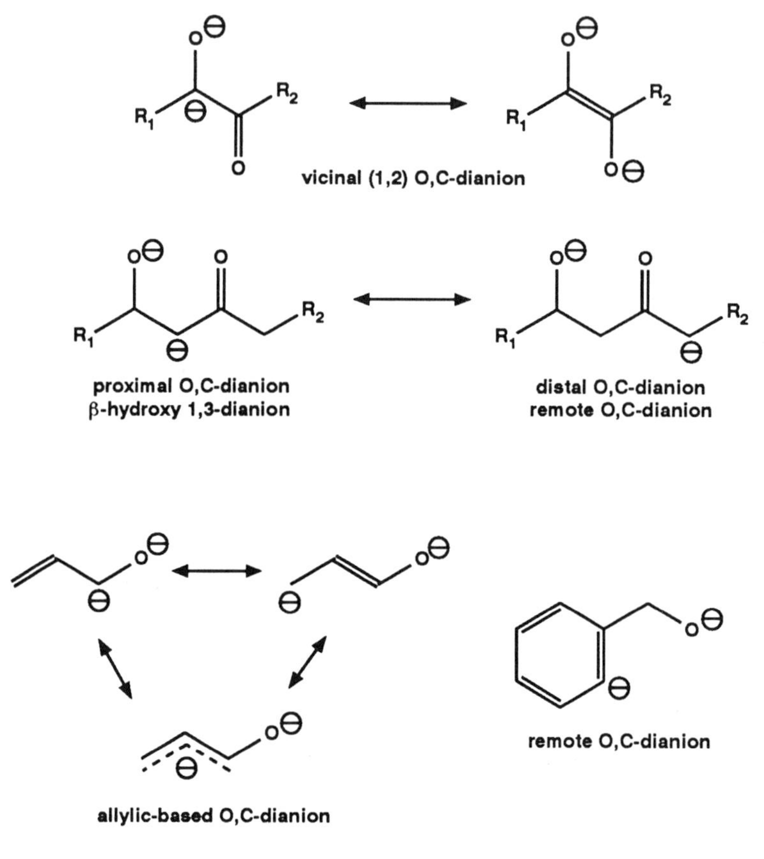

Figure 3.8. Representative hydroxyl-based dianions

Within the group of hydroxyl-based O,C-dianions, the β-hydroxyl carbanion-stabilized (1,3-O,C-dianions) have been most studied, although this is a relatively recent development. This increased usage emanates from investigations with enantioenriched β-hydroxycarbonyl (and related) compounds which, following bis-deprotonation, undergo stereoselective alkylations and condensations. The products thus obtained are aldols, and reaction with these dianions represent an alternative synthetic approach to stereoselective aldol condensations. O,C-Hydroxy (1,3)-dianions appear to have a three-fold benefit over the corresponding protected β-hydroxy-protected monoanions. First, direct enolate formation via the dianion of β-hydroxylcarbonyls avoids protection-deprotection sequences. Second, the protected hydroxyls (e.g., ethers or esters) are usually more prone to β-elimination pathways giving alkenes. Third, deprotonation of a chiral carbinol generally protects the asymmetric center from concomitant deprotonation, thereby avoiding epimerization.

Clever usage of the sulfonyl, sulfinyl, or related groups as substitutes for the carbonyl moiety has greatly extended the versatility of 1,3 β-hydroxy-dianions. Work in this area has progressed significantly over the past decade.

Other hydroxyl-based O,C-dianions include vicinal, allylic and propargylic alcohol-based, remote, and O,C-dianions that are prepared by methods other than bis-deprotonations.

3.3.1 Generation of hydroxy-based dianions

The low pKa of the hydroxyl OH permits ready access to this dianion class most commonly through metal hydride or organometallic bases. In cases where the hydroxy group deprotonation occurs at a remote site, judicious choice of the deprotonating base occurs only when compatability with the other functional group(s) present. Strong bases (e.g. RLi and R_2N-Li) are clearly needed to form the carbanion center of vicinal dianions. In general, there are no specific guidelines for the hydroxyl or C-H deprotonating base of O,C-dianions with the exception when either regio- or stereochemical control is desired. Virtually all extensive studies of bases have been conducted with the 1,3-O,C-dianions (Section 3.3.3), with particular attention given to thermodynamic versus kinetic enolate control, and the chelating capability of the cation.

A specific area of concern in the generation of hydroxyl-based O,C-dianions is the allylic species. Following O,C-bis-deprotonation, the carbanion may exist in more than one resonance-stabilized form, and strict control of the conditions is needed to provide unambiguous alkylation products. Further, several addition reactions of alkyllithium reagents to allylic alcohols have been reported, which must be considered when utilizing this dianion type. Since addition to the alkene moiety of an allylic alcohol results in the formation of a O,C-dianion but does not result from O,C-deprotonation, these intermediates and their chemistry are placed in Section 3.3.6, which includes hydroxyl-based O,C-dianions prepared via alternative pathways.

3.3.2 Reactions of hydroxyl-based 1,2-dianions (vicinal O,C-dianions)

Little exploration has been conducted on the direct generation of hydroxyl-based 1,2-dianions. These α-hydroxyl, vicinal-type dianions require a carbanion-stabilizing group (X; Fig. 3.9) that is capable of directing the C-H deprotonation toward the unfavorable α-site, which strictly limits the number of possible starting alcohol structures. Further

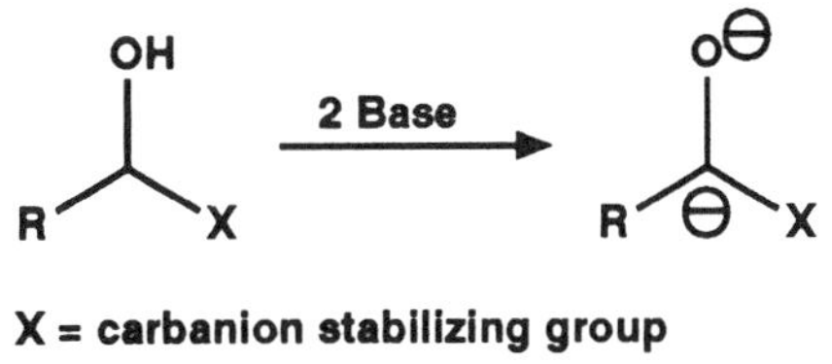

Figure 3.9. Vicinal O,C-dianions

discouragement in utilizing these intermediates stems from the fact that little potential for stereochemical extension is possible owing to enediolate resonance contributors. However, vicinal O,C-dianions have been used successfully in the preparation of α-hydroxyl α-disubstituted esters (Eq. 3.52).[182]

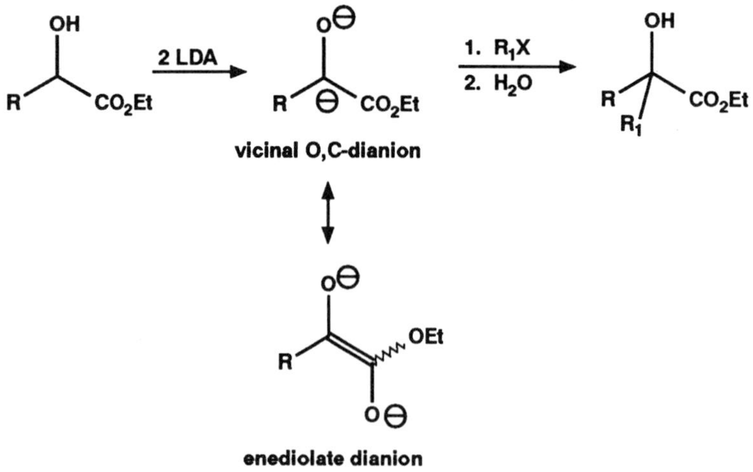

(3.52)

The cyclic adipoin, 2-hydroxycyclohexanone, has been bis-deprotonated using either LDA or Li-TMP to form the kinetic enolate, which was captured with TMS-Cl to furnish cyclohexene-2,3-diol bis(trimethylsilyl) ether in 73% yield. In contrast, if the adipoin dianion is generated under thermodynamic conditions (TEA, DMF, reflux) and reacted with excess TMS-Cl, the thermodynamic silyl enol ether predominates (Scheme 3.23).[183]

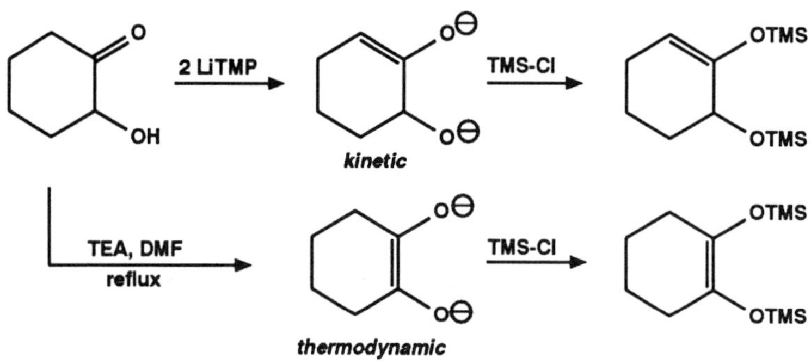

Scheme 3.23. Kinetic vs. thermodynamic generation of α-hydroxy O,C-dianions

Vicinal O,C-dianions also have been prepared from benzylic alcohols, but their use has primarily been relegated to titrating agents for alkyllithiums (Fig. 3.10).[184]

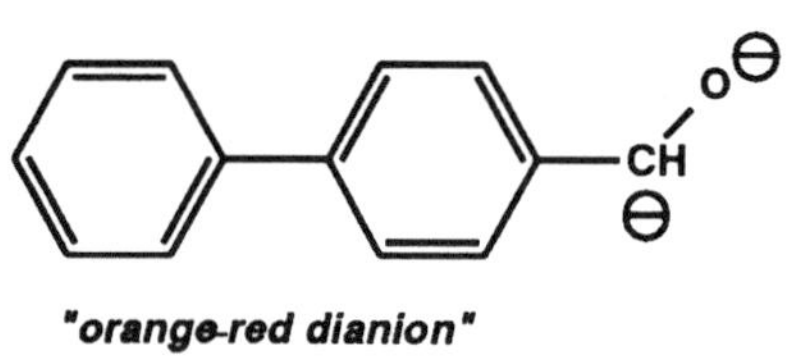

Figure 3.10. A vicinal O,C-dianion indicating reagent

It is important to note that the 1,2-O,C-dianions of the enediolate type also represent the putative intermediates of the acyloin condensation (Eq. 3.53). Therefore, the reactions

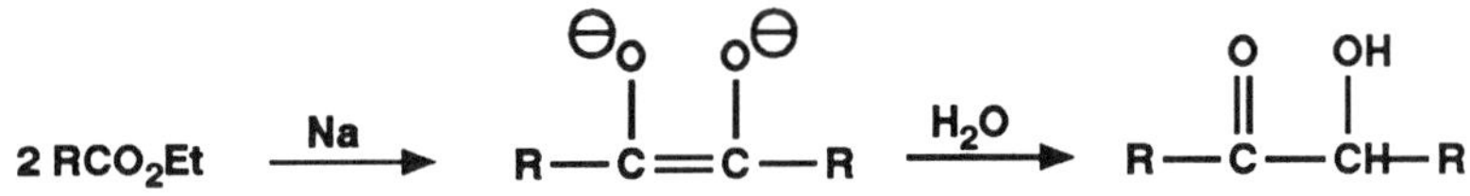

(3.53)

that have been performed on the acyloin enediolate dianion (alkyations, condensations, etc.) should be applicable to the same 1,2-dianion achieved through bis-deprotonation.[183] A comprehensive review on the acyloin condensation is available.[185]

3.3.3 Reactions of hydroxyl-based 1,3-dianions (β-hydroxyl-α-stabilized O,C-dianions)

Certain compounds that bear a hydroxyl group beta to a carbanion stabilizing moiety (X) can be bis-deprotonated to form the corresponding O,C-dianion (Eq. 3.54). A variety of

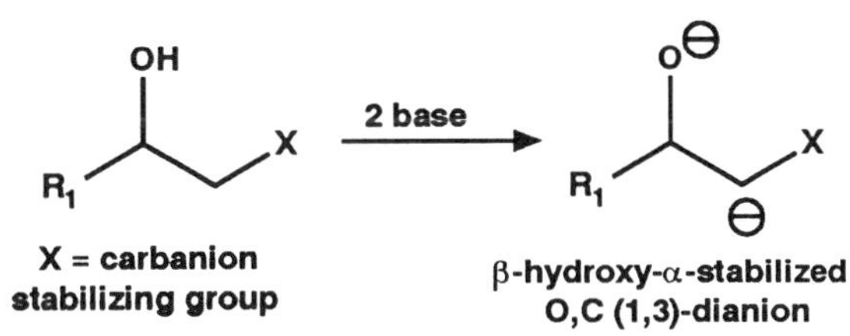

(3.54)

stabilizing groups have been employed to aid formation of the carbanionic site, the most common being carbonyl and sulfur-based types. Two significant developments in the use of these intermediates have greatly expanded their utility and versatility. First, Albizati and coworkers described precise protocols for the generation of either proximal or distal enolates of β-hydroxyketones (Eq. 3.55),[186] offering a novel regiochemical handle. The proximal and distal types fit the 1,3- and remote classes of O,C-dianions, respectively. Since these dianions are prepared from a single substrate, they shall be covered here with minor notation in Section 3.3.5.

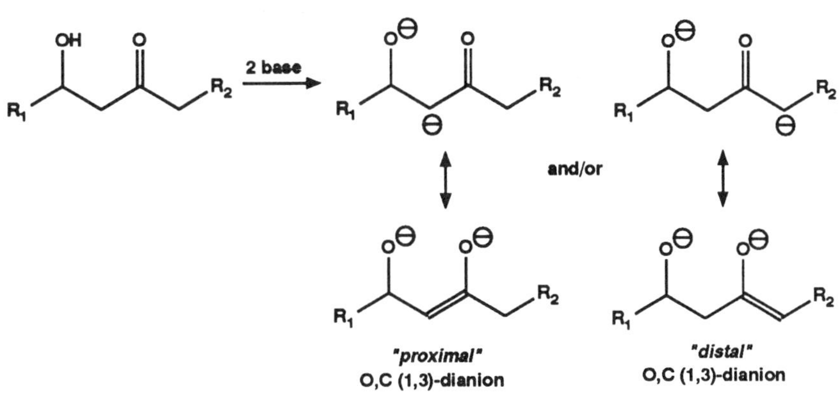

(3.55)

Second, as mentioned in Section 3.3., 1,3-hydroxyl-based dianions have been used in stereoselective syntheses. Enantioenriched β-hydroxycarbonyl- and sulfonyl compounds can be bis-deprotonated and reacted with electrophiles to give high diastereoselectivities. The products obtained differ in their stereochemical outcome, and can be used to complement similar condensations conducted with the enolates of hydroxyl-protected β-hydroxycarbonyls. The β-hydroxysulfinyl dianions also can be prepared with resolved carbon and sulfur stereocenters to further influence the product stereochemistry.

In addition to carbon and sulfur π-containing groups, researchers have also used other carbanion stabilizing moieties to direct 1,3- O,C-dianion formation. Each of these β-hydroxy O,C-dianions will be covered individually in the sections below.

β-Hydroxycarbonyl 1,3-dianions

Several distinct features of β-hydroxycarbonyl dianions have been explored; variation in the carbonyl component (ketone, ester, lactone, etc.), kinetic versus thermodynamic enolate generation (with certain β-hydroxyketones), and stereocontrol. Each of these features will be presented in this section with a focus upon their application to synthesis.

Over twenty years ago, Herrmann and Schlessinger showed that simple β-hydroxyesters could be bis-deprotonated with LDA to form the 1,3-O,C-dianion.[187] Reaction with alkyl halides gave excellent yields of the α-substituted β-hydroxyesters (Eq. 3.56). In this study,

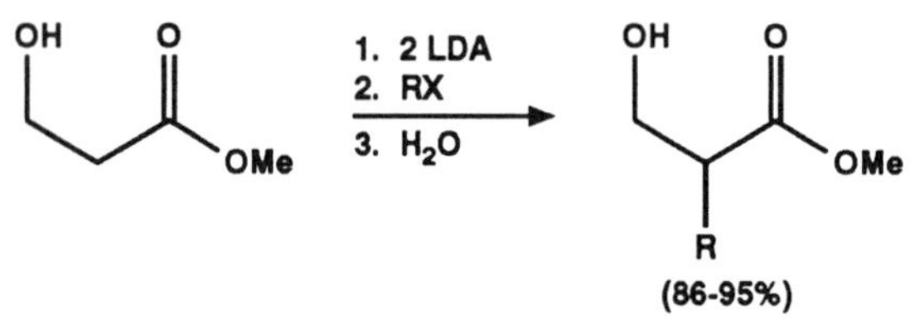

(3.56)

an unambiguous ester enolate forms, and this luxury is not generally extended to β-hydroxyketones, which can form complex enolates mixtures derived from regiochemical (1,3 versus 1,5 O,C-bis-deprotonations) and stereochemical (Z- or E-enolate geometry) ambiguities. This challenge was undertaken by Albizati and coworkers,[186,188,189] who were able to produce reliable experimental procedures for the generation of specific β-hydroxyketone O,C-dianion enolates. The β-oxidoenolate dianions generated with 2.2 LDA (from various β-hydroxyketones) led to exclusive formation of the kinetic ("distal") enolate when the degree of substitution was lower at the distal end. When the degree of substitution was equal and the enolate can express geometric preferences (e.g., Z- and E-enolates of 1-hydroxy-3-pentanone (1H3P) dianion), all four enolate isomers are observed under the experimental conditions (conditions for Scheme 3.24; a. 1H3P added to a mixture of LDA/TMS-Cl at -78°C; b. 1H3P added to the LDA solution and add TMS-Cl after 30 min. at -78°C; c. 1H3P added to LDA solution at -78°C, warm to 25°C for 15 min., recool to 0°C, and add TMS-Cl) with the proximal enolate favored by 2:1 at -78°C (Scheme 3.24).[188]

Aldehydes condense with the distal enolate dianion in good to excellent yield.[188]

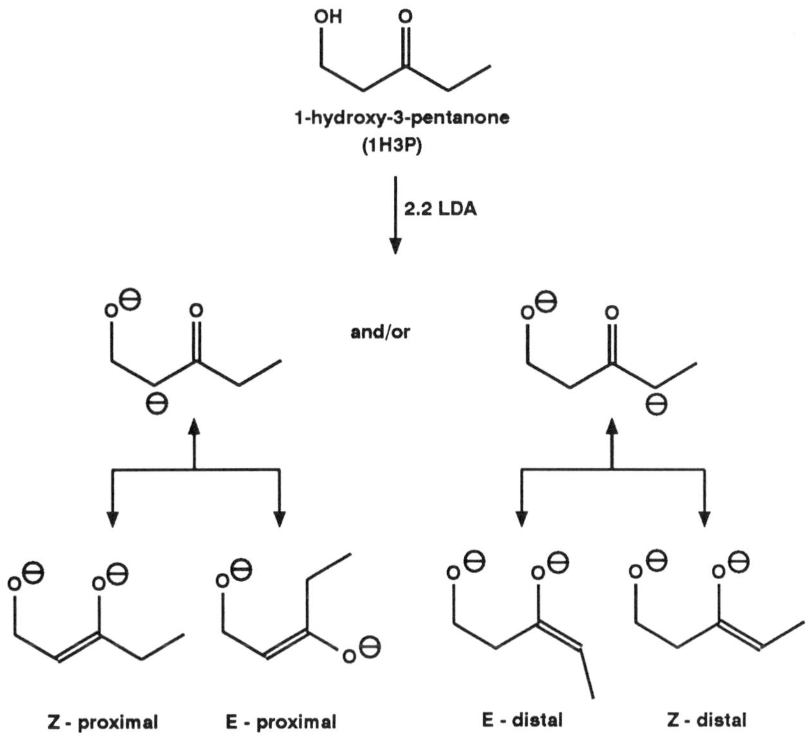

Conditions	Relative ratio			
a	2.0	2.3	1.1	1.0
b	<0.1	1.2	1.0	2.0
c	<0.1	1.0	7.0	15.7

Scheme 3.24. Proximal and distal dianions of 1-hydroxy-3-pentanone

In situ trapping of the distal enolate dianion with TMS-Cl gave the β-silyloxy silyl enol ether (predominant Z-geometry where applicable), which reacts with dioxenium cations (formed *in situ* from an orthoformate and $SnCl_4$) to afford substituted tetrahydropyran-4-ones in good yields (Eq. 3.57).[189]

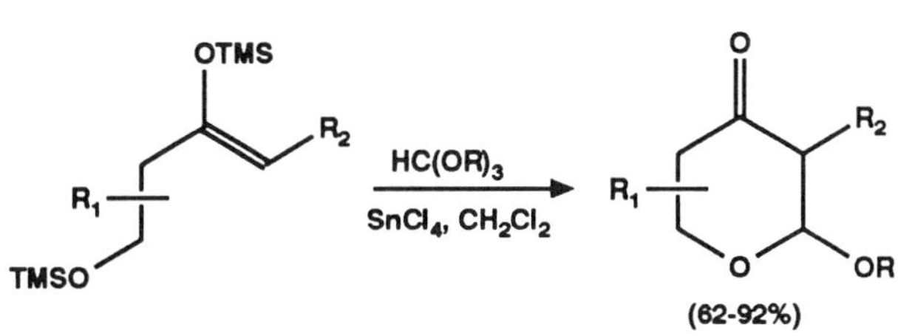

(3.57)

In a comprehensive study of distal/proximal β-hydroxyketone dianions, these workers also found that (by comparison to the monoanion β-alkoxy or β-silyloxy enolates) the β-hydroxy group can strongly direct the enolate regiochemistry toward the distal (kinetic) product, whereas the corresponding protected alcohols exert only a mild directing effect in favor of the proximal enolate.[186]

Relative easy access to enantioenriched β-hydroxyesters permits a study of stereoselective α-alkylations of the derived dianion. Frater and coworkers showed that (S)-ethyl-3-hydroxybutanoate forms the corresponding 1,3-dianion with 2 eq. of LDA at -50°C and undergoes alkylation to give the 2S,3S-diastereomers (Eq. 3.58).[190-192] The product stereochemistry was confirmed through parallel synthetic sequences including a one-pot, stepwise aldol condensation-dianion generation-alkylation process, and microbial reduction.[190] This work was extended to the synthesis of the elm bark beetle pheromone.[192]

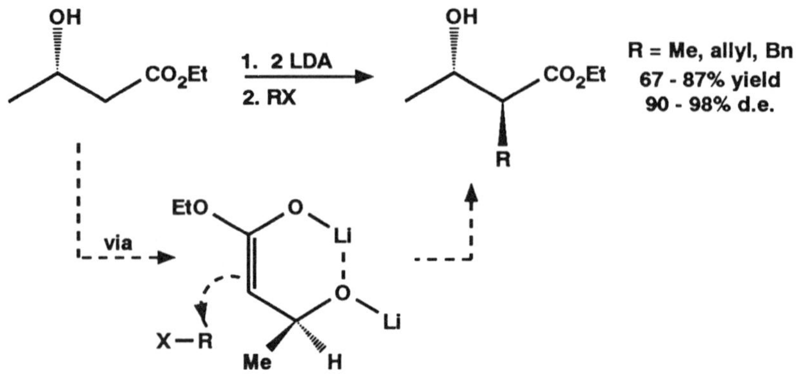

(3.58)

Concomitantly, Seebach and coworkers used diastereoselective functionalization at the α-position of enantioenriched β-hydroxyesters and diesters to access epoxides (via capture with I_2),[193] and γ-butyrolactams (via Michael addition to nitroolefins) (Scheme 3.25).[194]

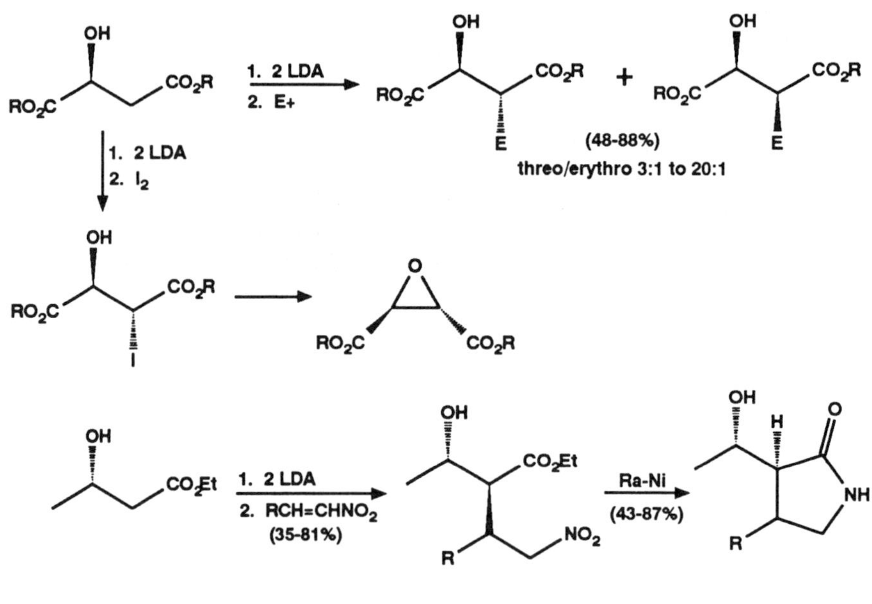

Scheme 3.25. Select transformations of β-hydroxyester dianions

Cyclic β-hydroxycarbonyl dianions have received considerable attention owing to both easy access of the cyclic precursor from readily available chiral starting materials, and better control of the alkylation stereochemistry.

Shieh and Prestwich were among the first to examine both alkylation and aldol condensations of enantioenriched β-hydroxy-γ-lactone dianions.[195] They found that the alkylation proceeded with excellent stereocontrol giving virtually exclusive trans-product, but the aldol condensations gave erythro:threo products ranging from 1:1 to 100:1 (Scheme 3.26). Methylation led to formation of the α-methyl-β-methoxy product from secondary reaction at the alkoxide. They also found that $ZnCl_2$ increased the yield of the reaction from 34-48% to 48-89%, but this enhancement came at the expense of the stereoselectivity.

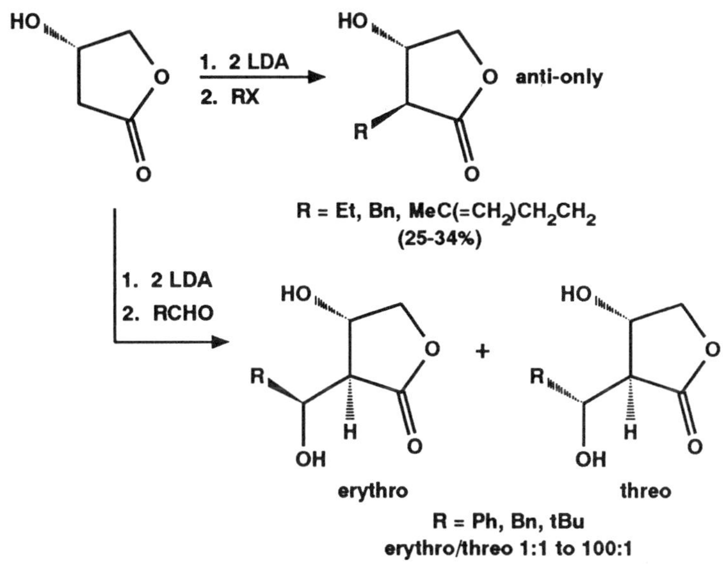

Scheme 3.26. Reactions of enantioenriched β-hydroxylactones

Chamberlin conducted studies with a more highly functionalized chiral hydroxy lactone, but found that alkylation of the corresponding dianion with MeI gave only methylation at the enolate position, again in high stereoselectivity (Eq. 3.59).[196]

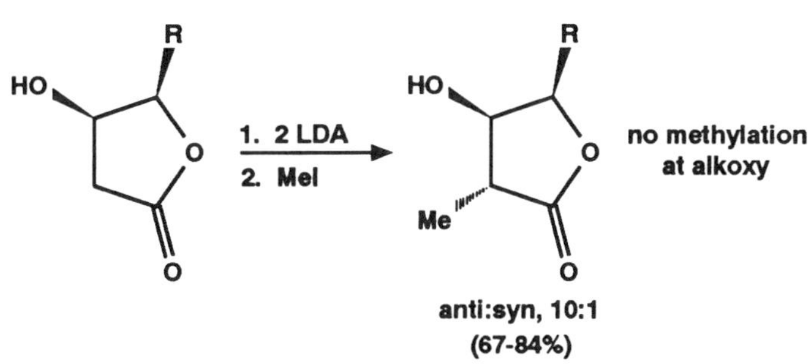

(3.59)

Chen and Jouille studied the alkylation (Eq. 3.60) and aldol condensation (Eq. 3.61) of ribonolactone derived dianions, and found that alkylation proceeded with good stereocontrol but condensation with aldehydes gave modest erythro:threo mixtures.[197]

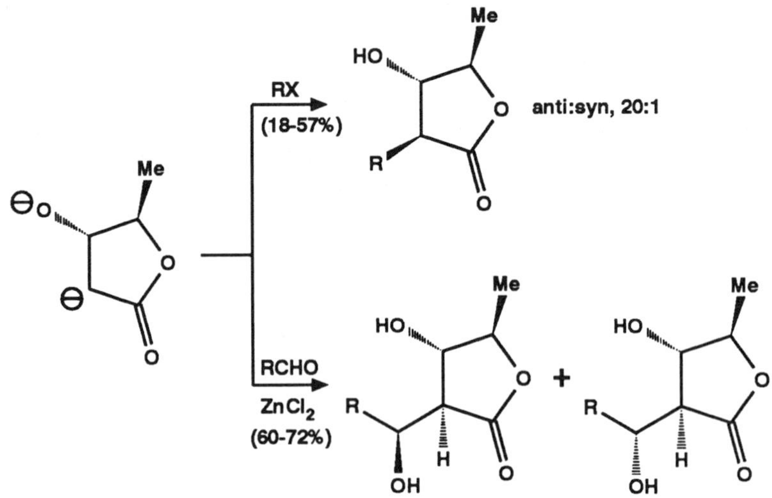

(3.60)

(3.61)

R			% Yield
$CH_3(CH_2)_{11}CH_2$-	1	1.5	60
Ar	3	1	75
HC≡C$(CH_2)_9CH_2$-	1	2	72

Cyclic β-hydroxy dianion systems have been extended to lactams. Work by the Speckamp[198] and Hutchins[199] groups showed that suitably substituted hydroxy lactams could be bis-deprotonated and reacted with electrophiles to give functionalization of the enolate position bearing exclusive trans stereochemistry (Eqs. 3.62 and 3.63).

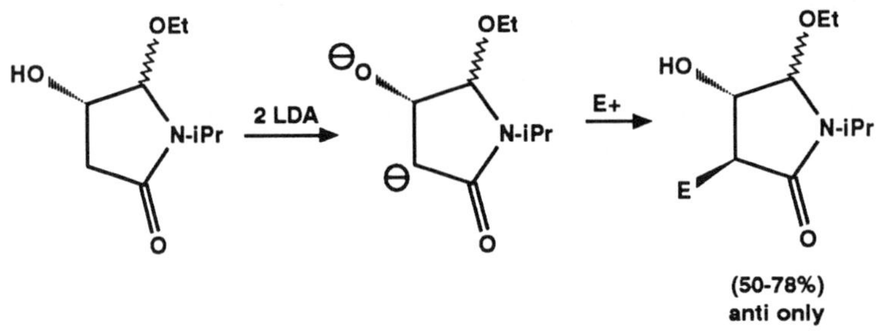

(3.62)

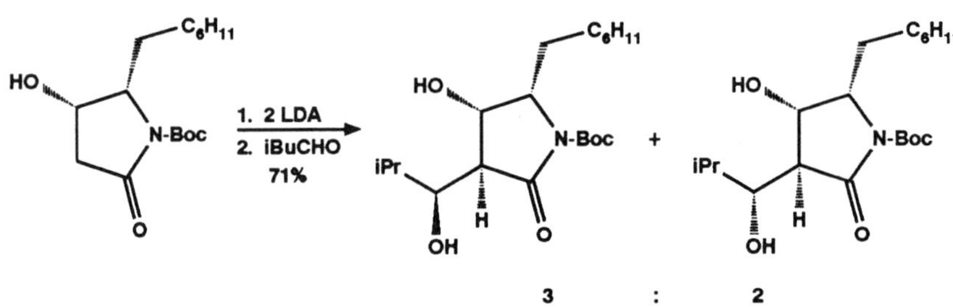

(3.63)

Further imaginative use of β-hydroxyester dianions was conceived by Kurth and co-workers. In particular, recognition of allyl ester enolates as Claisen rearrangement precursors set the stage for the generation of a β-hydroxy allyl ester 1,3-dianion at -78°C that, upon warming to 50°C, rearranged to set three contiguous stereocenters (Eq. 3.64).[200-202] The crotyl ester geometry (E or Z) correlated well with the resultant diastereoselectivity.

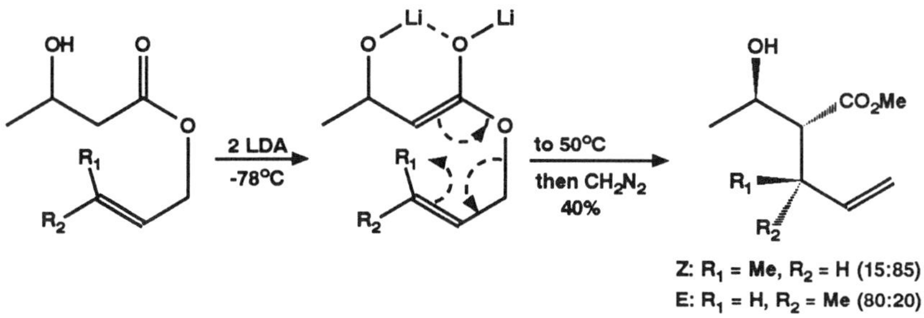

(3.64)

Similar Claisen rearrangement stereochemical outcomes were achieved by Normant, although via a different metal enolate complex.[203]

β-Hydroxysulfinyl and β-hydroxysulfonyl 1,3-dianions

Dianions derived from β-hydroxysulfoxides and β-hydroxysulfones (Fig. 3.11) also have been employed in synthesis with equal success to the carbonyl congeners.[5,181,204] However, three features of these sulfur-functional groups lead to variation in the generation and

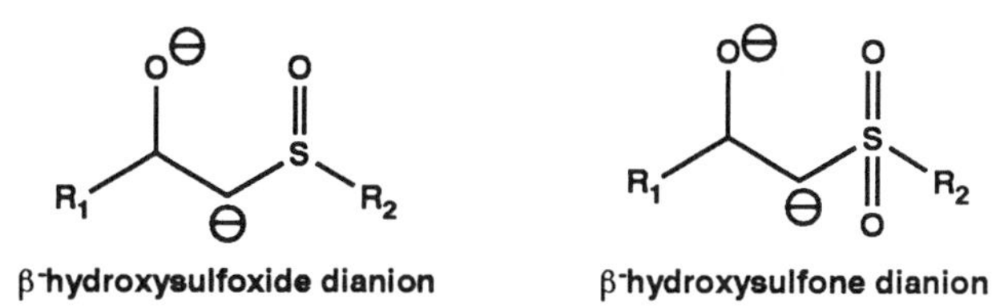

Fig. 3.11. Dianions of β-hydroxysulfoxides and β-hydroxysulfones

usage of the derived dianions. First, unlike the carbonyl-based stabilizing groups, sulfoxides and sulfones are not reactive toward alkyllithium bases. Second, sulfoxides may be asymmetric at the sulfur atom, which can greatly affect the stereochemical outcome of alkylation and condensation reactions through unique chelation orientations. Last, sulfur-based carbanion-stabilizing groups can be excised from the product or converted into another functional group (e.g. Pummerer reaction). In this section, the major uses of β-hydroxy-sulfoxide and -sulfone dianions will be presented. The major methods for their preparation has been very recently covered.[204] Owing to the fact that the sulfone moiety is achiral, this simpler dianion class will be covered first.

2-Phenylsulfonyl ethanol and its substituted analogues can undergo iterative dianion formation and reaction with alkyl halides and ICH_2CO_2Na to furnish 4-substituted furanones following cyclization in excellent yield (Eq. 3.65).[205] A reversal in the synthetic order using methyl phenylsulfone as a starting material affords the 5-substituted furanones (Eq. 3.66).[205]

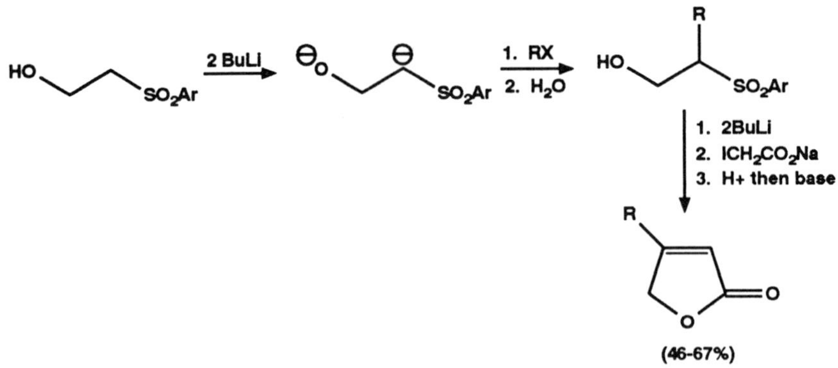

(3.65)

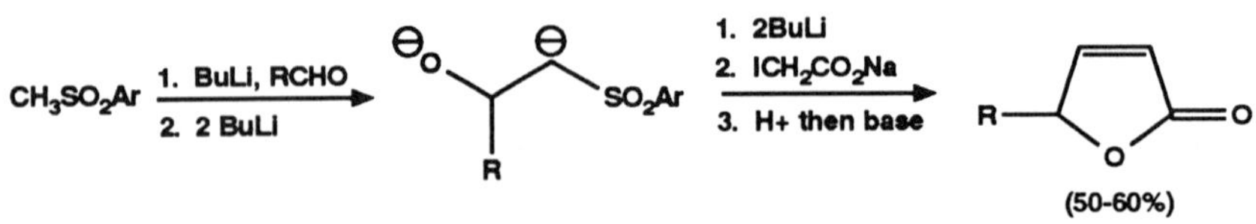

(3.66)

Kozikowski and coworkers were among the first to explore the reactions of asymmetric β-hydroxysulfone dianions.[206] In order to study the reaction stereoselectivity, enantioenriched (S)-β-hydroxysulfones were prepared from the corresponding β-ketosulfones via bakers yeast reduction. Formation of the hydroxysulfone dianion with 2 BuLi and reaction with alkyl halides led to good conversion to products with modest stereoselectivity (Eq. 3.67).

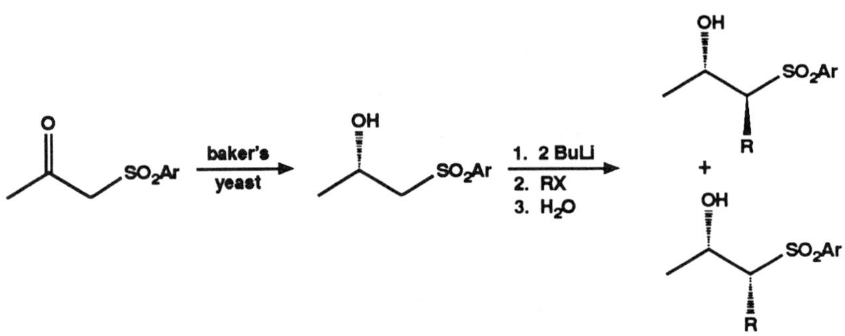

(3.67)

The products of alkylation with allyl bromide were converted to a common butenolide in order to assess the stereochemical outcome (Eq. 3.68).

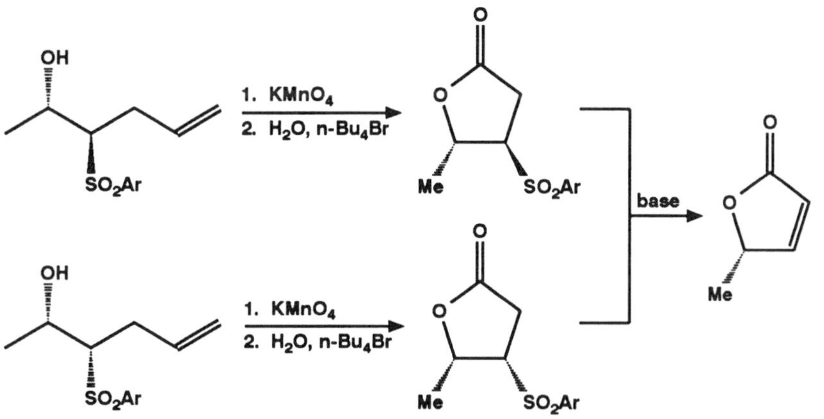

(3.68)

Further exploration of this system was conducted by Tanikaga and co-workers who reacted (S)-β-hydroxysulfone dianions with aldehydes to furnish the aldols corresponding to deprotonation only at the pro-*R* position (Scheme 3.27).[207] This observed stereoselectivity is in contrast to the alkylation results.[206] The sulfone moiety was reductively eliminated from the aldol products through first conversion to the tosylate leading to allylic alcohols and acrylonitriles in 100% e.e. (Scheme 3.27).

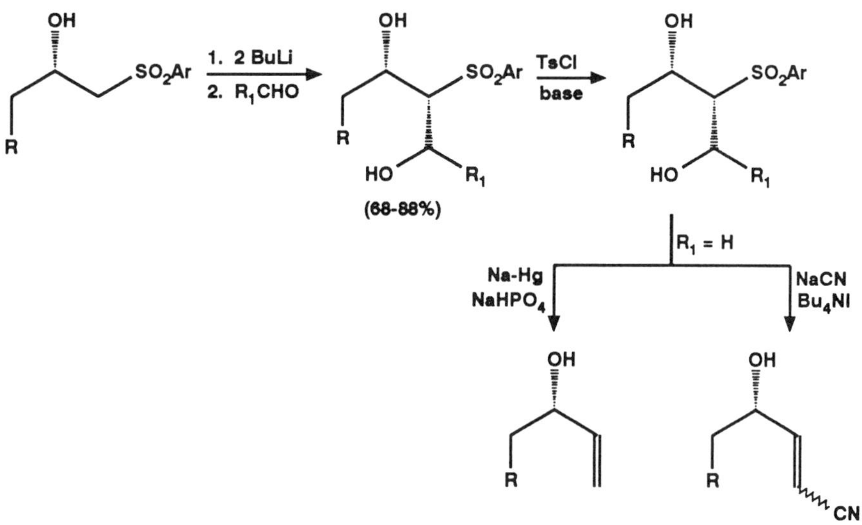

Scheme 3.27. Aldol condensation of (S)-β-hydroxysulfones

In a subsequent report,[208] the condensation and alkylation of related enantioenriched β-hydroxysulfone dianions was re-examined. It was found that the reaction of the dianion of (E)-2-hydroxy-1-phenylsulfonylpropane with electrophiles gives the more sterically demanding *erythro* products (50-100% d.e. except methyl) (Eq. 3.69) (Table 3.8).

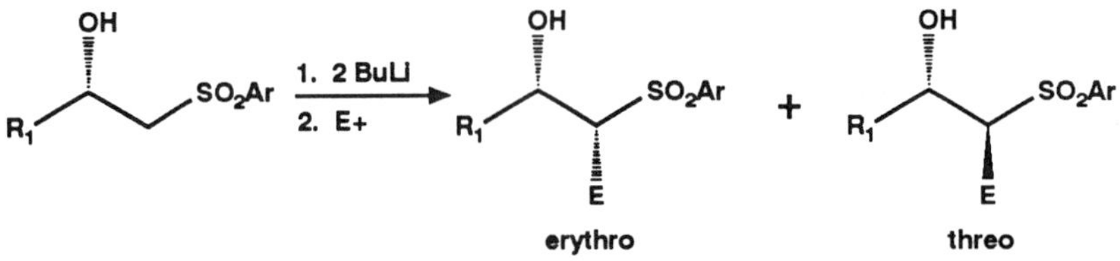

(3.69)

Table 3.8
Stereochemistry of β-Hydroxysulfone Dianions Reactions (Eq. 3.69)

R^1	E	erythro	threo	Yield (%)
Me	MeI	51	49	98
Me	$(MeO)_2SO_2$	83	17	96
Me	CH_2=$CHCH_2Br$	73	27	83
Me	$C_{11}H_{23}CHO$	80	20	73
Bu	MeI	78	22	72
Bu	CH_2=$CHCH_2Br$	80	20	78
$PhCH_2$	MeI	85	15	62
$PhCH_2$	CH_2=$CHCH_2Br$	98	2	69

Dianions of β-hydroxysulfones linked to sugars have been generated and reacted with electrophiles to provide good yields of products (Eq. 3.70).[209] The stereoselectivity was not very high, ranging from 58:42 to 67:33.

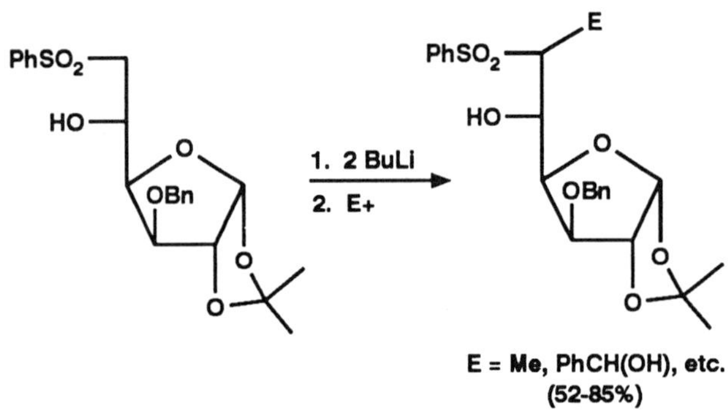

(3.70)

Prior to the acyclic studies, the dianion of either cis- or trans-2-phenylsulfonyl cyclopentanol was reacted with electrophiles to give products that have the sulfone and alkoxy moiety in the anti disposition (Eq. 3.71).[210]

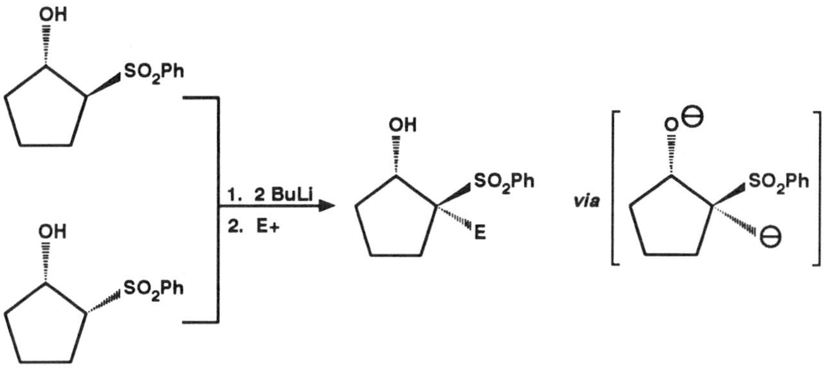

(3.71)

As early as 1977, β-hydroxysulfoxide dianions were shown to be useful intermediates in carbon-carbon bond formation.[211] Prenyl bromide was reacted with a β-hydroxysulfoxide dianion to afford an alkylated product that was subjected to a β-elimination in route to linalool (Eq. 3.72). Thus, the β-hydroxysulfoxide moiety was used as a masked alkene.

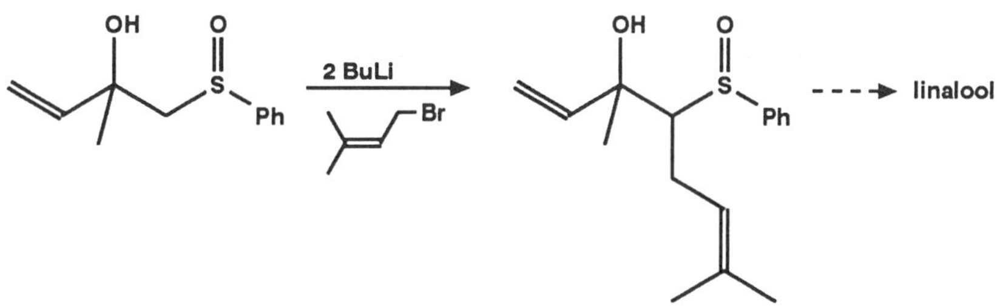

(3.72)

Tanikaga and coworkers provided some of the first stereochemical investigations of β-hydroxysulfoxide dianions.[212-214] Their first report[212] contrasted the [stereochemical] reaction outcomes between structurally related β-hydroxysulfoxide and β-hydroxysulfone dianions in order to assess the impact of oxidation state and asymmetry at sulfur upon the erythro/threo preferences. Thus, they compared the three possible dianion species of simple β-hydroxy-phenylsulfonyl and -phenylsulfinyl toward alkylation (Scheme 3.28). The stereoisomeric sulfoxides were prepared by reduction of the corresponding ketosulfoxides using diisobutylaluminum hydride (DIBAL-H).

In general, alkylations of the sulfoxide dianions occurred stereospecifically with noticeable increases in the predominant isomer as the bulk of the R-group increased. The S_S-β-hydroxysulfoxide dianion gave threo/erythro ratios ranging from 78:22 to 98:2, whereas the R_S-β-hydroxysulfoxide dianion gave threo/erythro ratios ranging from 89:11 to 99:1. The corresponding sulfone dianion gave a greater quantitity of the erythro product (1:1 to 85:15). This reversal in stereochemical preference was attributed to strong intramolecular chelation by the more polar sulfinyl group.[212]

The R_S- and S_S-1-phenylsulfinyl-2(S)-propanols (R=Me; Scheme 3.28) were converted into their respective dianions and reacted with n-octyl iodide and n-decanal.[213] Interestingly, alkylation led to the expected threo selectivity (R_S yields 66:34; S_S yields 96:4), whereas condensation gave increased erythro selectivity (R_S yields 19:81; S_S yields 51:49).

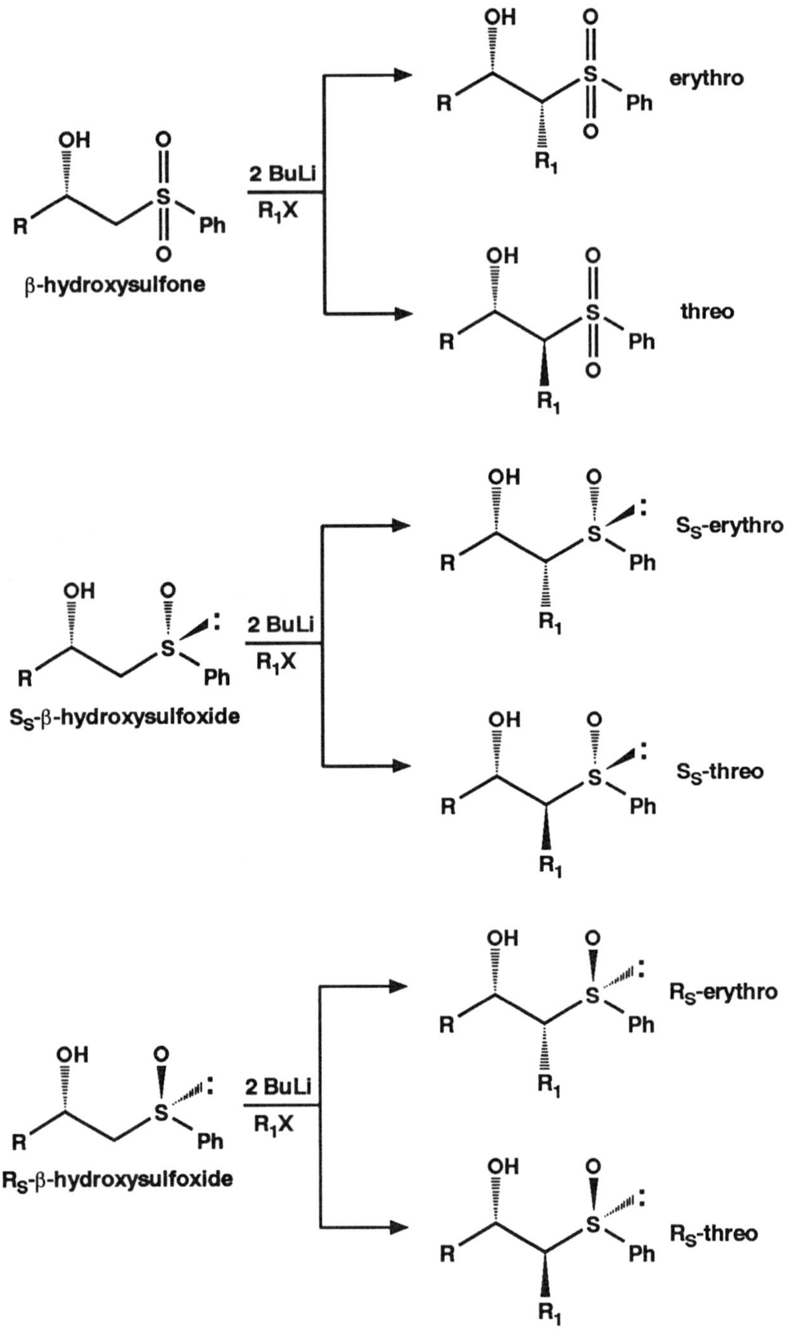

Scheme 3.28. Possible stereochemical outcomes following the alkylation of a β-hydroxysulfone and β-hydroxysulfoxides

Follow-up studies showed that certain β-hydroxysulfoxide dianions (e.g. R=Me) are capable of configurational interconversion (Eq. 3.73), although significant decomposition occurs prior to complete equilibration.[214] This study would suggest that the unsubstituted species (R=H) could undergo relatively easy interconversion.

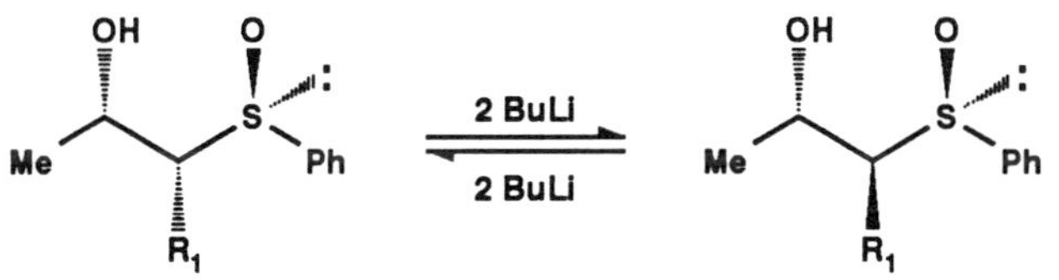

(3.73)

Miscellaneous β-hydroxy-1,3-dianions

Several studies have been conducted with 1,3-(O,C) dianions using carbanion stabilizing groups other than carbonyls and oxidized sulfur groups. Sulfur-stabilizing groups (e.g., sulfides, dithianes, and dithioesters) have most served this purpose.

Bis-1,3-diphenylthio-2-propanol reacts with 2 BuLi to form the β-hydroxy dianion, which collapses via intramolecular expulsion of the 3-phenylthio group to a cyclopropane (Eq. 3.74).[215]

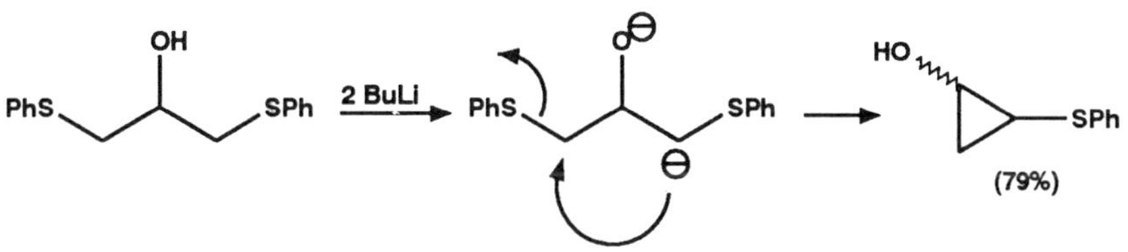

(3.74)

Paulsen and coworkers used the dianion of 1,3-dithiane-2-methanol in route to the synthesis of pillarose (Eq. 3.75).[216,217] The dithiane methanol dianion can be considered as a masked α-hydroxyacetaldehyde reagent.

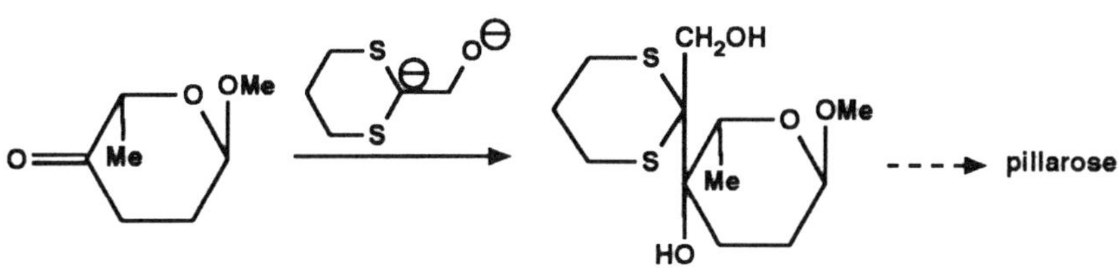

(3.75)

Further use of this reagent as a building block toward chain-elongated sugars was explored.[218] For example, the addition of the dithiane methanol dianion to glyceraldehyde gives the diastereomeric erythrose analogues (Eq. 3.76).

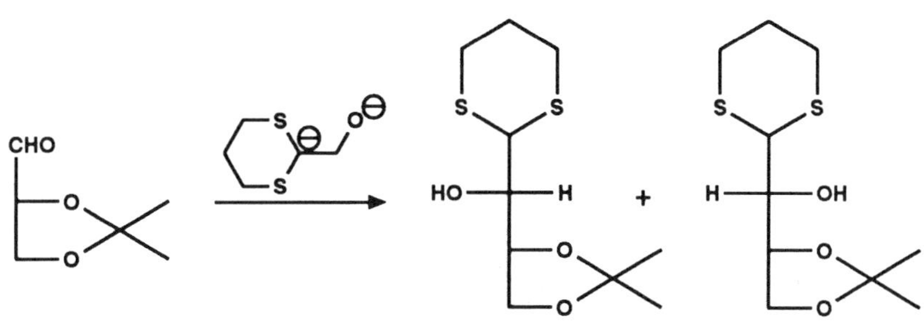

(3.76)

Dithioesters bearing a β-hydroxyl group may be bis-deprotonated using 2 LDA, but alkylation with allyl bromide occurs at sulfur to form the dithioacetal.[219] These products undergo Claisen rearrangement with excellent stereocontrol offering an interesting route to the enantioenriched α-functionalized products (Eq. 3.77) (Table 3.9).

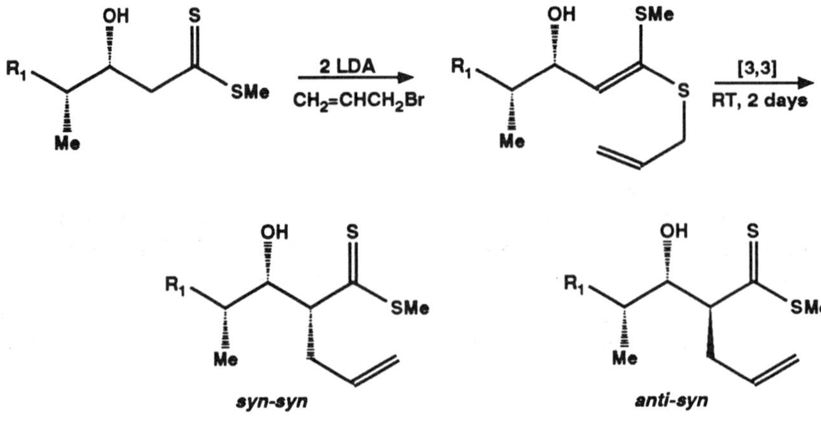

(3.77)

Table 3.9
Claisen Rearrangement of β-Hydroxydithioacetals (Eq. 3.77)

alkene	R_1	syn-syn	anti-syn	Yield (%)
E	tBu	94	6	47
Z	tBu	85	15	54
E	Ph	99	1	91
Z	Ph	99	1	92

Alkynes also have been used to stabilize the carbanion site of β-hydroxy dianions. Bis-deprotonation of a terminal homopropargylic alcohol with 2 LDA leads to the O,C-dianion that reacts with MeI to form predominantly the erythro product in an overall yield of 65% (Eq. 3.78).[220]

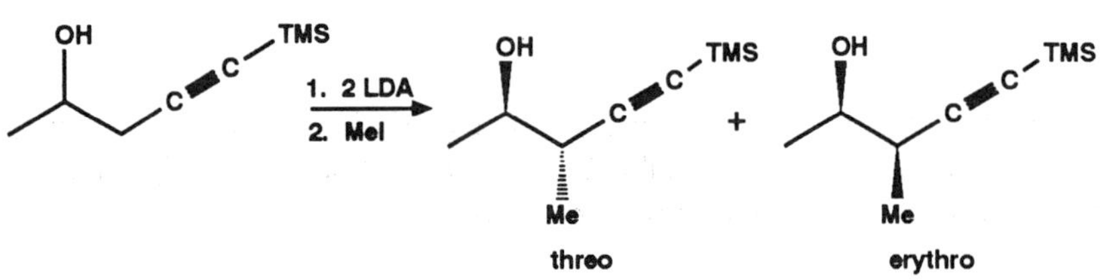

(3.78)

3.3.4 Reactions of allylic and propargylic alcohol-based O,C-dianions

The alkene and alkyne groups have provided functional anchors at which a second site of deprotonation can occur following ROH deprotonation. The resultant O,C-dianions may be categorized as allylic-, vinylic-, or propargylic-based. Allylic alcohols may undergo bis-deprotonation at the hydroxyl and at an allylic position to give a resonance-stabilized O,C-dianion. Subsequent reaction with electrophiles, therefore, is possible at more than one site, although most of the allylic-based O,C-dianions react regioselectively. Conversely, suitably substituted allylic alcohols may form O,C-dianions where the carbanion site is localized at a vinylic position, and is usually accessed via metal-metal or halogen-metal exchange.

Propargylic alcohols undergo O,C-deprotonation at the alkyne-CH (1,4) site, or at the propargylic (1,2) site. This section will explore the generation and utility of π-system O,C-dianion derived from allylic and propargylic alcohols, and related applications of these intermediates.

Allylic site deprotonation of allylic alcohols leading to O,C-dianions

There are two major reactions schemes that can be conducted between allylic alcohols and organolithium reagents; O,C-bis-deprotonation followed by reaction with an electrophile, or hydroxyl deprotonation followed by addition of the alkyllithium to the alkene moiety. The two reactions are, on occasion, competitive, frequently leading to a complex array of products. As such, only in cases where a carbanion-stabilizing group controls the regiochemistry of a deprotonation does the reaction of an allylic alcohol with 2 equiv. of RLi lead to useful outcomes. Since the addition reactions were initially discovered, these will be considered first.

Allylic alcohol combines with organolithium reagents to give 2-substituted-1-propanols via the 1,4 O,C-dianion (Scheme 3.29).[221] TMEDA (0.2 equiv.) increased the yields.

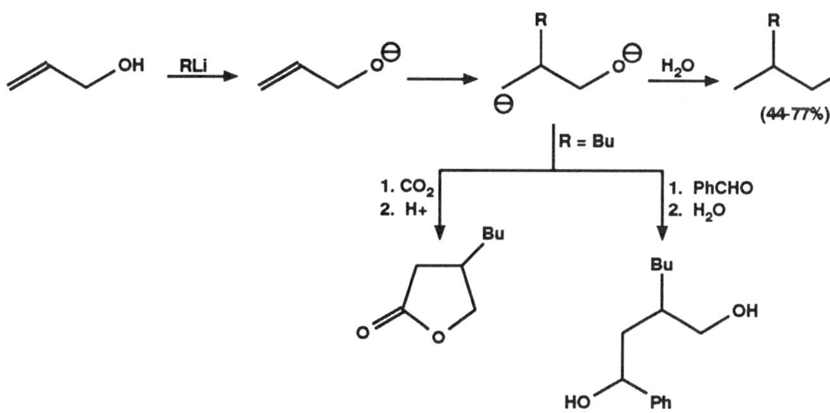

Scheme 3.29. Addition of alkyllithium reagents to allylic alcohol

Quenching the 1,4-dianion with benzaldehyde gave the diol, whereas reaction with carbon dioxide led to a 23% yield of the γ-butyrolactone.

The reaction of 3-buten-1-ol with BuLi or BnLi proceeds similarly to allyl alcohol (Eq. 3.79), whereas olefin formation predominates when 2-cyclohexenol reacts with alkyllithiums (Eq. 3.80).[221]

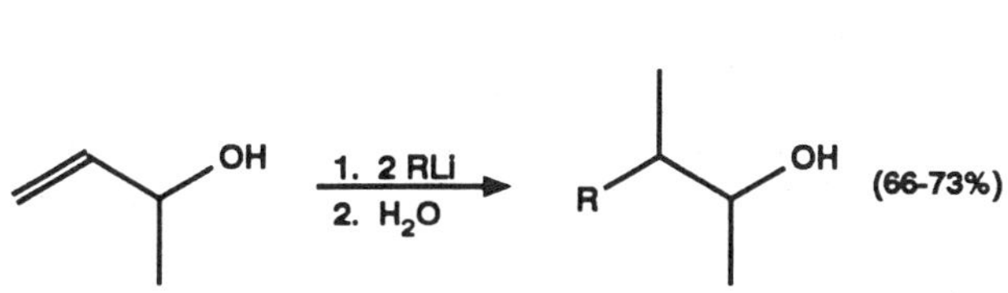

(3.79)

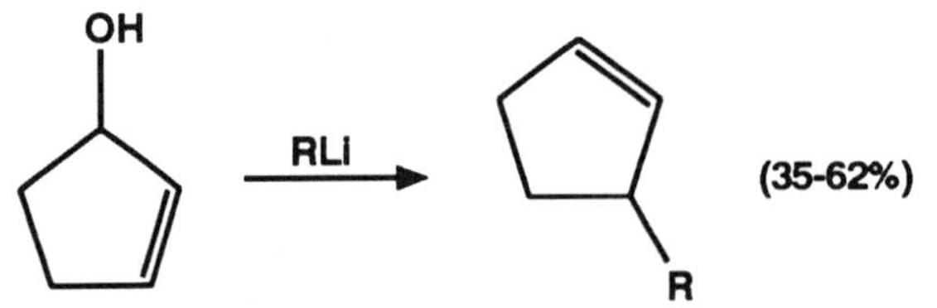

(3.80)

Dimmel and coworkers showed that α-vinyl benzylalcohol reacts with organolithiums, but can give different products than those indicated above.[222,223] For example, 2 equiv. of BuLi reacts with α-vinyl benzylalcohol to give propiophenone after proton abstraction from the solvent (THF or DME) and acid quench (Eq. 3.81). In hexane and TMEDA, mono- and di-addition products were formed, the corresponding benzylic alcohol and the β-butylheptylbenzene (Eq. 3.82).

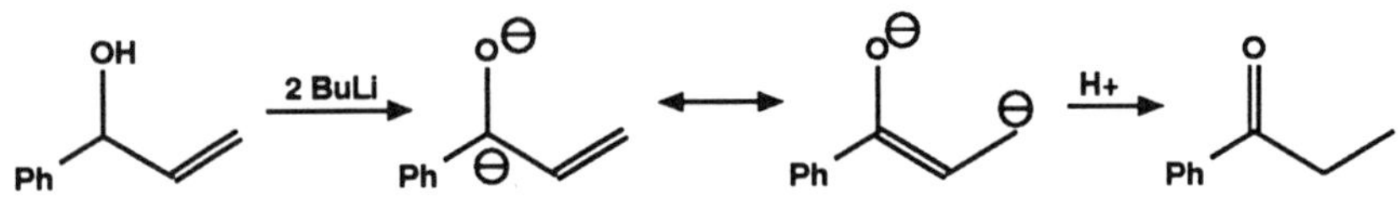

(3.81)

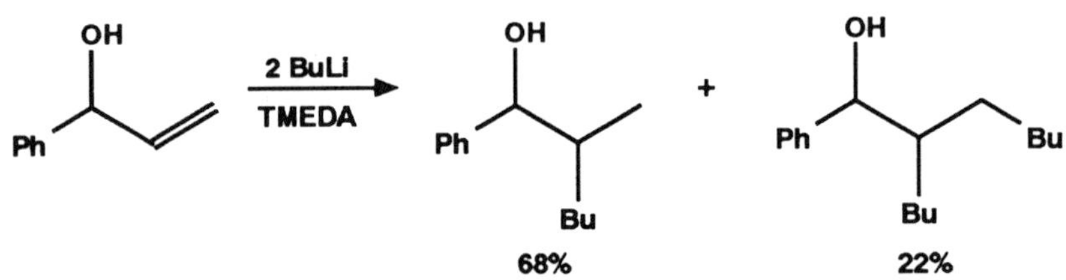

(3.82)

Interestingly, addition to form the benzyl alcohol proceeds stereospecifically giving only one pair of enantiomers.[222] The methyl-substituted analog, 2-phenyl-3-buten-2-ol gives similar products with BuLi, but under forcing conditions (overnight reflux, 3 equiv. BuLi) gave *cis*- (15%) and *trans*-2-phenyl-2-octene (10%).[223]

Addition reactions continued to thwart allylic alcohol dianion formation until Carlson, in a series of studies,[224-226] showed that reaction of methallyl or crotyl alcohol with tBuOK/BuLi complex (hexane, 0°C) leads to dianion formation, and no addition product. The dianions derived from methallyl[225,226] and crotyl[226] alcohol are formed regiospecifically, leading to 3- and 4-substituted products, respectively. The dianion of methallyl alcohol reacts with aldehydes or ketones[224] to give 1,4-diols that undergo oxidation at the allylic alcohol carbon to form α-methylene-γ-lactones (Eq. 3.83). Reaction with epoxides, followed by MnO_2 oxidation gives α-methylene-δ-lactones (Eq. 3.84).

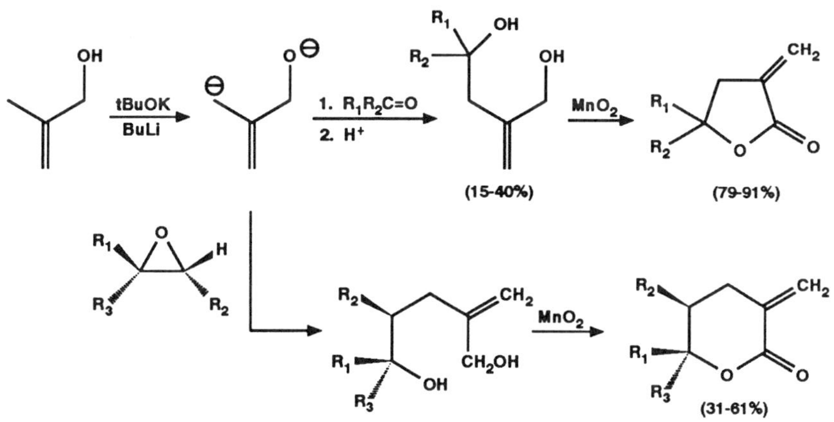

(3.83)

(3.84)

The same reaction conditions led to formation of an E/Z-mixture of crotyl dianions,[226] which may be successfully reacted with carbonyls (Eq. 3.85). However, this dianion exists in both resonance-stabilized forms (1,5 and 1,3), with the 1,5-dianion dominanting the reaction. Triethylaluminum was found to alter the regiochemistry.

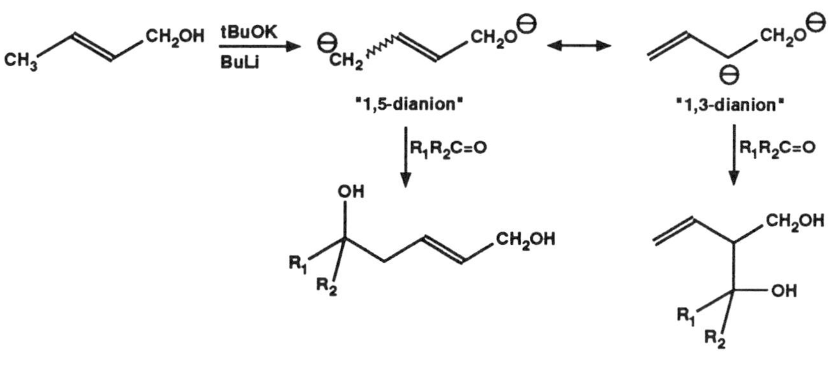

(3.85)

Methallyl alcohol reacts with BuLi-TMEDA in hexane to give the vinylic-based dianion (following section), whereas the allylic-based dianion is formed in ether solvents.[227-230] These differences were exploited by Trost in the bis-silylation of methallyl alcohol,[227,228] and by Julia and coworkers in an extensive investigation of ketone homoenolate dianions.[229,230] The ketone homoenolates give predominant β-alkylated products (Eq. 3.86).[230]

A cyclic variation of an allylic-based O,C-dianion was generated and explored in the construction of 14-deoxyisoamijiol (Eq. 3.87).[231] Unfortunately, the desired silylation at the C-6 position did not occur presumably owing to possible formation of the favored 6-membered chelate structure.

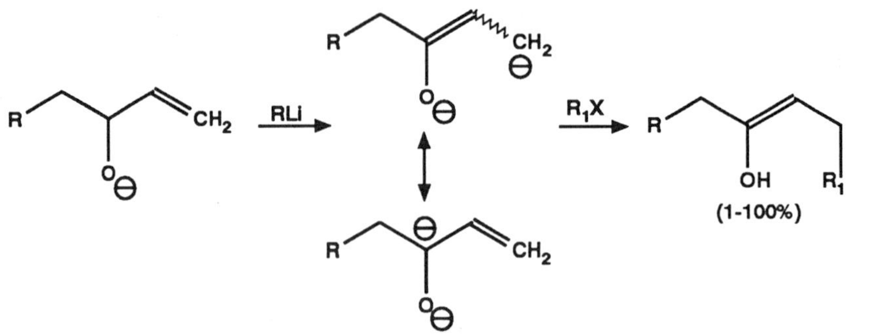

(3.86)

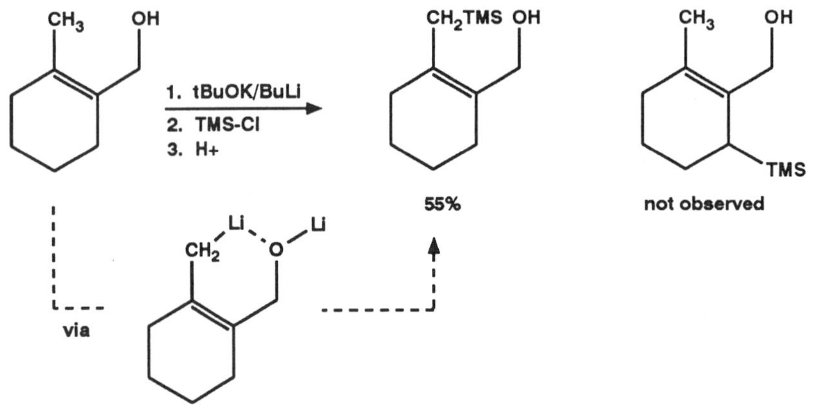

(3.87)

Allylic-based O,C-dianions also have been prepared using a carbanion-stabilizing group. Scheme 3.30 shows specific uses of a sulfone[232] and heterocyclic amine[233] for this purpose.

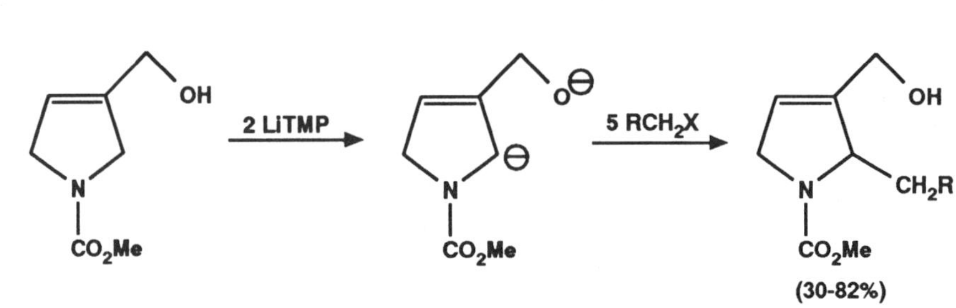

Scheme 3.30. Heteroatom-stabilized allylic alcohol dianions

Sturtz and coworkers discovered an interesting entry to allylic-alcohol based dianions that are further stabilized by a heteroatom. Reaction of an allylic phosphate ester with base induces a transposition to the α-alkoxyphosphonate, which can be further deprotonated to the allylic dianion (Eq. 3.88).[234] Electrophiles react predominantly at the γ-site.

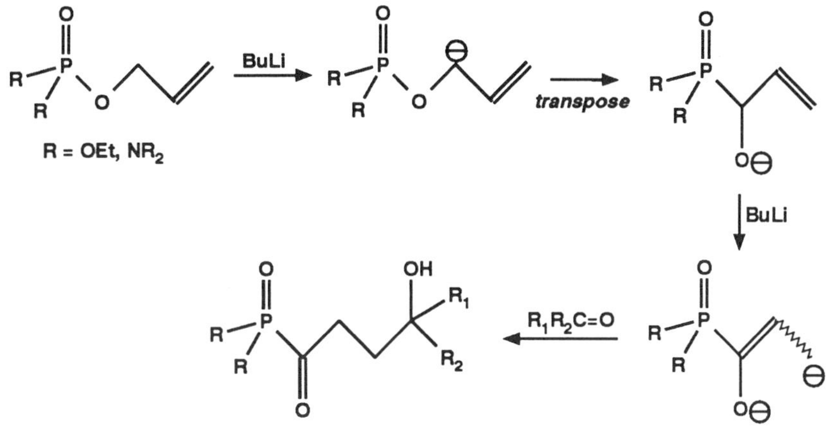

(3.88)

Vinylic site deprotonation of allylic alcohols leading to O,C-dianions

Allylic alcohols may be converted to O,C-dianions in which the carbanion site is positioned on the alkene (vinylic). Because of the greater acidity of the allylic position(s), vinylic O,C-dianions must be generated through halogen or metal exchange reactions. Schlosser and Corey showed that 2-bromoallyl alcohol could be converted to the corresponding dianion, and condensed with carbonyls to give good yields of condensation products (Eq. 3.89).[235,236]

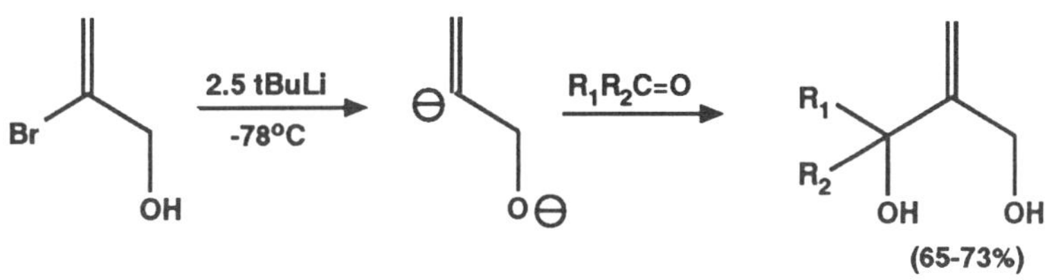

(3.89)

O,C-Dianions bearing a vinylic carbanion site may be prepared by deprotonation followed by metal-halogen exchange on the allylic alcohol or by sequential treatment of an α-chloro α,β-unsaturated aldehydes with Grignard reagents and lithium powder.[237] These vinylic O,C-dianions undergo reactions with a variety of electrophiles (Eq. 3.90). Generation of a related vinylic-based O,C-dianion will be covered in the following section.

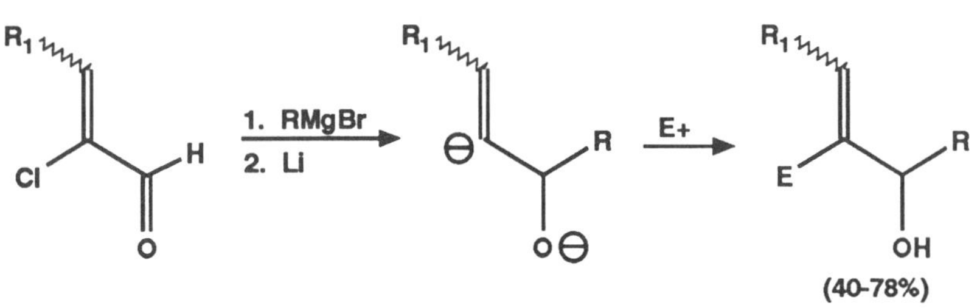

(3.90)

Tin-lithium exchange has been used as an interesting entry to these dianions, which are useful reagents in tetrahydrofuran synthesis (Scheme 3.31).[238]

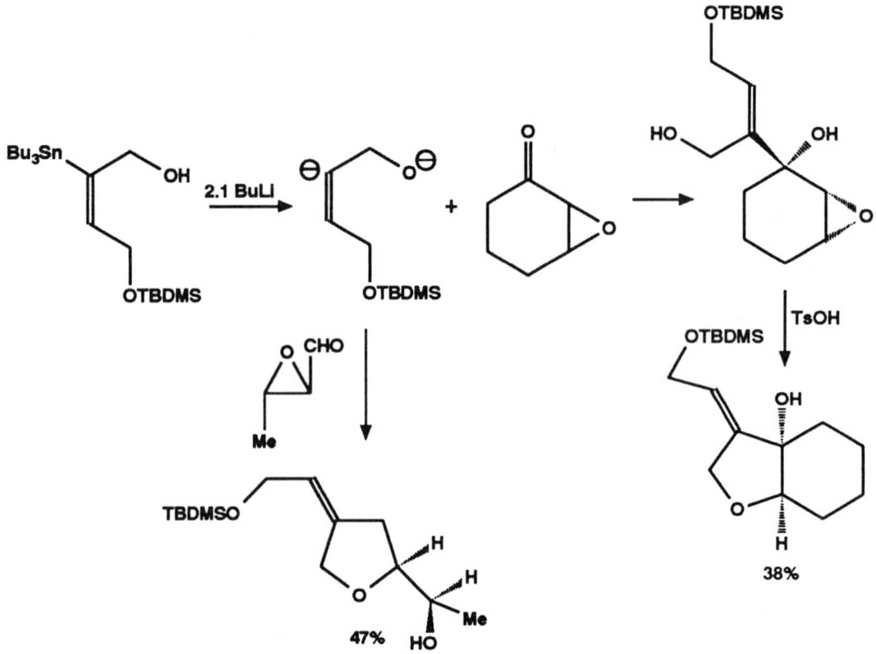

Scheme 3.31. Use of vinylic-based O,C-dianions in tetrahydrofuran synthesis

Propargylic alcohol-based dianions

Propargylic alcohols undergo deprotonation or addition reactions with organolithium reagents depending upon the substituent appended.

3-Phenylpropargylic alcohol reacts with 2-5 eq. of BuLi to furnish an excellent yield of E-β-n-butyl-cinnamyl alcohol (Eq. 3.91),[239] or the γ-lactone depending upon the workup method.

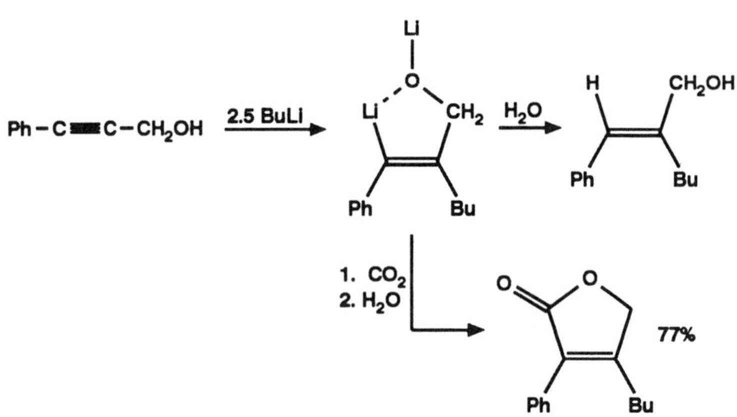

(3.91)

Propargyl alcohol has been bis-deprotonated to afford the alkyne-based (remote) O,C-dianion, which reacts with enones regioselectively 1,2,[240] or alkyl halides[241] in good yield (Eq. 3.92).

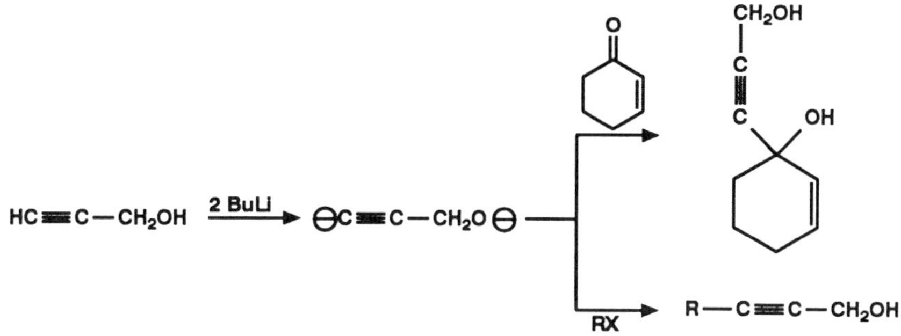

(3.92)

3.3.5 Reactions of remote, hydroxyl-based dianions

There are several reports that have investigated remote hydroxy-based O,C-dianions including the distal β-hydroxyketone dianions previously covered in Section 3.3.3. A few noteworthy examples are assembled here, and their utility in carbon-carbon bond formation. In general, a specific carbanion-stabilizing or chelating group is installed or used to direct the second deprotonation toward the remote site. The sulfone group has gained the most attention in this regard because, as mentioned prior, sulfone groups serve to both direct a carbanion toward the α-position but is also capable of being removed under mild conditions.

Julia and coworkers C-alkylated a range of remote dianions using the sulfone stabilizing group (Eq. 3.93).[242] Similarly, the optically active camphor derivative is converted to the putative dianion with three equivalents of BuLi and methylated at the remote α-sulfone position to give a 96:4 diastereomer mixture (Eq. 3.94).[243,244]

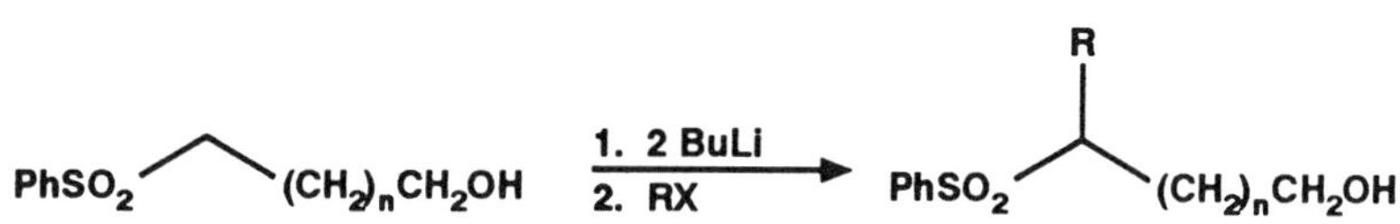

(3.93)

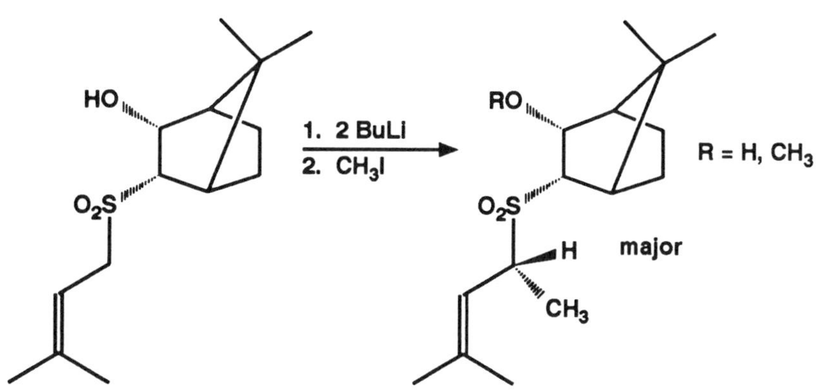

(3.94)

Tin-lithium exchange was used to prepare both acyclic (Eq. 3.95)[245] and cyclic (Eq. 3.96)[246] remote hydroxy-based O,C-dianions.

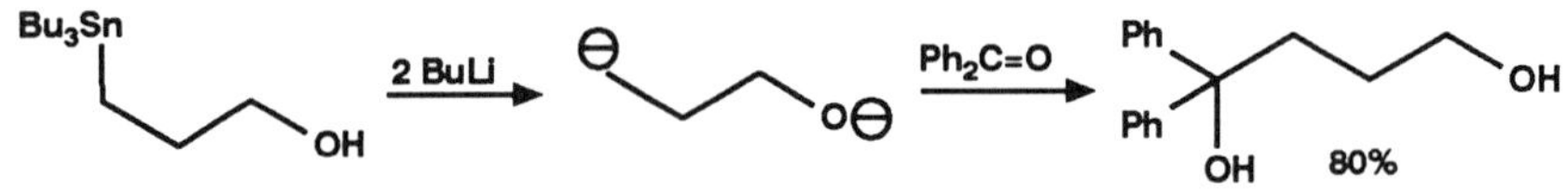

(3.95)

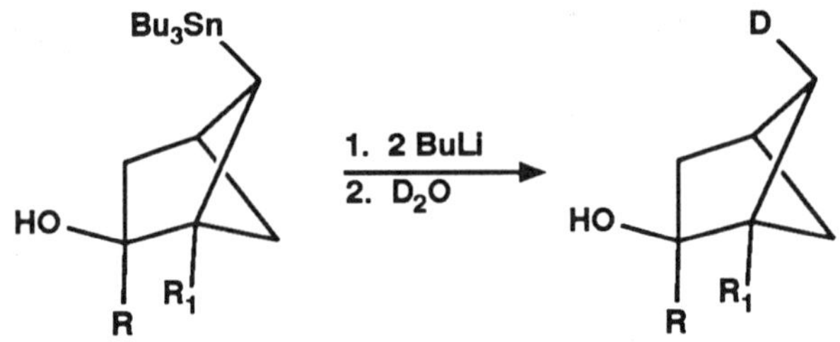

(3.96)

Alkene (allylic-type)[247] and carbonyl[248] groups also have been used to generate remote dianions (Eq. 3.97 and 3.98).

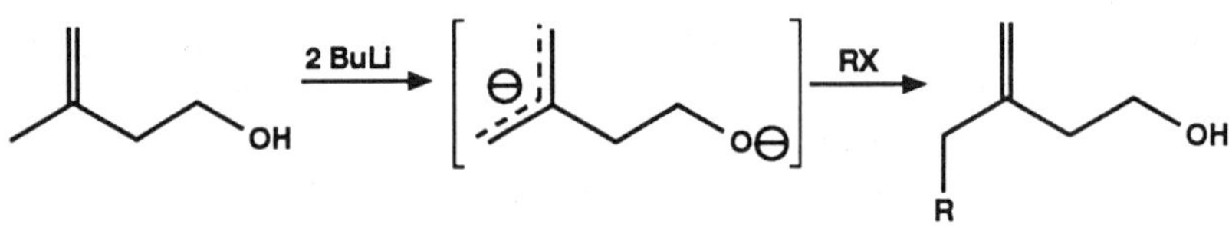

(3.97)

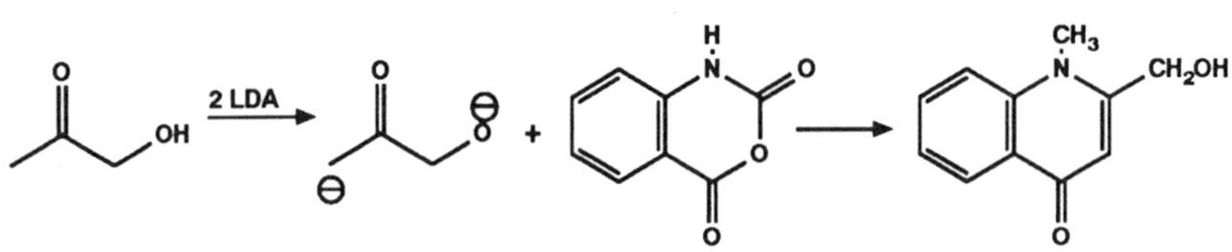

(3.98)

3.3.6 Preparation and reactions of O,C-dianions not derived through initial deprotonation of a hydroxyl group

Because the alkoxide group may be easily accessed via methods other than deprotonation, e.g. organometallic addition to carbonyls or epoxides, concomitant formation of various hydroxyl-based O,C-dianions is possible. Such O,C-dianions, therefore, would not be formed through traditional O,C-bis-deprotonation. As a result, a separate section has been generated to accommodate these species. As noted earlier, Barluenga[237] and coworkers showed that certain allylic alcohol-based dianions could be obtained via first addition of an organometallic to an α,β-unsaturated carbonyl, followed by halogen-metal exchange.

Addition of (poly)chloro organolithiums to aldehydes gives alkoxides, which upon subsequent treatment with BuLi eliminate LiCl and transpose to α-chloroketones (Eq. 3.99).[249]

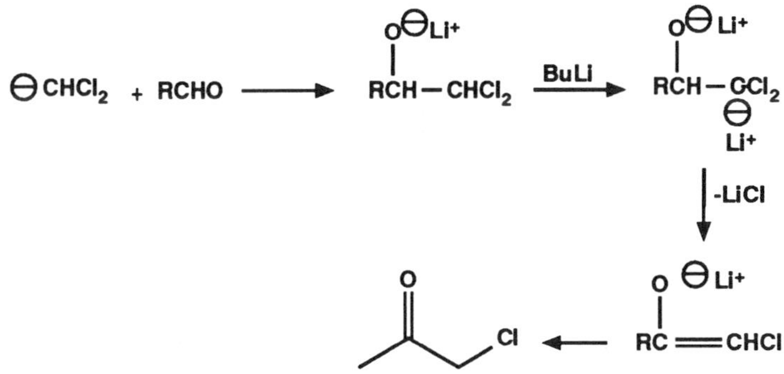

(3.99)

Epoxides are excellent precursors to hydroxyl-based O,C-dianions. Crandall and coworkers furnished the groundwork for epoxide and organolithium reactions.[250,251] The Fuchs groups set up a clever epoxide system, which reacts with organolithium reagents leading to ring opening. Addition of a second equivalent of RLi deprotonates an α-sulfone CH to form the O,C-dianion. Alkylation and aqueous workup gives the disubstituted β-hydroxysulfone, which is oxidized to the ketone, and treated with base to induce elimination to the enone (Eq. 3.100).[252,253]

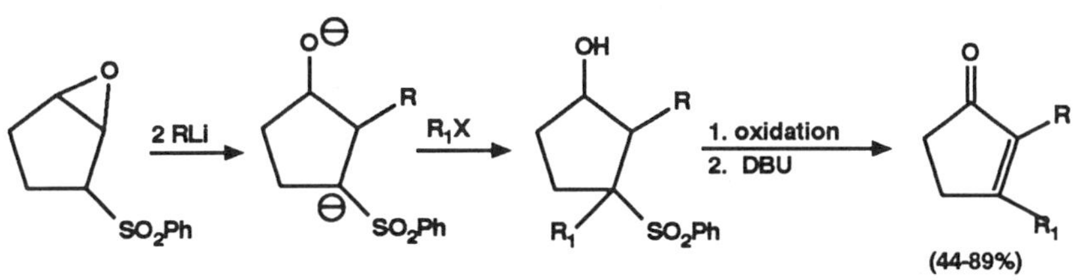

(3.100)

Epoxides bearing an α-tosyl group undergo successive ring opening and vinylic metalation with MeLi.[254] Reaction to form the 3-substituted-3-tosyl allylic alcohols proceeds in good yield following reaction with electrophiles (Eq. 3.101).

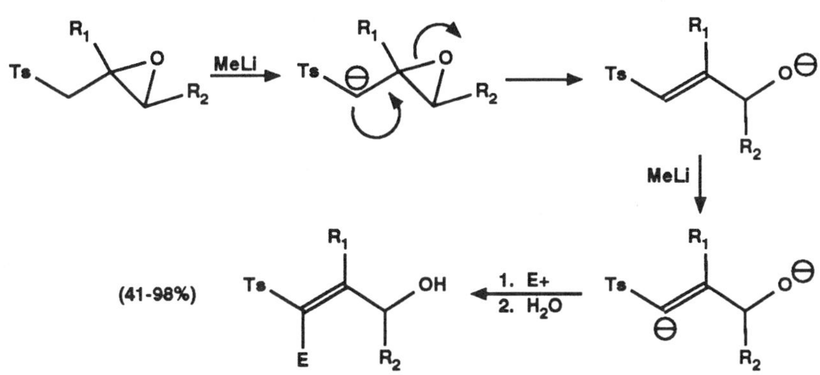

(3.101)

Allylphenyl ether undergoes a 1,3-shift following reaction with excess RLi to give a hydroxyl-based O,C-dianion that results in a propiophenone upon workup.[255]

Certain bis-stannyl derivatives undergo preliminary condensation with carbonyls to furnish the tin alkoxide. The second tin moiety can be metal-exchanged for lithium giving the tin/lithio O,C-dianion, which reacts with CO_2 to furnish lactones after cyclization (3.102).[256]

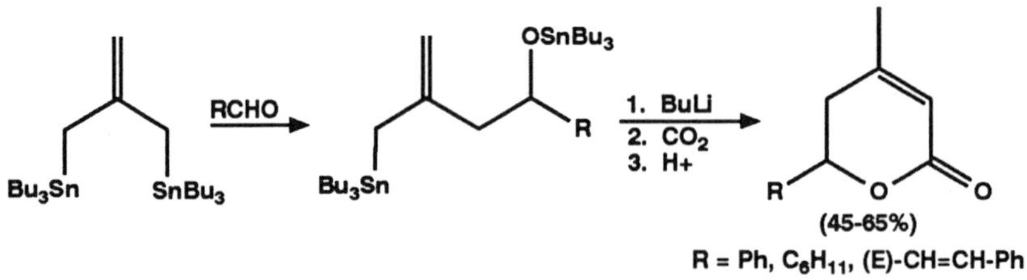

(3.102)

Comins and Brown reacted aromatic aldehydes with various lithium amides to form the α-amino alkoxides that both protect the carbonyl group from further reaction with organolithiums and serve as an ortho-directing group (Eq. 3.103).[257]

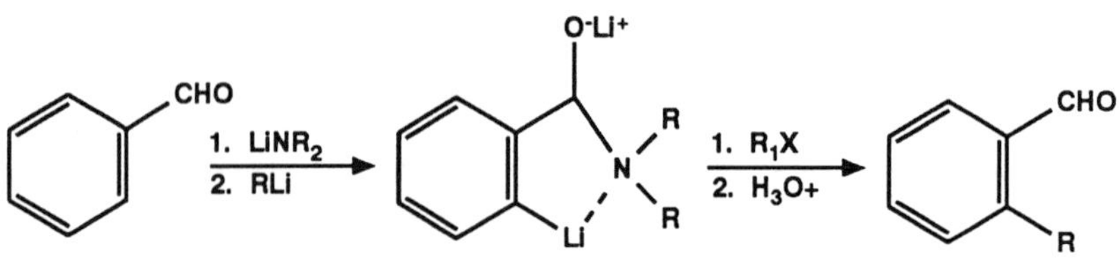

(3.103)

Reductive lithiation has recently provided entry to a variety of hydroxyl-based dianions (Eq. 3.104, 3.105 and 3.106).[258-260]

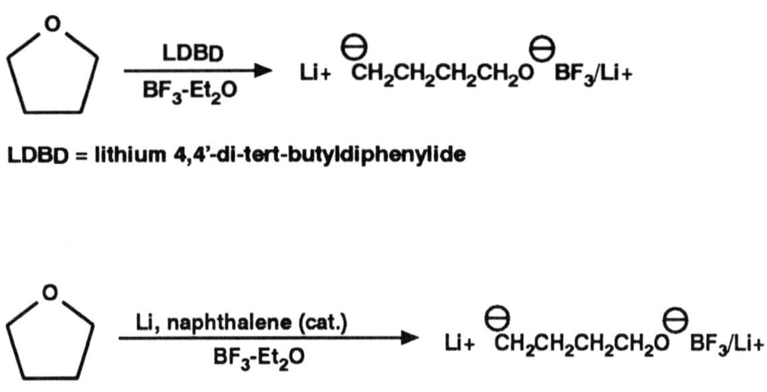

(3.104)

(3.105)

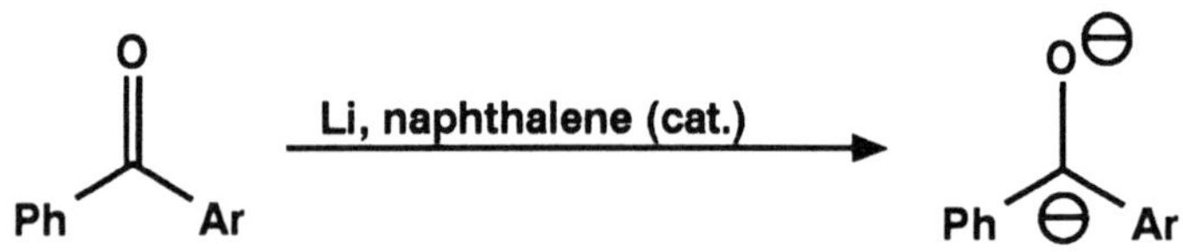

(3.106)

3.4 PHENOL- AND CRESOL-BASED DIANIONS

An interesting class of O,C-dianions may be generated by initial deprotonation at a phenolic-OH moiety. Although only a scant number of reports have appeared that use these intermediates, their range of utility has been adequately demonstrated. Several carbanion sub-types of phenol-based dianions have been investigated including those that undergo metalation, lateral deprotonation (typically a benzylic position), or enolate formation.

3.4.1 Generation of phenol-based dianions

Owing to the relatively low pKa of the phenol-OH group, bases of modest strength can be used in the first deprotonation. In some studies, the preformed salt,[261] has been used in conjunction with BuLi for the subsequent carbanion formation, although most investigators prefer the use of 2 BuLi or mixtures of t-BuOK and BuLi to access these dianions. Phenol-based O,C-dianions may be prepared using metalation and lateral deprotonation for the carbanion site, but these species generally require alkyllithium bases (with or without TMEDA as additive). Remote enolate formation following the phenol deprotonation has been accomplished with t-BuOK.

3.4.2 Reactions of phenol-based dianions

Gilman and co-workers first reported that lithium phenoxide reacts sequentially with BuLi and carbon dioxide to afford salicylic acid via the ortho metalated dianion (Eq. 3.107).[261] Although a poor yield (0.7%) of the product was obtained, this early study showed that dianions of this type could be formed regiospecifically.

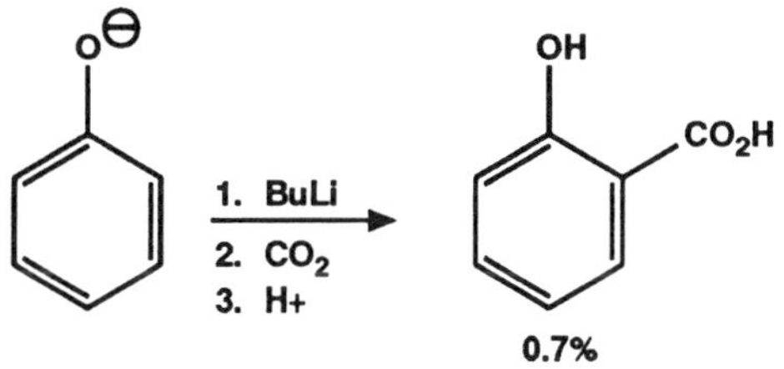

(3.107)

Similar reaction conditions were performed on β-naphthol to yield 2-hydroxy-3-naphthoic acid in 7% yield.[261] No metalation at the 1-position was found. The investigators did show, however, that phenols that bear alkoxy substituents also form the dianion when treated with strong bases. Little subsequent work on the ring metalation of unsubstituted phenols and naphthols followed. Advances in ring metalation of substituted phenols have been reported especially when ortho-directing, chelating groups are present.[262,263] Dimethylamino vanillylamine undergoes reaction with 2 equiv. of BuLi to form the ring-metalated phenol O,C-dianion (Eq. 3.108).[262] Interestingly, dimethylamino isovanillylamine gives no dianion formation under the same conditions. The authors indicate that the position and presence of the alkoxy moiety is an important determinant for successful dianion formation.

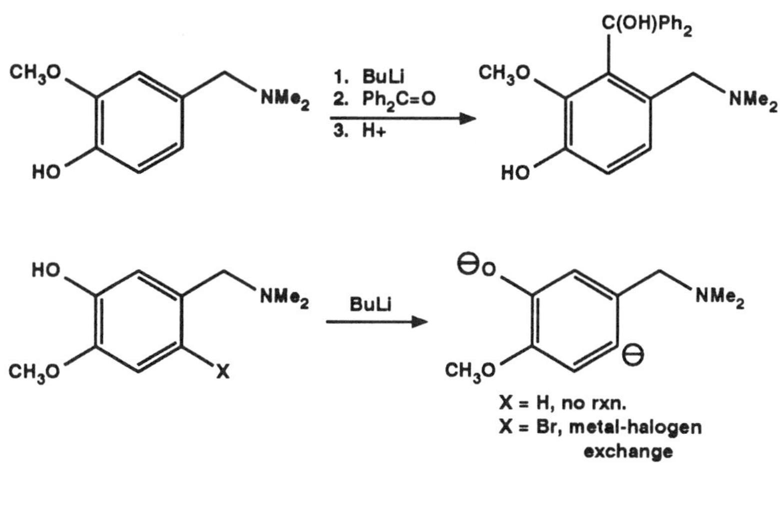

(3.108)

As an alternative to ring metalation, the corresponding bromophenol analogues were prepared and converted to the dianion via metal-halogen exchange (Eq. 3.108).[262]

Salicylaldehyde was converted to the 1,3-dimethylimidazolidine, and treated with base to afford the O,C-ring metalated dianion in high conversion (Eq. 3.109).[263] Reaction with electrophiles gave good yields of the 6-substituted products following removal of the chelating group (40-95%). Some unusual product outcomes were also obtained via this process. Reaction of the 6-metalated dianion with benzaldehyde gave the benzophenone product rather than the secondary alcohol, and reaction with allyl bromide led to bromination rather than alkylation. Successful reaction of this dianion with acetaldehyde was accomplished only when 4 eq. of $CeCl_3$ was added.

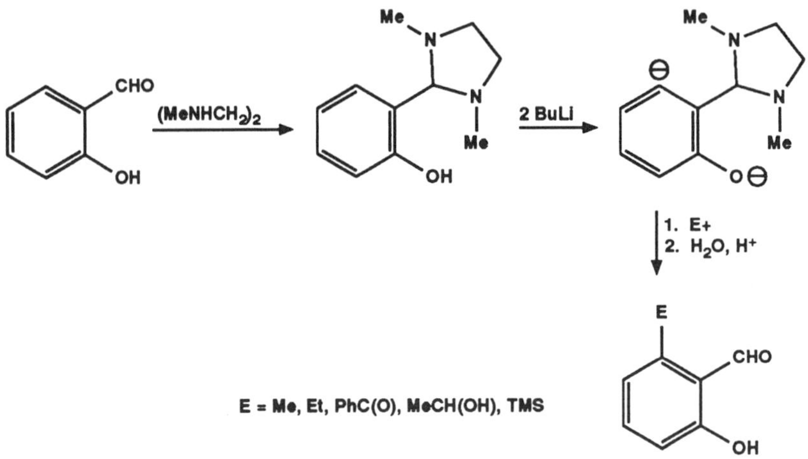

(3.109)

Bates and Siahaan conducted an extensive study of the formation and reaction of the o-, m-, and p-cresol dianions.[264] They found that the extent of benzylic deprotonation was linked to the regioisomer under consideration. Using 4 equivalents of t-BuOK/BuLi as base, the o-cresol, m-cresol, and p-cresol dianions were formed in 85, 95, and 42%, respectively, as determined by their reaction with electrophiles (Scheme 3.32). Further indication of the difficulty of formation of the p-cresol dianion was evidenced in the experimental conditions. Whereas the o- and m-cresol dianion formation was optimum after 3 h reflux in hexane/heptane, formation of the p-cresol dianion required 20 h at reflux.

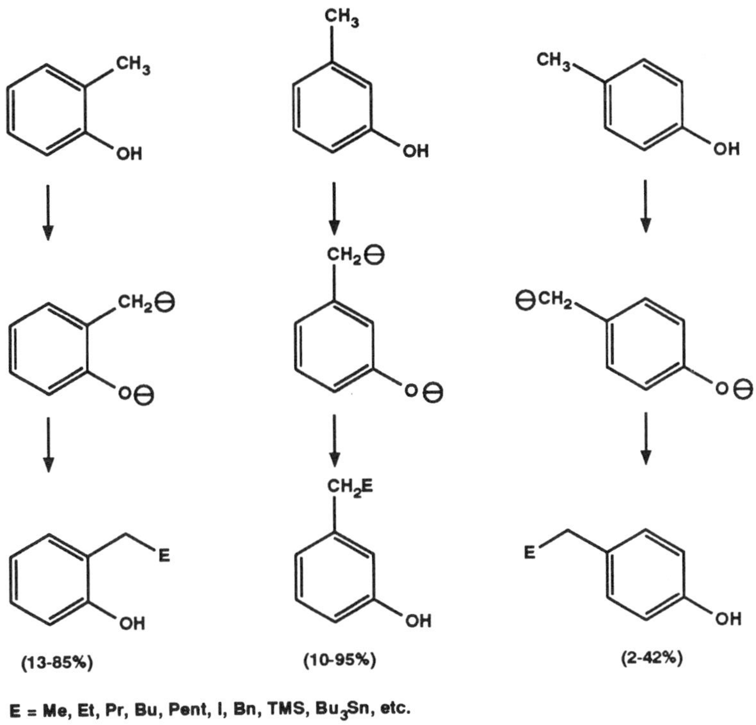

Scheme 3.32. Formation and reactions of the cresol dianions

The addition of a second methyl group on the phenol ring does not adversely affect dianion formation. Kende and coworkers formed the dianion of 3,5-dimethylphenol using KH and BuLi, and sequentially reacted it with allyl chloride and MOM-Cl in 56% yield (Eq. 3.110).[265]

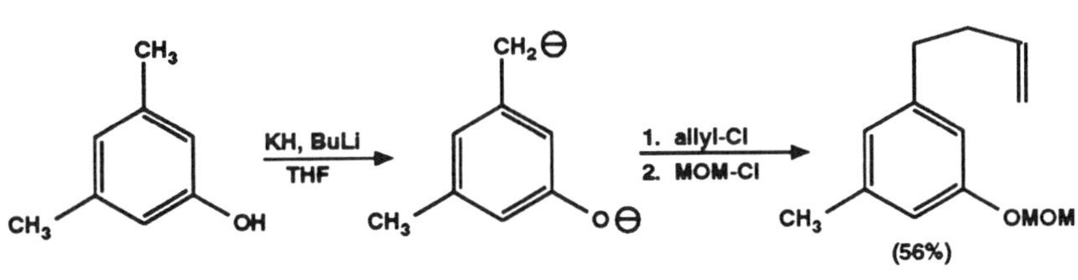

(3.110)

In related studies, benzylic O,C-dianions, 2-(4-hydroxybenzyl)pyridine was bis-deprotonated and acylated, although the yield was modest.[266]

When ring-metalation or lateral protonation is not desired, a few reports have shown that a remote carbanion site can be formed following phenol-OH deprotonation. 4-Hydroxy-2-methylindanone gives O,C-(enolate) dianion formation with 2 KNH_2 which, following reaction with lithium bromoacetate, gives the indanoyl acetic acid product (Eq. 3.111).[267] The enolate-based dianion of o-hydroxy acetophenone was prepared by reaction with 3 LiHMDS, and reacted with dimethylcarbonate to give the oxoester (Eq. 3.112).[268] This dilithio species did not react with alkylnitrates, but use of the mixed lithio/potassio salt did react with propylnitrate in 45% yield.

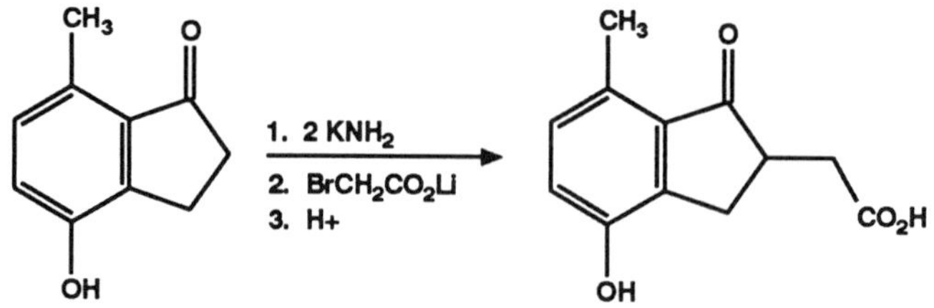

(3.111)

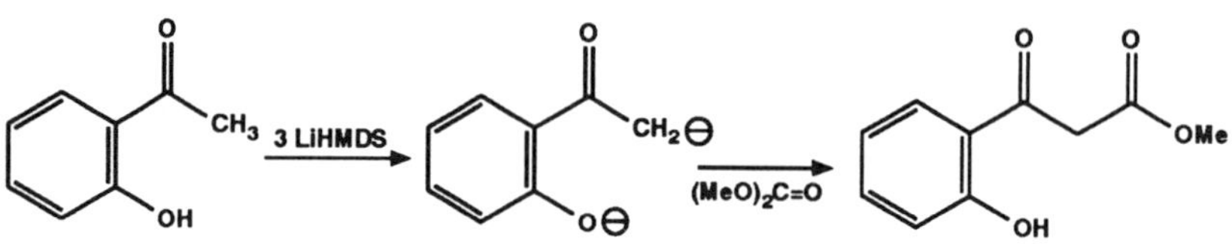

(3.112)

Although not resulting from an O,C-deprotonation sequence, Bates and coworkers found that alkyl groups migrate from oxygen to carbon in alkyl aryl ethers which have been metalated in the benzylic position. The intermediates proposed in this novel discovery were the putative phenol benzylic dianions emerging by way of homolytic cleavage of the alkyl-oxygen bond.[269]

3.5 OXIME-BASED DIANIONS

Ketoximes, and to a lesser extent aldoximes, have been converted to O,C-dianions and reacted successfully with electrophiles. Oxime-based dianions have been used for regiospecific α-functionalization which, following subsequent reaction or various workup procedures, can afford the substituted oxime, the corresponding carbonyl, or can be directly converted to novel heterocycles or nitriles. Oxime dianions are more stable than the corresponding oxime methyl ether α-monoanions, which decompose at 0°C. Two major classes of oxime dianions have been explored; the oxime 1,4-dianions (Fig. 3.12), and the remote oxime dianions, although only scant examples exist for the latter. Prior to a discussion of the utility of oxime dianions, it is worthwhile to remind the reader that certain oximes may exist as a mixture of *syn* and *anti* (or E and Z) forms. The stereochemistry of oxime dianion formation is an important feature, and should be reviewed prior to use of these intermediates.

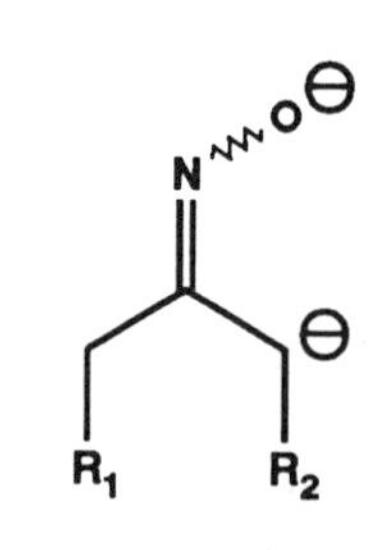

FIGURE 3.12. Structure of oxime 1,4-dianion

3.5.1 Generation of oxime-based dianions

Owing to the stability of oximes, alkyllithiums are the preferred bases for conversion into dianions. In certain circumstances, LDA has been employed with equal success. Early studies showed that metal amides also were capable of converting oximes to the respective dianions, although lower yields of alkylated products were obtained as compared to alkyllithium base use.[270] Several investigators of aldoxime dianions suggest the use of TMEDA as additive to suppress addition of the alkyllithium reagent to the oxime C=N bond.

3.5.2 Reactions of oxime 1,4-dianions

Several sub-classes of oxime 1,4-dianions have been explored, namely, aldoximes that give only one possible dianion, ketoximes where one α-position is blocked (e.g. acetophenone oxime), symmetric ketoximes, and unsymmetric ketoximes. In cases where the oxime is formed from an aldehyde or an unsymmetrical ketone, stereoisomeric (syn/anti) oximes result. Separation of the stereoisomers prior to dianion formation may be preferred (*vide infra*).

Syn/Anti isomerism and oxime dianion formation

The most intriguing aspect of the reaction of oximes with alkyllithium reagents is that only the *syn*- or Z-isomer of the oxime will form the dianion, even when forcing conditions are employed (Eq. 3.113).

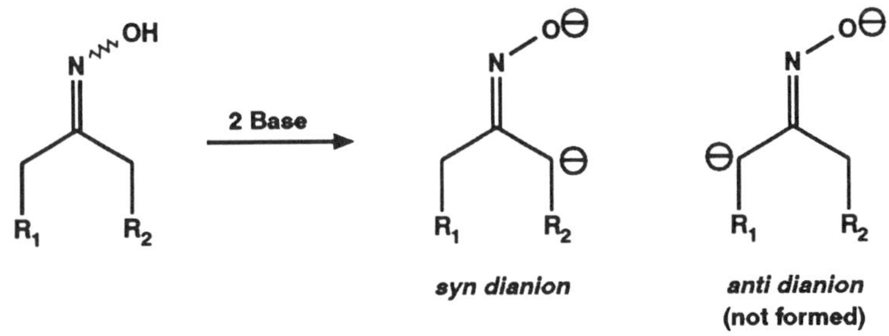

(3.113)

The dominant feature in this (regio)stereochemical bias for the *syn*-dianion has been ascribed to intramolecular chelation (Fig. 3.13), which is not possible for the *anti*-form.[271]

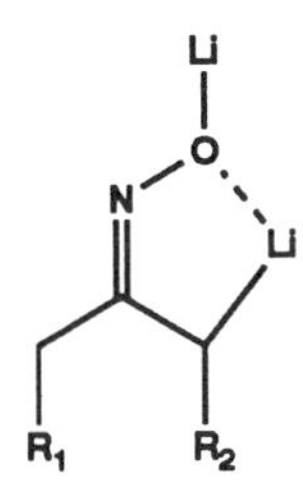

Figure 3.13. Chelated *syn*-oxime form of the dianion

Experimental support for this premise is found in virtually all examples where oxime dianion stereoisomerism is possible, but certain investigations have exploited this aspect. Kofron and Yeh[271] converted a 72:28 (anti/syn) mixture of 2-butanone oximes to the corresponding dianions, which react with Bn-Cl to give a 73:27 mixture of alkylated products (Eq. 3.114). This result confirmed that each oxime stereoisomer gave sole regiochemical alkylation, and also indicated that no change in the oxime configuration occurred during formation of the dianion despite the fact that an α-nitroso, carbanion resonance structure is possible.

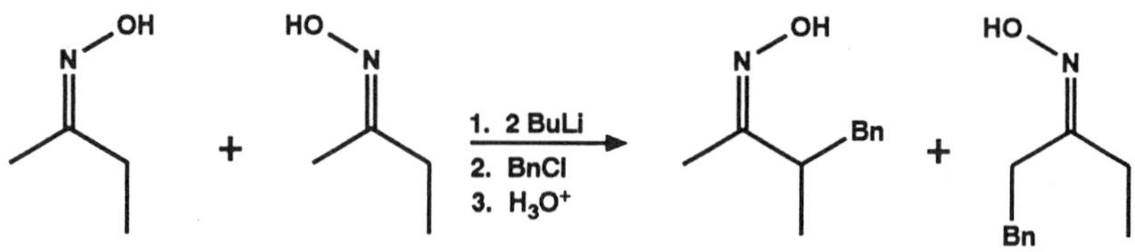

(3.114)

28 : 72 27 : 73

Jung and coworkers[272] reacted the acetoxime dianion with D_2O or CH_3I to give exclusive formation of the corresponding *syn*-products (Eq. 3.115) as evidenced by NMR.

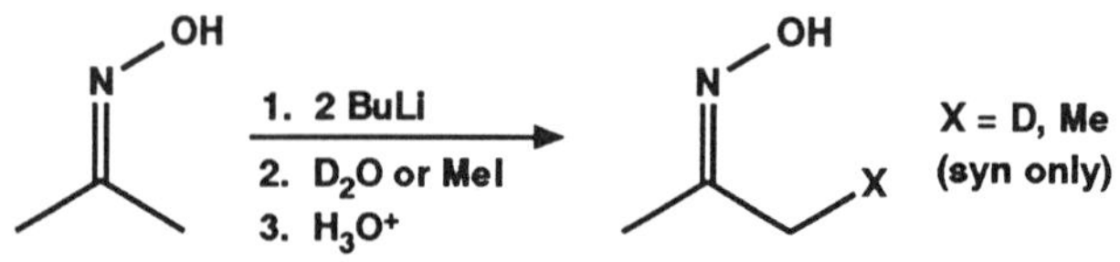

(3.115)

Aldoxime dianions also prefer intramolecular chelation although no regiochemical ambiguity is possible. Hassner and Naumann[273] reacted a 56:44 *anti:syn* mixture of propanal oximes with 5 equiv. of BuLi (hexanes, 0°C), and found that reaction with D_2O led to incorporation of deuterium only for the *syn*-oxime. Changes in the experimental conditions (solvent, HMPA, cation, etc.) had no effect on the outcome. Further indirect evidence for intramolecular chelation was first reported by Kofron[271] who observed that treatment of the oxime solution with 1 eq. of BuLi gave a "milky" heterogeneous solution. Upon addition of the second equivalent of BuLi, the reaction turned to clear yellow.

Recently, Glaser and Streitweiser conducted *ab initio* studies on the acetaldoxime dianion in an effort to clarify the experimental data.[274] Their results indicated that the dimetalated derivatives showed a preference for the syn-configured isomer supporting the prior experimental results. However, the anti-isomer was found to be thermodynamically favored by 2.2 kcal-mol^{-1} for the isolated dianion. The *ab initio* results further showed that chelation by the heteroatoms is more important than the interaction between the metal cation and the methylene carbanion center. The energy difference in favor of the syn-isomer was 4.4 and 8.3 kcal-mol^{-1} for the dilithio-and disodio-species, respectively.

Reactions of oxime 1,4-dianions

Successful alkylations, sulfenylations, and condensations of oxime 1,4-dianions have been reported (Eq. 3.116) (Table 3.10). In general, the yields range from good to excellent.

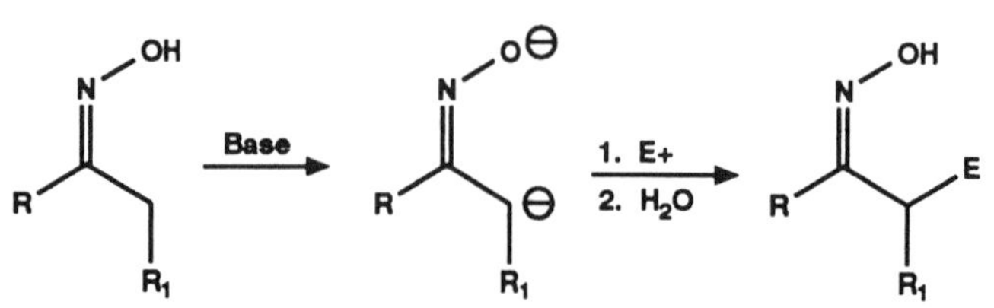

(3.116)

Table 3.10
Reaction of Oxime 1,4-Dianions with Electrophiles

R	R_1	Base	E	Yield (%)	Ref.
H	H	2 BuLi	Bn-	100	275
H	H	2 BuLi	Bn-	96	275
Me	H	2 BuLi	D-	100	272
Me	H	2 BuLi	Me-	73	272
Me	H	2 BuLi	Bn-	72[a]	271
Me	H	2 BuLi	Bu-	63[a]	271
Me	Me	2 BuLi	Me, nPr, iPr-	46-81	272
Me	Me (allyl, Bn)	2 LDA	PhS-	70-75	276
H	nBu	2 BuLi	PhCH(OH)-	82	273
H	nBu	2 BuLi	$R_2R_3C(OH)$-	32-99	273
H	n-Hex	2 BuLi	$Et_2C(OH)$-	99	273
H	n-Hex	2 BuLi	$Me_2CH_2C(OH)$-	90	273
Ph	H	2 BuLi	Bn-	68	270
Ph	H	2 KNH_2	Bn-	46	270
$-(CH_2)_n-$ (n=3,4,5)		2 LDA	PhS-	60-94	276
$-(CH_2)_3-$		2 BuLi	Bu, allyl, Bn-	54-64	271
$-(CH_2)_4-$		2 BuLi	Bu, allyl, Bn-	24-77	271
$-(CH_2)_4-$		2 BuLi	D-	100	272
$-(CH_2)_4-$		2 BuLi	Me-	74	272

[a] Isolated yield after hydrolysis to carbonyl

As indicated by Table 3.10, both acyclic and cyclic oxime 1,4-dianions undergo successful reaction sequences. Facile conversion of the oxime back to the original carbonyl suggests expanded versatility of the oxime dianions as masked carbonyl groups capable of stereo- and regioselective bias. Reduction to the amine also provides unique entry in further functional group management. In addition, Hiroi and coworkers[276] expanded the utility of oxime dianions using a Beckman fragmentation following α-sulfenylation of a cyclic oxime dianion to yield nitriles (Eq. 3.117).

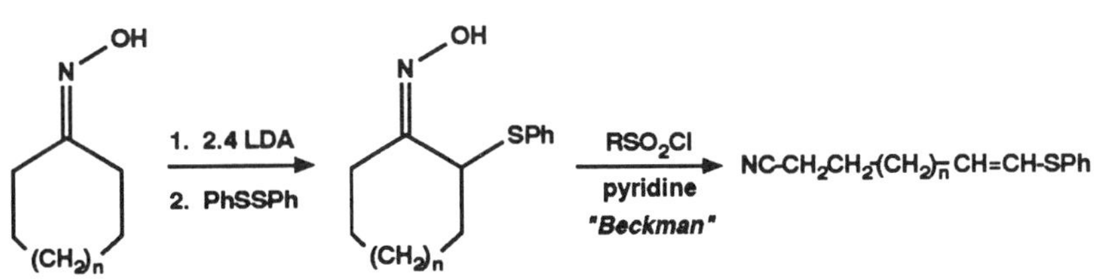

(3.117)

Detailed studies of alicyclic and heterocyclic oxime dianions were conducted by Lyle and coworkers, who showed that alkylations of these compounds occurred with remarkable conformational stereospecificity.[277,278] For example, when the dianion of 4-t-butylcyclohexanone oxime was reacted with methyl iodide, only the syn-oxime of *trans*-2-methyl-4-t-butylcyclohexanone oxime was isolated (Eq. 3.118).[277] Similarly, enantioenriched (Z)-(+)-1-methyl-2,6-diphenyl-4-piperidinone oxime was alkylated with marginal loss in optical purity.[278]

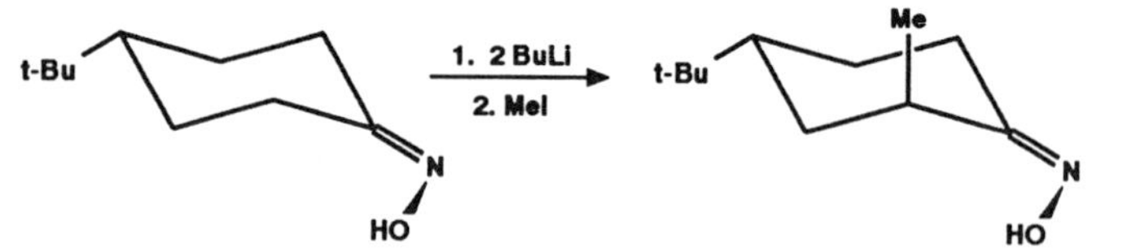

(3.118)

In contrast, (Z)-aldoxime dianions give 1:1 mixtures of β-hydroxy aldoximes following condensation with ketones, but this poor diasteroselectivity improves to 3:1 when the dianions are reacted with aldehydes.[273]

Clearly, the greatest utility of oxime dianions has been in the preparation of heterocycles, particularly isoxazoles, isoxazolines, and quinolines.[279-284] In most instances, the oxime 1,4-dianions used were acetophenone-based oximes that avoid regiochemical ambiguities. Examples of heterocycle synthesis from oxime 1,4-dianions are shown in Scheme 3.34. Isoxazoles are produced by reaction of oxime dianions with esters,[279,282,283] whereas isoxazolines are produced by reaction with aldehydes and ketones.[281] Clever usage of ortho amino-substituted benzoate esters in the reaction with oxime dianions led to quinolines.[284] In several of these studies, parallel investigations with hydrazone dianions produced the corresponding pyrazole heterocycles. For further studies with hydrazone dianions, see Chapter 4.

Oxime dianions also can furnish other heterocyclic systems. For example, oxime 1,4-dianions have been successfully condensed with aldimines to give 2-hydroxylamino azetidines (Eq. 3.119).[284] In this study, acetophenone type or symmetrical ketoximes were employed.

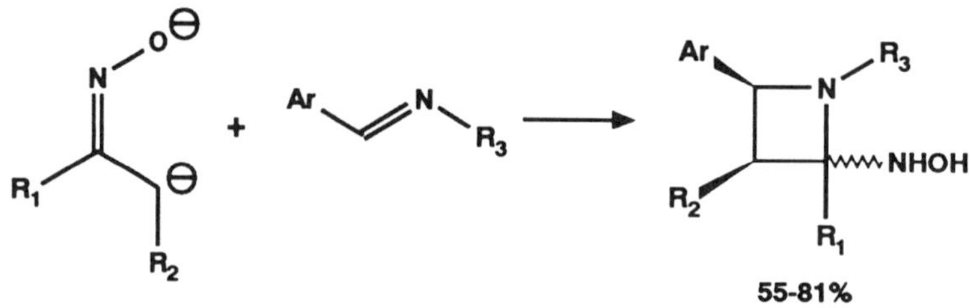

(3.119)

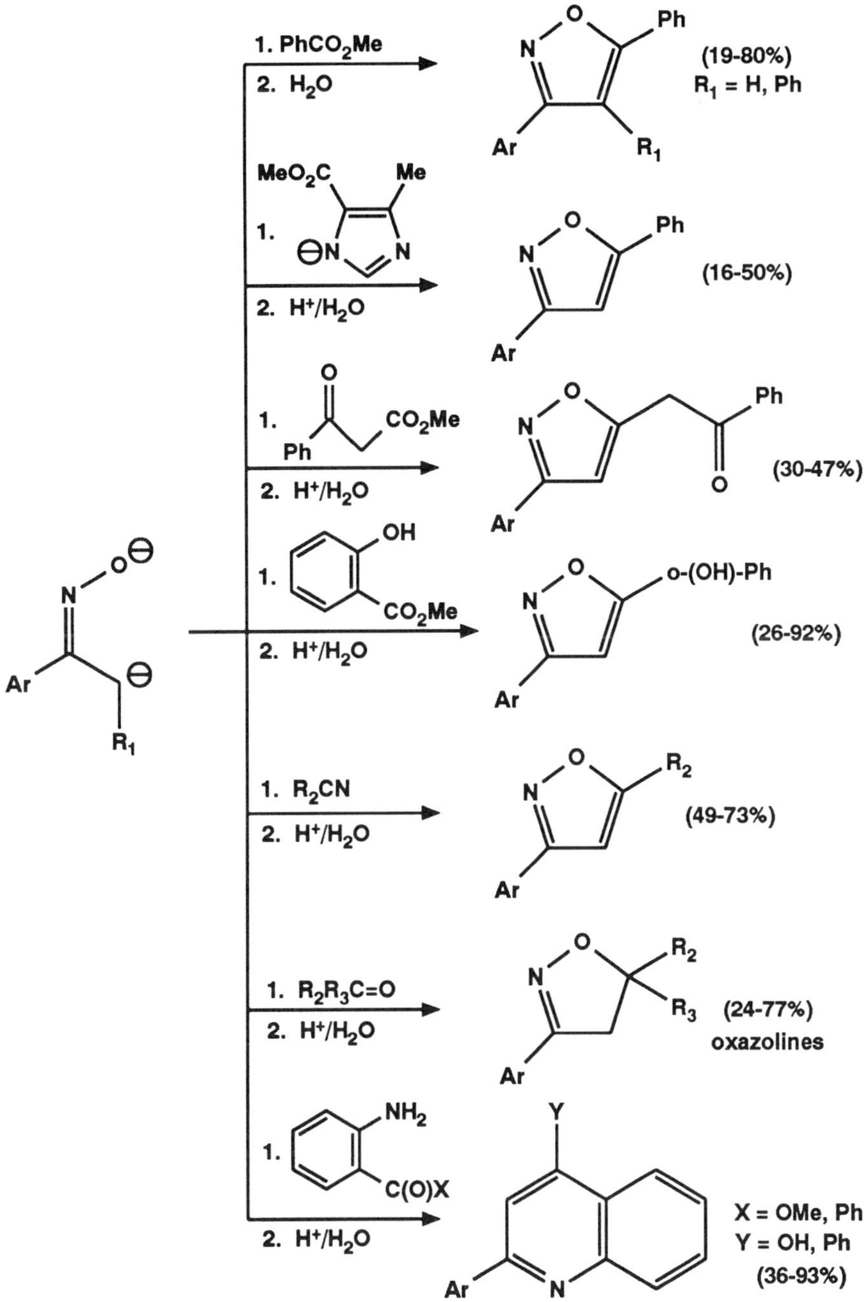

Scheme 3.34. Preparation of isoxazoles, isoxazolines and quinolines from oxime dianions

Yamamoto and coworkers also used oximes of cyclic ketones and acetophenones to prepare isoxazoles and ring-fused pyridine N-oxides (Eq. 3.120).[286]

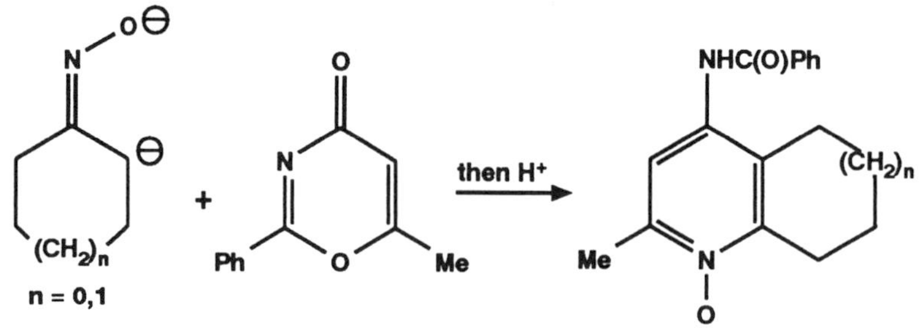

(3.120)

3.5.3 Reactions of remote oxime-based dianions

There have been very few studies of oxime dianions other than those of the 1,4-type. Two examples will demonstrate that oxime dianions are more versatile than the literature suggests.

Almost synchronous with the first reports of oxime dianions, an oxime, remote-benzylic dianion was prepared and alkylated in modest yield in route to acridizinium salts (Eq. 3.121).[287]

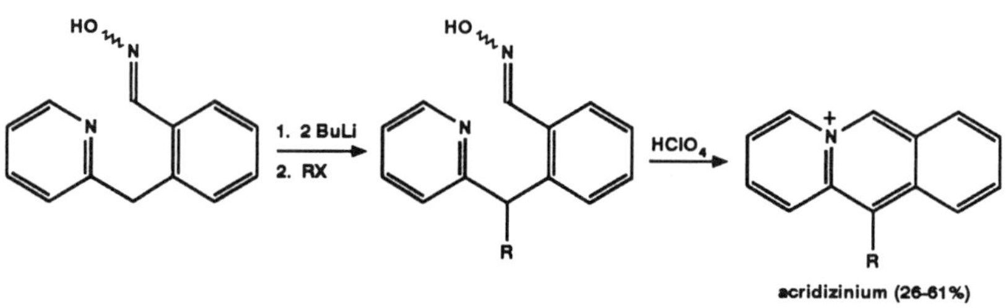

(3.121)

Of more general utility is the report by Ager,[288] who showed that 2-furfural oxime (the anti form) is metalated exclusively at the C-5 position, which undergoes both alkylation and condensation reactions, although ketones condense poorly with this dianion (Eq. 3.122). Interestingly, furan 2-carboxylic acid forms the 3-metalated O,C-dianion (Section 3.2.4).

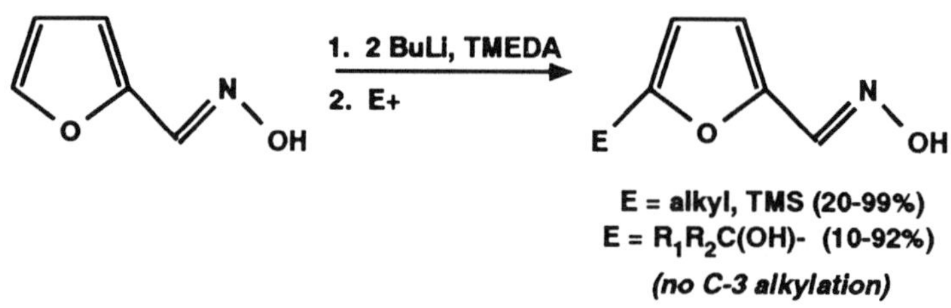

(3.122)

3.6 EXPERIMENTAL PROCEDURES

4′,5′-Dihydro-3β-hydroxy-(17R)-spiro[androst-5-ene-17,2′(3′)-furan]-5′-one (Eq. 3.4) - addition of the acetic acid dianion to steroidal spiroepoxides to form spirolactones (from Creger).[21]

To a solution of diisopropylamine (100 mmol) in 100-200 mL of THF was added n-BuLi in hexane (100 mmol) at a temperature below 10°C. After 10 min at 0°C, acetic acid (50 mmol) in a small volume of THF was added, and the reaction mixture warmed to 30-35°C to complete the metalation. A solution of (17S)-spiro[androst-5-ene-17,2′-oxira]-3β-ol (10 mmol) in 50 mL THF was added, and the mixture heated to reflux for 18 h. After the reaction had cooled, water (100 mL) and ethanol were added, followed by excess 3N HCl. After stirring at room temperature for 0.5 h, the phases were separated and the organic solvents evaporated. The residue was taken up in chloroform-ether and the solution washed with 2 N NaOH, brine, dried ($MgSO_4$), and evaporated. Silica gel chromatography gave 55% of the product lactone. mp = 180-187°C.

***Anti*-2-butyl-3-hydroxy-4-methyl-5-phenyl-4-pentenoic acid** (Eq. 3.14) - stereoselective preparation of γ,δ-unsaturated-β-hydroxy carboxylic acids from the 1,2-addition of an α-carboxylate dianion to an α,β-unsaturated aldehyde (from Mulzer et al.).[43]

A mixture of 288 mL ether, 48 mL THF, diisopropylamine (1 mole; 139 mL) and 63 mL α-methylstyrene were placed in a 2 L flask equipped with a reflux condenser and mechanical stirrer. Finely cut Li-wire was introduced (6.72 g; 1 mole) in small pieces which gave an exothermic reaction bringing the mixture to reflux. Reflux was maintained by external heating until all lithium had disappeared, and a clear solution of LDA formed. The LDA solution was cooled to -50°C and treated dropwise with hexanoic acid (500 mmol) in 240 mL of THF. A colorless precipitate of the dianion formed immediately, and the reaction was stirred at room temperature for 1.0 h to complete dianion formation. The mixture was recooled to -50°C, and α-methylcinnamaldehyde (71 g) was introduced via syringe. After 20 min, the cooling bath was removed and the mixture hydrolyzed with 100 mL water, extracted with ether, acidified with 2 N H_2SO_4, and extracted again with ether. The second extract was dried over magnesium sulfate, and evaporated to afford a 6:1 anti:syn ratio of condensation products as determined by NMR. The mixture was dissolved in ether-pentane and refrigerated, whereupon the anti-isomer crystallized over a period of a few days (42%). mp = 93-94°C.

2,3-Bis[(1,3-benzodioxol-5-yl)methyl]butanedioic acid (Eq. 3.21) - oxidative coupling of α-carboxylate dianions using 0.5 equiv. of I_2 (from Belletire and Fry).[62]

To a solution of diisopropylamine (8.2 mmol) in 40 mL THF at -25°C was added BuLi (8.2 mmol). The reaction was stirred for 10 min below -20°, then at room temperature for 20 min, recooled to -78°C, and to it was added rapidly a THF (22 mL) solution of 3,4-(methylenedioxy)-hydrocinnamic acid (4 mmol). The clear, greenish-yellow solution was stirred below -50°C for 1 h, and then at room temperature for 1.5 h. The dianion solution was then chilled to -62°C, and to it was added I_2 (2.0 mmol) in THF (10 mL) over 1 min. The resultant deep yellow solution was allowed to react for 21 h. The THF was removed and the brownish residue acidified with 20 mL of 1 N HCl and extracted with EtOAc (2 x 75 mL). The organic layers were combined and washed with $NaHSO_3$, water, and brine to give a quantitative yield of product. Recrystallization from EtOAc gives analytically pure material. mp = 190-191°C.

2-Hydroperoxy-2-methyl-tetradecanoic acid (Eq. 3.23) - preferential formation of 2-(hydroperoxy)carboxylic acids by the reaction of O_2 with α-carboxylate dianions (from Konen, Silbert, and Pfeffer).[66]

Diisopropylamine (50 mmol) was dissolved in THF (150 mL) and chilled to -10°C. BuLi (50 mmol) and 2-methyltetradecanoic acid were added in sequence while the temperature was maintained below 0°C. Dianion formation was completed by heating to 50°C for 2 h, and then recooled to ambient temperature. The dianion solution was added dropwise (the tip of an addition funnel was placed below the surface of the O_2 solution) to an oxygen-saturated diethyl ether solution over a period of 1-2 h at -78°C. The mixture was acidified with cold, dilute HCl, extracted with ether, dried and evaporated to afford a mixture of products whose composition by GC ratio was 12% unreacted starting acid, 68% 2-hydroperoxy-2-methyltetradecanoic acid, and 20% 2-hydroxy-2-methyltetradecanoic acid. Yield of desired peroxide product was 66%.

Ethyl butanoate (Eq. 3.25) - reaction of an α-monoethyl malonate dianion with alkyl halide as an alternative to the malonic acid synthesis (from McMurry and Musser).[71]

Isopropylcyclohexylamine (29.4 mmol) was dissolved in 10 mL THF at -78°C, and BuLi (29.4 mmol) was added. Monoethyl malonate (14.7 mmol) in THF (10 mL) was added, and the reaction permitted to warm to 0°C to form the dianion. After 15 min, HMPA (4 mL) was added, followed by ethyl iodide (14.7 mmol) in THF (5 mL). The reaction mixture was allowed to warm to room temperature and was stirred for 2 h to effect alkylation. The reaction was brought to reflux (68°C) overnight to induce decarboxylation. After cooling, the reaction was poured into water, and extracted with ether to give ethyl butanoate (60%).

Methyl-4-[(benzyloxycarbonyl)amino]-3-oxobutanoate (Eq. 3.26) - reaction of the magnesio-dianion of methyl hydrogen malonate with carbonyl imidazolides (from Moyer, Feldman, and Rapoport).[73]

To a solution of CBZ-glycine (10 mmol) in THF was added N,N′-carbonyldiimidazole (12 mmol), and the resulting solution stirred for 12 h at room temperature. Treatment of hydrogen methyl malonate (15 mmol) with isopropyl magnesium bromide (30 mmol) at 0°C for 0.5 h, then at room temperature for 0.5 h, and finally at 40°C for 0.5 h generated the dianion as its magnesium chelate. To this solution at 0°C was added the imidazolide solution, and a gummy precipitate began to form immediately. After warming to room temperature and stirring for 4 h, the reaction was poured onto ice-cold 1 M H_3PO_4. Extraction with ethyl acetate was followed by washing the combined organic phases with saturated NaCl, and drying over $MgSO_4$. Evaporation of the solvent gave the crude β-ketoester in 71% yield. mp = 53.5 - 55°C.

General procedure for the reaction between α-(phenylthio)carboxylic acid dianions and epoxides to give lactones (Scheme 3.6) - (Iwai, Kosugi, Uda, and Kawai).[89]

To a solution of LDA (20 mmol) in THF-ether was added dropwise a solution of (phenylthio)carboxylic acid (10 mmol) in THF (7 mL) at 0°C. The mixture was stirred at 0-4°C for 1 h to effect formation of the dianion. To this solution was added a slight excess of the epoxide (10.5 mmol) in THF (approx. 3 mL) at -60°C, the reaction stirred at this temperature for 1 h, and gradually warmed to room temperature within a period of 15 h. Aqueous NH_4Cl was added to the mixture and the solvent removed under reduced pressure. The residue was poured into a mixture of ether and dilute H_2SO_4, and the water layer extracted with ether. The combined organic layers were washed with water, brine, and dried. After removal of solvent, the residue was heated with benzene (approx. 10 mL) containing a trace of H_2SO_4 using a water-separator for 3 h. Removal of solvent gave a mixture of *cis*- and trans-γ-substituted-α-phenylthio-γ-butyrolactones in 70-94% yield.

Methyl (E)-2,6-heptadienoate (Scheme 3.8) - α-allylation of the crotonic acid lithium dienolate with allyl bromide (from Katzenellenbogen and Crumrine).[113]

To LDA (20 mmol) in THF at -78°C was slowly added a solution of crotonic acid (10 mmol) in 10-15 mL of THF. The solution was stirred for 0.5 h at 0°C to form a clear yellow solution of the dianion, which was brought to -78°C and treated with cuprous iodide (20 mmol). The resultant slurry was stirred rapidly (-78°C) leading to a pale to bright yellow color if the copper dienolate formation was successful. Allyl bromide was added with stirring at -78°C, and continued overnight permitting the reaction to come to room temperature. Aqueous workup using NaOH to adjust the aqueous to pH = 12 caused the copper salts to precipitate, which were removed by filtration. Acidification to pH = 3 and extraction with ether gave a yellow oil that was esterified with diazomethane to give the α-allylated product in 97% yield.

Methyl (E)-2-methyl-5-hydroxytetradec-2-enoate (Eq. 3.36) - reaction at the γ-site of the tiglic acid dianion with decanal, and subsequent conversion to the methyl ester (from Cardillo et al.).[116]

To a suspension of NaH (30 mmol) in dry THF (50 mL) was added tiglic acid (30 mmol) at 0°C in THF (30 mL). The mixture was stirred at 40°C for 3 h, cooled to 0°C and reacted dropwise with LDA (30 mmol; formed in 30 mL THF). The solution was stirred for 0.5 h, heated at 40°C for 1.5 h, and recooled to -78°C, whereupon HMPA (10 mL) was added. Decanal (30 mmol) was then added slowly over 1 h as a THF solution (30 mL), and maintained at -78°C overnight with stirring. The reaction was brought to room temperature, stirred for 36 h, quenched with ice water and 2 N HCl and extracted with ether. The ether extracts were washed with water, dried with $MgSO_4$, and the solvent removed. The crude product was esterified with diazomethane in ether and purified using silica gel chromatography to give 70% yield of the product.

5-[Hydroxy(phenyl)methyl]furan-2-carboxylic acid (Scheme 3.14) - formation of the ring metalated (5-position) furan-2-carboxylic acid dianion and condensation with benzaldehyde (from Knight and Nott).[145]

A solution of LDA (5 mmol) in THF (10 mL) was generated at -10°C and then chilled to -78°C, whereupon a solution of furan-2-carboxylic acid (2.5 mmol) in THF (5 mL) was added. The mixture was stirred at -78°C for 0.5 h and benzaldehyde (2.5 mmol) was added. After the addition of benzaldehyde, the ice bath was removed and the solution stirred an additional 0.5 h. A standard workup gave the 5-functionalized product in 80-85% yield.

2-Ethyl-4-methyl-3-furoic acid (Scheme 3.15) - formation of the lateral deprotonated 2,4-dimethyl-3-furanoic acid dianion and alkylation with methyl iodide (from Tada).[147]

A 4.2 mmol solution of LDA was prepared from BuLi (4.2 mmol in hexane) and diisopropylamine (4.2 mmol) in THF (15 mL) at 0°C. The solution was cooled to -78°C, then 2,4-dimethyl-3-furanoic acid (2 mmol) was added and the reaction stirred for 0.5 h to afford the dianion as an orange solution. Methyl iodide (2.2 mmol) was added and the reaction stirred at 0°C for 1 h dissipating the color. Water was added to the mixture, acidified with HCl, and extracted with ether to afford a 96% yield after sublimation (100°C, 3 mm Hg). mp = 69.5 - 70.5°C.

General methods for the preparation of distal and proximal aldolate dianions (Scheme 3.24) - formation of 1,3-(O,C)- or 1,5-(O,C)-dianions derived from β-hydroxyketones as determined by trapping with TMS-Cl (from Albizati et al).[186]

Preparation of distal β-silyloxy silyl enol ethers. A solution of LDA (2.2 equiv.) was prepared by addition of 2.0 mL of BuLi (2.5 M, hexanes, 2.2 equiv.) to a solution of 0.75

mL of diisopropylamine (5.35 mmol, 2.36 equiv.) in 5 mL THF at -78°C. After 15 min, the β-hydroxyketone (2.27 mmol, 1.0 equiv.) was added dropwise, and after an additional 15 min, the reaction was warmed to 25°C and stirred 15 min more. The solution of the aldolate dianion was then cooled to 0°C, and silylated by the rapid addition of TMS-Cl (2.5 equiv.). The reaction mixture was then warmed to room temperature, quenched after 15 min by the addition of 75 mL of saturated $NaHCO_3$, and extracted twice with 25 mL portions of CH_2Cl_2. The organic extracts were combined, washed with aqueous $NaHCO_3$, filtered and dried over sodium sulfate. Removal of the solvent gave the β-silyloxy silyl enol ethers in 70-93% yield.

Preparation of enforced proximal aldolate dianions. Lithium hexamethyldisilazide (2.2 equiv.) was generated at -78°C by the addition of 2.2 equiv. of BuLi to a THF solution of 2.4 equiv. of hexamethyldisilazane. After the solution was stirred for 10 min, 1.0 equiv. of β-hydroxyketone (as a 1.0 M solution in THF) was added dropwise. After 15 min the dianion was generated, and TMS-Cl (2-5 equiv.) was added rapidly. The silylation was conducted for 15 min, then quenched with an ice-cold solution of saturated $NaHCO_3$. The mixture was extracted thrice with CH_2Cl_2, and the combined extracts dried over Na_2SO_4. Purification was accomplished by chromatography on silica to give the enforced proximal silyl enol ether (1,3-bis-trimethylsilyl-1-phenyl-1-propene analogs) in 20-93% yield.

Ethyl-(2S,3S)-2-allyl-3-hydroxybutyrate (Eq. 3.58) - formation of the dianion from a chiral, β-hydroxy ester, and subsequent stereoselective alkylation (from Frater, Muller and Gunther).[190]

To LDA (100 mmol) in 30 mL THF and 70 mL ether at -50°C was added ethyl-(S)-3-hydroxybutyrate (41.5 mmol) in one portion while letting the temperature rise to 0°C. The mixture was stirred for 2 min, and allyl bromide (45 mmol) in 20 mL HMPTA was added quickly. The temperature rose to 25°C, and briefly to 38°C, and permitted to return to room temperature over 0.5 h. The reaction mixture was worked up with water, 2 N H_2SO_4, and extracted with ether. The crude product was chromatographed on silica gel using hexane:ether (1:1) to furnish 75.5% of a 95:5 mixture of (2S,3S)- and (2R,3S)-2-allyl-3-hydroxybutyrates.

(2R, 3R, 4R)-(+/-)-3-Methoxycarbonyl)-4-methyl-5-hexen-2-ol (Eq. 3.64) - bis-deprotonation of a β-hydroxycarbonyl compound, and example of the dianionic Claisen rearrangement (from Kurth and Yu).[201]

To a solution of LDA (38 mmol; THF, 34 mL; hexane 24 mL) was added (Z)-2-butenyl-3-hydroxybutanoate (17.3 mmol) in THF (7 mL) at -78°C. After 0.5 h, the mixture was allowed to stir for 6 h at room temperature, and then warmed to 50°C for an additional 6 h. The reaction mixture was then poured into 150 mL of 5% NaOH, and extracted with ether (2 x 20 mL), acidified with conc. HCl at 0°C, and successively extracted with methylene chloride. The organic extracts were combined, dried over Na_2SO_4, filtered and concentrated to an oil. The crude carboxylic acid product was esterified with diazomethane to give an inseparable mixture of rearranged β-hydroxy esters in 44% yield. The stereoisomeric ratio was 81:19.

3-Phenylsulfonylhex-5-en-2-ol (Eq. 3.69; Table 3.8) - formation of a β-hydroxysulfone dianion and alkylation with allyl bromide to give the sterically more crowded *erythro* isomer (from Tanikaga, Hosoya and Kaji).[208]

Under an argon atmosphere, BuLi (2.2 mmol) was added dropwise to a solution of 1-phenylsulfonylpropan-2-ol (1 mmol) in THF at -78°C followed by TMEDA (2.0 mmol). The mixture was stirred for 0.5 h, and allyl bromide (1 mmol) was introduced dropwise. The resultant solution was stirred for 0.5 h at -78°C, allowed to warm to 20°C over 2 h, and quenched by the addition of saturated NH_4Cl. The organic layer was extracted twice with ethyl acetate, dried (Na_2SO_4), the solvents removed under reduced pressure, and the product 3-phenylsulfonylhex-5-en-2-ol isolated as a viscous oil following column chromatography in 82% yield existing as an 85:15 erythro:threo mixture.

5,5-Diethyl-3-methylene-butyrolactone (Eq. 3.83) - reaction of the dianion of methallyl alcohol with a carbonyl to form a 1,4-diol, and oxidation to the butyrolactone (from Carlson).[224]

To a suspension of 95-99% tBuOK (21 mmol) in hexane (10 mL) was rapidly added methallyl alcohol (10 mmol) via syringe. n-Butyllithium (21 mmol) in hexane was added over a ten minute period giving a gold color, which was stirred for 15 min at 0°C. 3-Pentanone was added (10 mmol) in hexane (3 mL) causing a rapid loss in color. Water (10 mL) was added, and the mixture extracted with ether resulting in a 40% yield of 2-methylene-4-ethyl-1,4-hexanediol. The diol was converted to the butyrolactone by stirring with active magnesium dioxide (10-fold wt of diol) in methylene chloride at room temperature for 2 h. Filtration and evaporation of the solvent provides 5,5-diethyl-3-methylene-butyrolactone in 86% yield.

2-[(Trimethylsilyl)methyl]phenol, 3-[(trimethylsilyl)methyl]-phenol, and 4-[(trimethylsilyl)methyl]-phenol (Scheme 3.32) - preparation of the *o*-cresol, *m*-cresol or *p*-cresol dianion and reaction with trimethylsilylchloride (from Bates and Siahaan).[264]

Argon was bubbled through 130 mL of heptane for 0.25 h, and t-BuOK (40 mmol) and n-BuLi (40 mmol) were added. After stirring for 0.25 h, *o*-cresol (10 mmol) was added resulting in a bright yellow solution. The mixture was brought to reflux for 3 h (turning the mixture to brown), recooled to 0°C, and the dianion salt filtered and washed with pentane (100 mL). The dianion was taken up in 0°C THF (150 mL), trimethylsilylchloride (TMS-Cl) added (20 mmol), and the reaction stirred for 20-30 h at 25 °C. Standard workup afforded 2-[(trimethylsilyl)methyl]phenol in 40% yield.

The *m*-cresol dianion was prepared in the same manner as *o*-cresol dianion except hexane was used as solvent, and refluxing for 3 h gave a yellowish brown color. Reaction with TMS-Cl gave an 88% yield of 3-[(trimethylsilyl)methyl]phenol.

The *p*-cresol dianion was prepared in the same manner as the *o*-cresol dianion except reflux for 20 h was required for complete conversion. Reaction with TMS-Cl gave a 42% yield of 4-[(trimethylsilyl)methyl]phenol.

2-(1′-hydroxyethyl)cyclohexanone oxime (Eq. 114, in part) - preparation of the *syn*-dianion derived from cyclohexanone oxime and condensation with acetaldehyde (from Kofron and Yeh).[271]

Cyclohexanone oxime (50 mmol) was dissolved in THF (70 mL) and reacted with n-BuLi (100 mmol) using a dry-ice bath to form a yellow solution of the 1,4-dianion. To this mixture was added acetaldehyde (50 mmol) in THF (40 mL), the ice bath removed and the reaction allowed to warm to room temperature. Saturated brine was added and the aqueous solution extracted with 80 mL of a 2:1 mixture of ether and acetone. The organic extracts were combined, dried over $MgSO_4$, and evaporated to afford the substituted oxime. The crude oxime could be converted to 2-ethylidenecyclohexanone following hydrolysis of the oxime, and dehydration of the aldol hydroxyl using oxalic acid (46% overall).

5-Phenacyl-3-phenylisoxazole (Scheme 3.34) - formation of the acetophenone oxime dianion and reaction with ethyl benzoylacetate to form isoxazoles (from Livingston, Chick, Shealy and Beam).[282]

A solution of LDA (44 mmol) in 25-30 mL of THF was generated and stirred for 20 min at 0°C. Acetophenone oxime (10 mmol) dissolved in 25-30 mL of THF was added over 5 min period, then stirred an additional 0.75 h. Ethyl benzoylacetate (11 mmol) in 25-30 mL THF was added over 5 min, the mixture stirred for 0.5 h, then neutralized by the addition of 100 mL of 3 N HCl. The flask was fitted with a reflux condenser, heated to reflux for 1 h, cooled, and carefully neutralized with solid $NaHCO_3$. The layers were separated, and the aqueous layer extracted thrice with 75 mL portions of ether. The resultant precipitate was recrystallized from ethanol and water to give 5-phenacyl-3-phenylisoxazole in 30% yield. mp = 87°C.

REFERENCES

1. **Angelo, B. and Normant, H.** Dianions fonctionnels et leurs applications en synthese organique, *Isv. Khim.* **8**, 5 (1975).
2. **Ivanov, D., Vassilev, G. and Panayotev, I.** Syntheses and reactions of organolithium reagents derived from weakly acidic carbon-hydrogen compounds, *Synthesis* 83 (1975).
3. **Kaiser, E.M., Petty, J.D. and Knutson, P.L.A.** Di- and polyalkali metal derivatives of heterofunctionally substituted molecules, *Synthesis* 509 (1979).
4. **Petragnani, N. and Yonashiro, M.** The reactions of dianions of carboxylic acid and ester enolates, *Synthesis* 521 (1982).
5. **Thompson, C.M. and Green, D.L.C.** Recent advances in dianion chemistry, *Tetrahedron* **47**, 4223 (1991).
6. **Renaud, P. and Fox, M.A.** An electrochemical characterization of dianions: dilithiated carboxylic acids, *J. Am. Chem. Soc.* **110**, 5705 (1988) and references therein.
7. **Jorgenson, M.J.** Preparation of ketones from the reaction of organolithium reagents with carboxylic acids, *Org. React.* **18**, 1 (1971).
8. **Creger, P.L.** Metalated carboxylic acids I. Alkylation, *J. Am. Chem. Soc.* **89**, 2500 (1967).
9. **Pfeffer, P.E., Silbert, L.S. and Chirinko, J.M.** α-Anions of carboxylic acids II. Formation and alkylation of α-metalated aliphatic acids, *J. Org. Chem.* **37**, 451 (1972).
10. **Creger, P.L.** Metalated carboxylic acids II. Monoalkylation of metalated toluic acids and dimethylbenzoic acids, *J. Am. Chem. Soc.* **92**, 1396 (1970).
11. **Ferraz, H.M.C. and Ribiero, C.M.R.** Thallium (III) salts-induced lactonizations of unsaturated carboxylic acids, *Synth. Commun.* **22**, 399 (1992).
12. **Schlosser, M. and Heinz, G.** Fluoroorganic synthesis. I. Fluorolysis of organometallic compounds by perchloryl fluoride, *Chem. Ber.* **102**, 1944 (1969).
13. **Jahngen, E.G., Phillips, D., Kobelski, R.J. and Demko, D.M.** Dimerization of cyclopropanecarboxylic acid dianion and thermal decarboxylative rearrangement of the dimer to 2-cyclopropyl-4,5-dihydrofuran, *J. Org. Chem.* **48**, 2472 (1983).
14. **Belletire, J.L., Spletzer, E.G. and Pinhas, A.R.** Oxidative coupling of carboxylic acid dianions, *Tetrahedron Lett.* **25**, 5969 (1984).
15. **Petragnani, N. and Yonashiro, M.** The reactions of lithium α-lithiocarboxylates (dianions of carboxylic acids) or α-lithiocarboxylic esters (ester enolates), with α-halocarboxylic esters or lithium α-halocarboxylates, *Synthesis* 710 (1980).
16. **Pfeffer, P.E. and Silbert, L.S.** α-Anions of carboxylic acids I. Effect of hexamethylphosphoramide on metalation and alkylation, *J. Org. Chem.* **35**, 262 (1970).
17. **Ireland, R.E., McKenzie, T.C. and Trust, R.I.** Total synthesis of terpenes XXII. Polyene cyclization approach to tetradecahydropicene derivatives for pentacyclic triterpene synthesis, *J. Org. Chem.* **40**, 1007 (1975).
18. **Morishita, S., Saito, T., Hirai, Y., Shoji, M., Mishima, Y. and Kawakami, M.** Synthesis and hypolipidemic activity of 2-substituted isobutyric acid derivatives, *J. Med. Chem.* **31**, 1205 (1988).
19. **Reich, H.J. and Renga, J.M.** Enolate alkylation with bromomethyl sulfides. Synthesis of α-methylene ketones and carboxylic acids, *J. Chem. Soc. Chem. Commun.* 135 (1974).
20. **Krafft, G.A. and Katzenellenbogen, J.A.** Synthesis of halo enol lactones. Mechanism-based inactivators of serine proteases, *J. Am. Chem. Soc.* **103**, 5459 (1981).

21. **Creger, P.L.** Metalated carboxylic acids. IV. Reactions of metalated carboxylic acids with epoxides. Substituted steroidal spiro γ-lactones from spiro β-epoxides, *J. Org. Chem.* **37**, 1907 (1972).
22. **Kok, P., De Clercq, P.J. and Vandewalle, M.E.** (+/-)-Carpesiolin: Total synthesis and structure determination, *J. Org. Chem.* **44**, 4553 (1979).
23. **Stamm, H. and Weiss, R.** Reactions of aziridines: 35. α-aryl-γ-amidobutyric acids (N-protected α-aryl-GABA's) by amidoethylation of dianions of α-arylacetic acids, *Synthesis* 395 (1986).
24. **Kuo, Y-N., Yahner, J.A. and Ainsworth, C.** Reactions of dianions of carboxylic acids with esters and α,β-unsaturated esters, nitriles, and aldehydes, *J. Am. Chem. Soc.* **93**, 6321 (1971).
25. **Mulzer, J., Bruentrup, G., Hartz, G., Kuhl, U., Blaschek, U. and Bohrer, G.** Additionen von Carbonsaure Dianionen an α,β-ungesattigte Carbonylverbindungen - Steuerung der 1,2-/1,4-Regioselectivitat durch sterische Subtituenteneffekte, *Chem. Ber.* **114**, 3701 (1981).
26. **Miyashita, M., Yamaguchi, R. and Yoshikoshi, A.** Michael reaction of conjugated nitro olefins with carboxylic acid dianions and with ester enolates: new synthesis of γ-keto acids and γ-keto esters, *J. Org. Chem.* **49**, 2857 (1984).
27. **Mulzer, J. and Bruentrup, G.** Decarboxylative dehydration of β-hydroxycarboxylic acids according to the redox-condensation scheme; a new olefin synthesis, *Angew. Chem. Int. Ed. Engl.* **89**, 265 (1977).
28. **Adam, W., Baeza, J. and Liu, J.** Stereospecific introduction of doubles bonds via thermolysis of β-lactones, *J. Am. Chem. Soc.* **94**, 2000 (1972).
29. **Mulzer, J., Segner, J. and Bruentrup, G.** Stereoselective synthesis of threo-3-hydroxycarboxylic acids. Stereochemistry of an aldol type addition under kinetic and thermodynamic control, *Tetrahedron Lett.* 4651 (1977).
30. **Moersch, G.W. and Burkett, A.R.** Synthesis of β-hydroxy acids using α-lithiated carboxylic acid salts, *J. Org. Chem.* **36**, 1149 (1971).
31. **Krapcho, A.P. and Jahngen, E.G.E.** General synthetic route to cycloalkylidenecycloalkanes. Reactions of α-anions of cycloalkanecarboxylic acid salts with cycloalkanones, *J. Org. Chem.* **39**, 1650 (1974).
32. **Watanabe, S., Suga, K., Fujita, T. and Fujiyoshi, K.** Direct synthesis of β-hydroxyacids by lithium naphthalene and acetic acid, *Israel. J. Chem.* **8**, 731 (1970).
33. **Hara, S., Taguchi, H., Yamamoto, H. and Nozaki, H.** New synthesis of olefins from β-hydroxy carboxylic acids, *Tetrahedron Lett.* 1545 (1975).
34. **Krapcho, A.P. and Jahngen, E.G.E.** Specific introduction of an isopropylidene group in the synthesis of the monoterpene terpinolene and the sesquiterpene (+/-)-α-curcumene, *J. Org. Chem.* **39**, 1322 (1974).
35. **Kutney, J.P., McGrath, M.J., Young, R.N. and Worth, B.R.** The chemistry of thujone. II. Insect juvenile hormone analogues via acid dianion coupling. The β-lactone route, *Can. J. Chem.* **57**, 3145 (1979).
36. **Black, T.H. and Dubay, W.J.** A new synthesis of substituted spiro butyrolactones via dyotopic rearrangement, *Tetrahedron Lett.* **28**, 4787 (1987).
37. **Black, T.H. and Dubay, W.J.** Lactone synthesis via dyotopic rearrangement. Stereospecific construction of fused butyrolactones with three contiguous asymmetric centers, *Tetrahedron Lett.* **29**, 1747 (1988).
38. **Black, T.H., Hall, J.A. and Sheu, R.G.** Improved, stereospecific synthesis of highly substituted butyrolactones via dyotopic rearrangement, *J. Org. Chem.* **53**, 2371 (1988).
39. **Black, T.H. and Maluleka, S.L.** An efficient, highly regioselective synthesis of substituted (1-cyclohexenyl) acetic acid derivatives via ionization/elimination of β-lactones, *Tetrahedron Lett.* **30**, 531 (1989).
40. **Black, T.H. and Eisenbeis, S.** A novel synthesis of mono- and disubstituted (1-cycloalkenyl) acetic acid derivatives via ionization/elimination of β-lactones, *Synth. Commun.* **19**, 2243 (1989).
41. **Trost, B.M. and Fortunak, J.M.** A new diene synthesis via organopalladium chemistry, *J. Am. Chem. Soc.* **102**, 2841 (1980).
42. **Mulzer, J., Zippel, M., Bruentrup, G., Segner, J. and Finke, J.** Zur Stereochemie der Carbonsauredianion-Aldehyd-Addition unter kinetisch und thermodynamisch kontrollierten Bedingungen-Reindarstellung und Konfigurationszuordnung von 2,3-disubstituierten *threo-* und *erythro*-3-Hydroxycarbonsaure, *Leibigs Ann. Chem.* 1108 (1980).

43. **Mulzer, J., de Lasalle, P., Chucholowski, A., Blaschek, U., Bruentrup, G., Jibril, I. and Huttner, G.** Stereoselective additions to carboxylic dianions and β-lactone substituted ester enolates, *Tetrahedron* **40**, 2211 (1984).
44. **Pfeffer, P.E. and Silbert, L.S.** α-anions of carboxylic acids III. Preparation of aldehydes and nitroparaffins, *Tetrahedron Lett.* 699 (1970).
45. **Koch, G.K. and Kop, J.M.M.** Ethyl formate dianion - a new carbanion, *Tetrahedron Lett.* 603 (1974).
46. **Brocksom, T.J., Petragnani, N. and Rodrigues, R.** Ester enolates. New preparation of malonates, phosphonoacetates, and α-selenyl and sulfinyl esters, *J. Org. Chem.* **39**, 2114 (1974).
47. **Feiffers, S., Wynberg, H. and Strating, J.** α-Carboxylation of esters, *Tetrahedron Lett.* 3001 (1971).
48. **Krapcho, A.P., Jahngen, E.G.E. and Kashdan, D.S.** α-Carbalkoxylation of carboxylic acids. General synthetic route to monoesters of malonic acids, *Tetrahedron Lett.* 2721 (1974).
49. **Konen, D.A., Pfeffer, P.E. and Silbert, L.S.** α-Anions. VIII. Dithioesters and ketene mercaptals from aliphatic acids and esters, *Tetrahedron* **32**, 2507 (1976).
50. **Trost, B.M. and Tamaru, Y.** New synthetic reactions. Oxidative decarboxylation, *J. Am. Chem. Soc.* **97**, 3528 (1975).
51. **Trost, B.M. and Tamaru, Y.** 2-Methylthioacetic acid and diethylmalonate as acyl anion equivalents. Synthesis of juvabione, *Tetrahedron Lett.* 3797 (1975).
52. **Trost, B.M. and Tamaru, Y.** New synthetic reactions. Oxidative decarboxylation of α-methylthiocarboxylic acids, new approach to acyl anion and ketene synthons, *J. Am. Chem. Soc.* **99**, 3101 (1977).
53. **Trost, B.M.** α-Sulfinylated carbonyl compounds in organic synthesis, *Chem. Rev.* **78**, 363 (1978).
54. **Seebach, D. and Teschner, M.** Preparation of α-thiolated carbonyl compounds, *Chem. Ber.* **109**, 1601 (1976).
55. **Reich, H.J.** Functional group manipulation using organoselenium reagents, *Accts. Chem. Res.* **12**, 22 (1979).
56. **Paulmier, C.** *Selenium Reagents and Intermediates in Organic Synthesis*, Pergamon Press, Oxford, 1986.
57. **Kuo, Y-N., Chen, F., Ainsworth, C. and Bloomfield, J.J.** metal reductions of malonates and oxalates. A convenient decarboxylation route for disubstituted malonates and synthesis of ketene acetals, *J. Chem. Soc. Chem. Commun.* 136 (1971).
58. **Ainsworth, C. and Kuo, Y-N.** Ketene bis(trialkylsilyl)acetals. Synthesis, pyrolysis, and spectral studies, *J. Organomet. Chem.* **46**, 13 (1972).
59. **Larson, G.L., Cruz de Maldonado, V. and Berrios, R.R.** The methyldiphenylsilylation of carboxylic acid dianions, *Synth. Commun.* 1347 (1986).
60. **Belletire, J.L. and Fremont, S.L.** Oxidative coupling. II. The total synthesis of enterolactone, *Tetrahedron Lett.* **28**, 127 (1987).
61. **Belletire, J.L. and Spletzer, E.G.** Oxidative coupling. III. The Duco reaction, *Tetrahedron Lett.* **27**, 131 (1986).
62. **Belletire, J.L. and Fry, D.F.** Oxidative coupling of carboxylic acid dianions: the total synthesis of (+/-)-hinokinin and (+/-)-fomentaric acid, *J. Org. Chem.* **52**, 2549 (1987).
63. **Belletire, J.L. and Fry, D.F.** Total synthesis of (+/-)-wikstromol, *J. Org. Chem.* **53**, 4724 (1988).
64. **Belletire, J.L., Fremont, S.L. and Fry, D.F.** A key intermediate for the synthesis of antitumor lignan prototypes, *Synth. Commun.* **18**, 699 (1988).
65. **Bergman, J. and Pelcman, B.** Coupling of indoleacetic acid trianion or methyl indoleacetic acid dianion. A biomimetic approach to indolecarbazole alkaloids, *Tetrahedron Lett.* **28**, 4441 (1987).
66. **Konen, D.A., Silbert, L.S. and Pfeffer, P.E.** α-Anions. VII. Direct oxidation of enolate anions to 2-hydroperoxy- and 2-hydrocarboxylic acids and esters, *J. Org. Chem.* **40**, 3253 (1975).
67. **Wasserman, H. and Lipshutz, B.H.** Oxidative decarboxylation of acids via dianion oxygenation, *Tetrahedron Lett.* 4611 (1975).
68. **Jefford, C.W., Currie, J., Richardson, G.D. and Rossier, J-C.** The synthesis of 1,2,4-trioxan-5-ones, *Helv. Chim. Acta* **74**, 1239 (1991).

69. **Yamada, S.I., Oguri, T. and Shioiri, T.** α-Amination of carboxylic acids: new syntheses of α-amino acids, *J. Chem. Soc. Chem. Commun.* 623 (1972).
70. **Krapcho, A.P., Glynn, G.A. and Grenon, B.J.** The decarboethoxylation of geminal dicarboethoxy compounds, *Tetrahedron Lett.* 215 (1967).
71. **McMurry, J.E. and Musser, J.H.** A simple one-step alternative to the malonic acid synthesis, *J. Org. Chem.* **40**, 2556 (1975).
72. **Brown, R.F.C., Coulston, K.J., Eastwood, F.W. and Moffat, M.R.** Pyrolysis of quinoline-3,3-dicarboxylic anhydrides bearing 2-phenyl, 2-benzyl, and 2-o-tolyl substituents: formation of products of carbene insertion and addition, *Tetrahedron* **48**, 7763 (1992).
73. **Moyer, M.P., Feldman, P.L. and Rapoport, H.** Intramolecular N-H, OH, and S-H insertion reactions. Synthesis of heterocycles from α-diazo β-keto esters, *J. Org. Chem.* **50**, 5223 (1985).
74. **Padwa, A., Dean, D.C., Hertzog, D.L., Nadler, W.R. and Lin, Z.** Azomethine ylide generation via the dipole cascade, *Tetrahedron* **48**, 7565 (1992).
75. **Ihara, M., Takahashi, M., Taniguchi, N., Yasui, K., Niitsuma, H. and Fukumoto, K.** Stereoselective alkylation of dianions derived from chiral half esters of monosubstituted malonic acids: asymmetric synthesis of α-alkyl α-amino acids and key synthetic intermediates for *hunteria* and *Aspidosperma* indole alkaloids, *J. Chem. Soc. Perkin Trans. I.* 525 (1991).
76. **Tanaka, K. and Kaji, A.** The synthetic utility of sulfur-containing dianions, *Sulfur Rep.* **1**, 97 (1980).
77. **Oae, S.** *Organic Sulfur Chemistry: Structure and Mechanism*, CRC Press, Boca Raton, Fl., 1991 and references therein.
78. **Trost, B.M., Crimmin, M.J. and Butler, D.** Synthesis of thermodynamically less stable enol thioethers. An alternative oxidative decarboxylation of α-thio acids, *J. Org. Chem.* **43**, 4549 (1978).
79. **Tanaka, K., Terauchi, M. and Kaji, A.** Dianion of 2-(methylthio)-4-pentenoic acid. A useful reagent for the preparation of 5-methylene-2(5H)-furanones, *Chem. Lett.* 351 (1982).
80. **Tanaka, K., Terauchi, M. and Kaji, A.** Synthesis of (+/-)-dihydromahubanolide using 2-methylthio-4-pentenoic acid dianion, *Bull. Chem. Soc. Japan* **55**, 3935 (1982).
81. **Grieco, P.A. and Wang, C-L.J.** Phenylthioacetic acid dianion: latent, vinyl, oxiranyl, and acyl carbanion equivalents, *J. Chem. Soc. Chem. Commun.* 714 (1975).
82. **Iwai, K., Kawai, M. Kosugi, H. and Uda, H.** New synthetic method for γ-substituted δ-butenolides, *Chem. Lett.* 385 (1974).
83. **Boatman, R.J., Whitlock, B.J. and Whitlock, H.W. Jr.** An anionic equivalent of the Friedel-Crafts cycloacylation, *J. Am. Chem. Soc.* **99**, 4822 (1977).
84. **Bates, G.S.** New α-keto acid synthon: alkylation of the potassium dianion of bis(ethylthio)acetic acid, *J. Chem. Soc. Chem. Commun.* 161 (1979).
85. **Bates, G.S. and Ramasamy, S.** Dianions of glyoxylic acid thioketals: convenient α-keto acid equivalents, *Can. J. Chem.* **58**, 716 (1980).
86. **Nicolaou, K.C., Seitz, S.P., Sipio, W.J. and Blount, J.F.** Phenylseleno- and phenylsulfenolactonizations. Two highly efficient and synthetically useful cyclization procedures, *J. Am. Chem. Soc.* **101**, 3884 (1979).
87. **Iwai, K., Kosugi, H. and Uda, H.** A new synthetic method for β-substituted δ-butenolides, *Chem. Lett.* 1237 (1974).
88. **Iwai, K., Kosugi, H. and Uda, H.** Michael addition of α-phenylsulfinyl δ-butenolides, *Chem. Lett.* 981 (1975).
89. **Iwai, K., Kosugi, H., Uda, H. and Kawai, M.** New method for synthesis of various types of substituted 2(5H)-furanones, *Bull. Chem. Soc. Japan* **50**, 242 (1977).
90. **Donaubauer, J.R. and McMorris, T.C.** Synthesis of lepichlorin, an antibiotic metabolite of a fungus cultivated by ants, *Tetrahedron Lett.* **21**, 2771 (1980).
91. **Kosugi, H., Uda, H. and Yamigawa, S.** Vinylogous Pummerer rearrangement of methyl cyclopentylidene(phenylsulfinyl)acetate, *J. Chem. Soc. Chem. Commun.* 192 (1975).
92. **Yamagiwa, S., Hoshi, N., Sato, H., Kosugi, H. and Uda, H.** Reactions of the α-carbanion species of (phenylthio)acetic acid and its ester with carbonyl compounds, *J. Chem. Soc. Perkin Trans. I.* 214 (1978).
93. **Miyashita, M., Yamaguchi, R. and Yoshikoshi, A.** New synthesis of γ-keto acids from nitro olefins and carboxylic acid dianions, *Chem. Lett.* 1505 (1982).

94. **Miyashita, M., Kumazawa, T. and Yoshikoshi, A.** 1-Nitro-1-phenylthiopropene, a new reagent for 3-methylfuran synthesis, *J. Chem. Soc. Chem. Commun.* 362 (1978).
95. **Caron, G. and Lessard, J.** Synthesis of trisubstituted enol ethers and of methyl ketones from ketones, *Can. J. Chem.* **51**, 981 (1973).
96. **Adam, W. and Fick, H.H.** Direct α-ltihiation of phenoxyacetic acid and electrophilic substitution, *J. Org. Chem.* **43**, 772 (1978).
97. **Adam, W. and Fick, H.H.** Synthesis of phenyl enol ethers via decarboxylation of β-lactones derived from the lithium α-phenoxy-α-lithioacetate synthon, *J. Org. Chem.* **43**, 4574 (1978).
98. **Adam, W. and Encarnacion, L.** Benzyl enol ethers via decarboxylation of α-(benzyloxy)-β-lactones derived from the lithium α-(benzyloxy)-α-lithioacetate synthon, *Synthesis* 388 (1979).
99. **Neef, G. and Eder, U.** Methyl dimethoxy acetate as a carbonyl anion equivalent, *Tetrahedron Lett.* 2825 (1977).
100. **Raucher, S., Gustavson, L.M. and Stenkamp, R.E.** An unusual cyclization of a homoallyl oxyacetic acid dianion, *J. Org. Chem.* **52**, 2760 (1987).
101. **Johnson, C.R. and Bade, T.R.** Alkylation, acylation, and cyclopropanation reactions of α-halo carboxylic acid dianions, *Synthesis* 284 (1982).
102. **Johnson, C.R. and Bade, T.R.** Dianions derived from α-halo acids. The Darzens condensation revisited, *J. Org. Chem.* **47**, 1205 (1982).
103. **Black, T.H., McDermott, T.S. and Brown, G.A.** A facile 3-alkyl-butenolide annulation sequence, *Tetrahedron Lett.* **32**, 6501 (1991).
104. **Coutrot, P. and El Gadi, A.** Dichloroacetic acid dianion; preparation, alkylation and Darzens reaction with carbonyl compounds, *Synthesis* 115 (1984).
105. **Grieco, P.A., Wang, C-L.J. and Burke, S.D.** Trimethylsilylacetic acid dianion: application to organic synthesis, *J. Chem. Soc. Chem. Commun.* 537 (1975).
106. **Petragnani, N. and Carvalho-Ferraz, H.M.** 2-Phenylselenopropanoic acid: a useful synthon for α-methylene-γ-lactones, *Synthesis* 476 (1978).
107. **Koppel, G.A. and Kinnick, M.D.** Carboxyvinylation, a one-step synthesis of α,β-unsaturated acids, *Tetrahedron Lett.* 711 (1974).
108. **Wasserman, H. and Lipshutz, B.H.** A new synthesis of β-lactams. Oxygenation of dianions of azetidine carboxylic acids, *Tetrahedron Lett.* 4613 (1976).
109. **Pfeffer, P.E. and Silbert, L.S.** α-Anions. IV. Positional and stereochemical isomerization of 2-and 3-unsaturated carboxylic acid dianions, *J. Org. Chem.* **36**, 3290 (1971).
110. **Pfeffer, P.E., Silbert, L.S. and Kinsel, E.** A reexamination of unsaturated carboxylate dianion reactions. Evidence for α- and γ-substitutions in alkenoic acids, *Tetrahedron Lett.* 1163 (1973).
111. **Cainelli, G., Cardillo, G., Contento, M., Trapani, G. and Umani-Ronchi, A.** Synthesis of compounds containing the isoprene unit. Condensation of 3-methyl-3-butenoic and 3-methyl-2-butenoic acids with carbonyl compounds, *J. Chem. Soc. Perkin Trans. I.* 400 (1973).
112. **Cainelli, G., Cardillo, G. and Orena, M.** Synthesis of compounds containing the isoprene unit. Condensation of 3-methylbut-3-enoic and 3-methylbut-2-enoic acids with carbonyl compounds, *J. Chem. Soc. Perkin Trans. I.* 1537 (1979) and references therein.
113. **Katzenellenbogen, J.A. and Crumrine, A.L.** Selective γ-alkylation of dienolate anions derived from α,β-unsaturated acids. Applications to the synthesis of isoprenoid olefins, *J. Am. Chem. Soc.* **98**, 4925 (1976) and references therein.
114. **Savu, P.M. and Katzenellenbogen, J.A.** Selective γ-alkylation of copper enolates derived from α,β-unsaturated acids: factors affecting scope and regio- and stereoselectivity, *J. Org. Chem.* **46**, 239 (1981).
115. **Cardillo, G., Orena, M. and Sandri, S.** Synthesis of compounds containing the isoprene unit, *Tetrahedron* **32**, 107 (1976).
116. **Cardillo, G., Orena, M., Porzi, G. and Sandri, S.** Synthesis of malyngolide, an antibiotic from the marine blue-green algae *Lyngbya majuscula* gomont, *J. Org. Chem.* **46**, 2439 (1981).
117. **Bernasconi, S., Columbo, M., Jommi, G. and Sisti, M.** A convenient synthesis of perillenal, *Gazz. Chim. Ital.* **116**, 69 (1986).
118. **Bongini, A., Orena, M. and Sandri, S.** Structural assignment of the dianion of 3-methylbut-2-enoic acid, *J. Chem. Soc. Chem. Commun.* 50 (1986).
119. **Johnson, P.R. and White, J.D.** Condensation of crotonic and tiglic acid dianions with aldehydes and ketones, *J. Org. Chem.* **49**, 4424 (1984).

120. **Ballester, P., Costa, A., Raso, A.G. and Gomez-Solivellas, A.** Dienediolates from unsaturated carboxylic acids: Michael addition of dilithium buta-1,3-diene-1,1-diolate (from crotonic acid) to unsaturated ketones, *J. Chem. Soc. Perkin Trans. I.* 1711 (1988).
121. **Carlson, R.M. and Oyler, A.R.** A direct α-methylene lactone synthesis via an itaconic acid ester, *Tetrahedron Lett.* 4099 (1975).
122. **Carlson, R.M., Delane, J.C. and Liu, T-L.** α-Alkylidene-γ-lactone synthesis: α-methyl and α-benzylcinnamic acid dianion addition to ketones, *Synth. Commun.* **15**, 657 (1985).
123. **Aurell, M.J., Gil, S., Parra, M., Tortajada, A. and Mestres, R.** Iodine oxidative coupling of diene and triene-diolates of unsaturated carboxylic acids, *Tetrahedron* **47**, 1997 (1991).
124. **Aurell, M.J., Gil, S., Martinez, P.V., Parra, M., Tortajada, A. and Mestres, R.** Polyenolates of unsaturated carboxylic acids in synthesis. Synthesis of unsaturated α-amino acids and γ-hydrazino acids, *Synth. Commun.* **21**, 1833 (1991).
125. **Deshmukh, A.R., Tran, L. and Biehl, E.R.** Reaction of 2-butenoic acid dianion and its N-(4-methoxyphenyl)amide with methoxy-substituted arynes, *J. Org. Chem.* **57**, 667 (1992).
126. **Kejian, C., Sanner, M.A. and Carlson, R.M.** The synthesis of butenolide α-formyl acetals, *Synth. Commun.* **20**, 901 (1990).
127. **Allan, R.D. and Fong, J.** Synthesis of analogues of GABA. VIII. Selective α-alkylation and γ-halogenation of the dianion from α,β-unsaturated nipecotic acid derivatives, *Aust. J. Chem.* **36**, 601 (1983).
128. **Bachi, M.D., Epstein, J.W., Herzberg-Minzly, Y. and Lowenthal, H.J.E.** Synthesis of compounds related to gibberelic acid. III. Analogs of ring A of the gibberelins, *J. Org. Chem.* **34**, 126 (1969).
129. **Marshall, J.A. and Wuts, P.G.** Stereoselective synthesis of racemic occidentalol and related cis-fused hexahydronaphthalene from o-toluic acid, *J. Org. Chem.* **42**, 1794 (1977).
130. **Sipio, W.J.** 1,2-Disubstituted-2,5-cyclohexadienes: selective inhibitors of 5-lipoxegenase, *Tetrahedron Lett.* **26**, 2039 (1985).
131. **Ramanathan, H. and Subba Rao, G.S.R.** Synthesis based on cyclohexadienes: Part 6 - a new total synthesis of (+/-)-methyl acorate and (+/-)-methyl epi-acorate, *Indian J. Chem.* **30B**, 901 (1991).
132. **Vedejs, E. and Gapinski, D.M.** A method for γ-functionalization of tiglaldehyde via the lithio aldimine, *Tetrahedron Lett.* **22**, 4913 (1981).
133. **Pitzele, B.S., Baran, J.S. and Steinman, D.H.** γ-Alkylation of 2-butynoic acid. A route to controlled prenol homologation, *J. Org. Chem.* **40**, 269 (1975).
134. **Pitzele, B.S., Baran, J.S. and Steinman, D.H.** γ-Alkylation of α,β-unsaturated acids. A technique of regiospecific isoprenoid homologation, *Tetrahedron* **32**, 1347 (1976).
135. **Meyer, H.H.** Aldol-Reaktion des 2-Butinsauredianions. Eine einfache Variante der Synthese von (+/-)-Goniothalamin, *Tetrahedron Lett.* 3221 (1975).
136. **Meyer, H.H.** Umsetzung des 2-Butinsauredianions mit Aldehyden; Synthese einiger Naturstoffe mit 5,6-Dihydro-2H-pyran-2-on-Struktur, *Leibigs Ann. Chem.* 327 (1978).
137. **Carlson, R.M. and Oyler, A.R.** The propiolic acid dianion as an acyl acetate equivalent: The synthesis of (+/-)-pestalotin, *Tetrahedron Lett.* 2615 (1974).
138. **Carlson, R.M., Oyler, A.R. and Peterson, J.R.** Synthesis of substituted 5,6-dihydro-2H-pyran-2-ones. Propiolic acid dianion as a reactive three-carbon nucleophile, *J. Org. Chem.* **40**, 1610 (1975).
139. **Coke, J.L. and Richon, A.B.** Synthesis of optically active δ-n-hexadecalactone, the proposed pheromone from Vespa orientalis, *J. Org. Chem.* **41**, 3516 (1976).
140. **Micetich, R.G.** Lithiation of five-membered heteroaromatic compounds. Methyl substituted 1,2-azoles, oxadiazoles, and thiadiazoles, *Can. J. Chem.* **48**, 2006 (1970).
141. **Snieckus, V.** The directed ortho metalation reaction. Methodology, applications, synthetic links, and a non-aromatic ramification, *Pure Appl. Chem.* **62**, 2047 (1990).
142. **Snieckus, V.** Regioselective synthetic processes based on the aromatic directed metalation strategy, *Pure Appl. Chem.* **62**, 671 (1990).
143. **Belletire, J.L. and Spletzer, E.G.** Oxidative coupling. The toluic acid dianion system, *Synth. Commun.* **16**, 575 (1986).

144. **Epsztajn, J., Plotka, M.W. and Scianowski, J.** Application of organolithium and related reagents in synthesis. Part 11. Metalation of 2-methyl- and 4-methyl-nicotinic acids. A useful method for preparation of aza-coumarins, *Synth. Commun.* **22**, 1239 (1992).
145. **Knight, D.W. and Nott, A.P.** The generation and chemistry of dianions derived from furancarboxylic acids, *J. Chem. Soc. Perkin Trans. I.* 1125 (1981).
146. **Dell, C.P. and Knight, D.W.** A brief, stereoselective total synthesis of the guaiane sesquiterpene (+/-)-gnididione, *J. Chem. Soc. Chem. Commun.* 349 (1987).
147. **Tada, M.** Total synthesis of (+/-)-isoalantolactone and (+/-)-dihydrocallitrisin, *Chem. Lett.* 441 (1982).
148. **Tada, M., Sugimoto, Y. and Takahashi, T.** Total synthesis of furanoeremophilane and 14-norfuranoeudesmane derivatives by a route involving alkylation of 2,4-dimethyl-3-furoic acid, *Bull. Chem. Soc. Japan* **53**, 2966 (1980).
149. **Buttery, C.D., Knight, D.W. and Nott, A.P.** Methods for the homologation of benzofurancarboxylic acids using dianionic intermediates, *Tetrahedron Lett.* **23**, 4127 (1982).
150. **Buttery, C.D., Knight, D.W. and Nott, A.P.** The generation and synthetic utility of dianions derived from benzofurancarboxylic acids, *J. Chem. Soc. Perkin Trans. I.* 2839 (1984).
151. **Costa, A.M.B.S.R.C.S., Dean, F.M., Jones, M.A., Smith, D.A. and Varma, R.S.** β-Deprotonation by LDA. Vinyl carbanions from oxygen heterocycles in the synthesis of carboxylic acids in the benzofuran, flavone, and coumarin series and in the regiospecific acylation of 2,6-dimethylchromone, *J. Chem. Soc. Chem. Commun.* 1224 (1980).
152. **Bedford, S.B., Begley, M.J., Cornwall, P. and Knight, D.W.** A brief and mild method for the generation of benzofuran-2,3-xylylene in solution, *SynLett.* 627 (1991).
153. **Knight, D.W. and Nott, A.P.** Formation and reactivity of dianions derived from 2- and 3-thiophencarboxylic acids, *Tetrahedron Lett.* **21**, 5051 (1980).
154. **Knight, D.W. and Nott, A.P.** Generation and synthetic utility of dianions derived from thiophencarboxylic acids, *J. Chem. Soc. Perkin Trans. I.* 791 (1983).
155. **Gould, N.P. and Lee, T-J.** Dianions of methylated thiophene-2-carboxylic acids: their formation and utility, *J. Org. Chem.* **45**, 4528 (1980).
156. **Cuomo, J., Gee, S.K. and Hartzell, S.L.** "Synthesis of heterocyclic sulfonamides," In *Synthesis and Chemistry of Agrochemicals II* (Baker, D.R., Fenyes, J.G. and Moberg, W.K., Eds.), ACS Symposium Series 443, ACS, Washington, DC, 1991, pp. 62-73.
157. **Cornwall, P., Dell, C.P. and Knight, D.W.** New methods for the homologation of thiazoles and oxazoles by regiospecific lithiations of thiazole- and oxazole-carboxylic acid derivatives, *Tetrahedron Lett.* **28**, 3585 (1987).
158. **Cornwall, P., Dell, C.P. and Knight, D.W.** Regioselectivity in the lithiation of methyl-substituted thiazole- and oxazole-carboxylic acids and carboxamides: general methods for the elaboration of trisubstituted thiazoles and oxazoles, *J. Chem. Soc. Perkin Trans. I.* 2417 (1991).
159. **Natale, N.R., McKenna, J.I., Niou, C-S., Borth, M. and Hope, H.** Metalation of isoxazolyloxazolines, a facile route to functionally complex isoxazoles: utility, scope, and comparison to dianion methodology, *J. Org. Chem.* **50**, 5660 (1985).
160. **Kofron, W.G. and Wideman, L.G.** Specific synthesis and selective alkylation and condensation of monoesters of substituted succinic acids, *J. Org. Chem.* **37**, 555 (1972).
161. **Iwai, K., Kosugi, H., Miyasaki, A. and Uda, H.** Generation of 3-carbanion species from 3-phenylsulfinyl- and 3-phenylsulfonyl propionic acids, *Synth. Commun.* **6**, 357 (1976).
162. **Bravo, P., Carrera, P., Resnati, G. and Ticozzi, C.** Chiral homoenolate anion equivalents: synthesis of optically pure 5-substituted furan-2-(5H)-ones, *J. Chem. Soc. Chem. Commun.* 19 (1984).
163. **Reider, P.J. and Grabowski, E.J.J.** Total synthesis of thienamycin; a new approach from aspartic acid, *Tetrahedron Lett.* **23**, 2293 (1982).
164. **Katritzky, A.R. and Sengupta, S.** α-Lithio trimethylsilylmethyl lithium carbonate as methanol dianion synthon. A one pot synthesis of α-hydroxy ketones, *Tetrahedron Lett.* **28**, 1847 (1987).
165. **Shinkai, I., Reamer, R.A. and Sletzinger, M.** The synthesis of 3-(1'hydroxyethyl)-2-azetidinone-4-yl acetic acid via dianion chemistry - an important intermediate in thienamycin total synthesis, *Tetrahedron Lett.* **23**, 4899 (1982).
166. **Rama Rao, A.V., Singh, A.K. and Varaprasad, Ch.V.N.S.** A convenient diastereoselective total synthesis of andrimid, *Tetrahedron Lett.* **32**, 4393 (1991).

167. **Thompson, C.M.** Remote dianions I. Utility of 4-(Phenylsulfonyl)butanoic acid in the mild conversion of aldehydes and ketones to lactones, *Tetrahedron Lett.* **28**, 4243 (1987).
168. **Thompson, C.M. and Frick, J.A.** 4-(Phenylsulfonyl)butanoic acid. Preparation, dianion generation and application to four-carbon homologation, *J. Org. Chem.* **54**, 890 (1989).
169. **Thompson, C.M., Frick, J.A. and Woytowicz, C.E.** Remote dianions II. Synthesis of sulcatole, *Synth. Commun.* **18**, 889 (1988).
170. **Williams, K. and Thompson, C.M.** Preparation of oxepan-2-ones (ε-lactones) via the dianion of 4-(phenylsulfonyl)butanoic acid, *Synth. Commun.* **22**, 239 (1992).
171. **Thompson, C.M., Green, D.L.C. and Kubas, R.** Remote dianions III. Novel synthesis of substituted 2-piperidones from imines, *J. Org. Chem.* **53**, 5389 (1988).
172. **Green, D.L.C. and Thompson, C.M.** Remote dianions V. Synthesis of δ-coniceine, *Tetrahedron Lett.* **32**, 5051 (1991).
173. **Thompson, C.M.** A route to 1-azaspirans related to perhydrohistrionicotoxin, *Heterocycles* **34**, 979 (1991).
174. **Green, D.L.C.G.** Ph.D. Dissertation, Loyola University of Chicago, Chicago IL (1991).
175. **Kiddle, J.** Ph.D. Dissertation, Loyola University of Chicago, Chicago IL (1994).
176. **Bravo, P., De Vita, C. and Resnati, G.** α-Methylene and α-methyl-δ-lactones through the condensation of α-methylene-γ-sulfinylbutyric acid on carbonyl electrophiles, *Gazz. Chim. Ital.* **117**, 165 (1987).
177. **Murray, W.V. and Wachter, M.P.** Synthesis and properties of aryl-1,3-dioxo carboxylic acids, *J. Org. Chem.* **55**, 3424 (1990).
178. **Michel, P., O'Donnell, M., Biname, R., Hesbain-Frisque, A.M., Ghosez, L., DeClercq, J.P., Germain, G., Arte, E. and Van Meersche, M.** Vicinal alkylation of olefins. Regio- and stereoselective addition of [C_M + C_N] units to cyclopentadiene, *Tetrahedron Lett.* **21**, 2577 (1980).
179. **Julia, M. and Badet, B.** Syntheses a l'aide de sulfones. XI. Synthese d'acides carboxyliques, *Bull. Chim. Soc. Fr.* 525 (1976).
180. **Knight, D.W. and Nott, A.P.** Total synthesis of wyerone and dihydrowyerone, phytoalexins from the broad bean *vicia faba* L., using the dianion derived from 3-(2-furyl)-acrylic acid, *J. Chem. Soc. Perkin Trans. I.* 623 (1982).
181. **Tanaka, K. and Kaji, A.** The synthetic utility of sulfur-containing dianions, *Sulfur Rep.* **1**, 97 (1980).
182. **Ciochetto, L.J., Bergbreiter, D.E. and Newcomb, M.** Alkoxy enediolates, *J. Org. Chem.* **42**, 2948 (1977).
183. **Wilson, S.R., Walters, M.E. and Orbaugh, B.** The dianion of adipoin. A model study for the C-ring of phorbol, *J. Org. Chem.* **41**, 378 (1976).
184. **Juaristi, E., Martinez-Richa, A., Garcia-Rivera, A. and Cruz-Sanchez, J.S.** Use of 4-biphenylmethanol, 4-biphenylacetic acid, and 4-biphenylcarboxylic acid/triphenylmethane as indicators in the titration of lithium alkyls. Study of the dianion of 4-biphenylmethanol, *J. Org. Chem.* **48**, 2603 (1983).
185. **Bloomfield, J.J., Owsley, D.C. and Nelke, J.N.** The acyloin condensation, *Org. React.* **23**, 259 (1976).
186. **Martin, V.A., Murray, D.H., Pratt, N.E., Zhao, Y-B. and Albizati, K.F.** Chemistry of aldolate dianions. Effects of β-heteroatom substituents on ketone enolization, *J. Am. Chem. Soc.* **112**, 6965 (1990).
187. **Herrmann, J.L. and Schlessinger, R.H.** A novel method of preparing alpha-substituted hydacrylate and acrylate esters, *Tetrahedron Lett.* 2429 (1973).
188. **Martin, V.A. and Albizati, K.F.** Generation, chemical stability, and reactivity of aldolate dianions, *J. Org. Chem.* **53**, 5986 (1988).
189. **Martin, V.A., Perron, F. and Albizati, K.F.** Chemistry of aldolate dianions. β-Silyloxy silyl enol ethers as bifunctional nucleophilic equivalents for oxygen heterocycle synthesis, *Tetrahedron Lett.* **31**, 301 (1990).
190. **Frater, G., Muller, U. and Gunther, W.** The stereoselective α-alkylation of chiral β-hydroxy esters and some applications thereof, *Tetrahedron* **40**, 1269 (1984).
191. **Frater, G.** Stereospecificity of the α-alkylation of β-hydroxy carboxylic acid esters. Preliminary communication, *Helv. Chim. Acta* **62**, 2825 (1979).

192. **Frater, G.** Stereospecific synthesis of (+)-(3R,4R)-4-methyl-3-heptanol, the enantiomer of a pheromone of the smaller European elm bark beetle (Scolytus multistriatus), *Helv. Chim. Acta* **62**, 2829 (1979).
193. **Seebach, D. and Wasmuth, D.** Preparation of erythro-2-hydroxysuccinic acid derivatives from malic esters, *Helv. Chim. Acta* **63**, 197 (1980).
194. **Zuger, M., Weller, T. and Seebach, D.** 2,3-Disubstituted γ-butyrolactams from the Michael-additions of doubly deprotonated, optically active β-hydroxycarboxylates to nitroolefins, *Helv. Chim. Acta* **63**, 2005 (1980).
195. **Shieh, H-M. and Prestwich, G.D.** Stereoselective alkylation and aldol reactions of (S)-(-)-β-hydroxy-γ-butyrolactone dianion, *J. Org. Chem.* **46**, 4319 (1981).
196. **Chamberlain, A.R. and Dezube, M.** Stereoselective methylation of β-hydroxylacetone dianions, *Tetrahedron Lett.* **23**, 3055 (1982).
197. **Chen, S-Y. and Joullie, M.M.** Use of *D*-ribonolactone in organic synthesis. 2. Scope and utility, *J. Org. Chem.* **49**, 2168 (1984).
198. **Klaver, W.J., Hiemstra, H. and Speckamp, W.N.** Diastereoselective alkylation of the dianion of 5-ethoxy-(4S)-hydroxy-1-isopropyl-2-pyrrolidone: synthesis of enantiomerically pure azabicycles, *Tetrahedron Lett.* **28**, 1581 (1987).
199. **Wittenberger, S.J., Baker, W.R., Donner, G. and Hutchins, C.W.** The design and synthesis of cyclic renin inhibitors, *Tetrahedron Lett.* **32** 7655 (1991).
200. **Kurth, M.J. and Yu, C-H.** The dianion Claisen rearrangement of β-hydroxy esters, *Tetrahedron Lett.* **25**, 5003 (1984).
201. **Kurth, M.J. and Yu, C-H.** Acyclic stereocontrol through the dianionic Claisen rearrangement of β-hydroxy esters: synthesis of (+/-)-botryodiplodin, *J. Org. Chem.* **50**, 1840 (1985).
202. **Kurth, M.J. and Beard, R.L.** β-Hydroxy esters: (E)- versus (Z)-enolate geometry in dianionic Claisen rearrangements, *J. Org. Chem.* **53**, 4085 (1988).
203. **Marek, H. and Normant, J.F.** Acyclic stereocontrol through the metalla-Claisen rearrangement. Part II., *Tetrahedron Lett.* **32**, 5973 (1991).
204. **Simpkins, N.S.** *Sulphones in Organic Synthesis*, Pergamon Press, Oxford, U.K., 1993.
205. **Tanaka, K., Ootake, K., Imai, K., Tanaka, N. and Kaji, A.** Dianions of β-hydroxysulfones: new and general approach to selective synthesis of 2(5H)-furanones, *Chem. Lett.* 633 (1983).
206. **Kozikowski, A.P., Mugrage, B.B., Li, C.S. and Felder, L.** Chemistry of baker's yeast reductions: use of optically active (S)-(+)-1-(p-toluenesulfonyl)propan-2-ol and (S)-(-)-phenylsulfonyl propan-2-ol in synthesis, *Tetrahedron Lett.* **27**, 4817 (1986).
207. **Tanikaga, R., Hosoya, K. and Kaji, A.** Reactions of (2S)-1-arenesulfonyl-2-alkanol dianions with aldehydes. Application to the synthesis of enantiomerically pure (3S)-1-alken-3-ols and (2E,4S)-4-hydroxy-2-alkenenitriles, *Chem. Lett.* 829 (1987).
208. **Tanikaga, R., Hosoya, K. and Kaji, A.** Unusual stereoselection in the reaction of dianions derived from 1-phenylsulphonylalkan-2-ols with electrophilic reagents, *J. Chem. Soc. Perkin Trans. I.* 2397 (1988).
209. **Kim, K.S., Sohng, J-K. and Ha, B.H.** Synthesis of higher carbon sugars: preparation and reaction of β-hydroxysulfonyl sugars, *Tetrahedron Lett.* **29**, 2847 (1988).
210. **Rothberg, I., Sundoro, B., Balanikas, G. and Kirsch, S.** Preparation and stereochemistry of methylation of some cycloalkyl sulfones, *J. Org. Chem.* **48**, 4345 (1983).
211. **Nederlof, P.J.R., Moolenaar, M.J., de Waard, E.R. and Huisman, H.O.** Isoprenoid homologation, *Tetrahedron* **33**, 579 (1977).
212. **Tanikaga, R., Hamamura, K., Hosoya, K. and Kaji, A.** Stereochemistry in alkylations of β-hydroxysulfoxides and β-hydroxysulfones, *Tetrahedron Lett.* **28**, 3705 (1987).
213. **Tanikaga, R., Hamamura, K. and Kaji, A.** Reactivities of diastereomeric carbanions in dianions generated from β-hydroxysulfoxides, *Chem. Lett.* 977 (1988).
214. **Tanikaga, R., Hamamura, K., Hosoya, K. and Kaji, A.** Effect of intramolecular chelation on configurational interconversion of sulfur-stabilized carbanions, *J. Chem. Soc. Chem. Commun.* 817 (1988).
215. **Tanaka, K., Uneme, H., Matsui, S., Tanikaga, R., and Kaji, A.** A simple, highly versatile synthesis of cyclopropyl sulfides, *Chem. Lett.* 287 (1980).

216. **Paulsen, H., Roden, K., Sinnwell, V. and Koebernick, W.** C-C- Linkage of polyfunctional molecules by nucleophilic addition of dianions. Synthesis of pillarose, *Angew. Chem. Int. Ed. Engl.* **15**, 439 (1976).
217. **Paulsen, H., Roden, K., Sinnwell, V. and Koebernick, W.** Einfache Synthese der Pillarose, *Chem. Ber.* **110**, 2146 (1977).
218. **Paulsen, H., Roden, K., Sinnwell, V. and Luger, P.** Kettenverlangerungen zu langkettigen Sacchariden durch reaktionen mit dem Dianion des 1,3-dithian-2-methanols und verwandten Dianionen, *Leibigs Ann. Chem.* 2009 (1981).
219. **Beslin, P. and Perrio, S.** Asymmetric induction in the thio-Claisen rearrangement. Creation of three contiguous stereogenic centers from α-hydroxy ketene dithioacetals, *Tetrahedron* **48**, 4135 (1992).
220. **Bac, N.V. and Langlois, Y.** New strategy in the stereocontrolled synthesis of the spiroketal subunit of milbemycins, *Tetrahedron Lett.* **29**, 2819 (1988).
221. **Crandall, J.K. and Clark, A.C.** The reaction of organolithium reagents with allylic alcohols, *J. Org. Chem.* **37**, 4236 (1972).
222. **Dimmel, D.R. and Huang, S.** Asymmetric additions of organolithium reagents to allylic alcohols, *J. Org. Chem.* **38**, 2756 (1973).
223. **Dimmel, D.R. and O'Malley, J.P.** Alkyllithium additions to allylic alcohols, *J. Org. Chem.* **40**, 132 (1975).
224. **Carlson, R.M.** Methallyl alcohol dianion additions as the salient feature in a facile synthesis of α-methylene-γ-lactones, *Tetrahedron Lett.* 111 (1978).
225. **Carlson, R.M. and White, L.L.** The use of the alkali metal dianion of methallyl alcohol in the synthesis of α-methylene-δ-valerolactones, *Synth. Commun.* **13**, 237 (1983).
226. **Meyer, F.K., Drewett, J.G. and Carlson, R.M.** The crotyl alcohol dianion: addition reactions, *Synth. Commun.* **16**, 261 (1986).
227. **Trost, B.M. and Chan, D.M.T.** Palladium-mediated cycloaddition approach to cyclopentanoids. Introduction and initial studies, *J. Am. Chem. Soc.* **105**, 2315 (1983).
228. **Trost, B.M., Chan, D.M.T. and Nanninga, T.N.** Silylation of 2-methyl-2-propen-1-ol dianion: 2-(hydroxymethyl)allyltrimethylsilane, *Org. Synth.* **62**, 58 (1984).
229. **Cuvigny, T., Julia, M. and Rolando, C.** Stereselective γ-cis-vinylic metalation of tertiary allylic alcohols, *J. Chem. Soc. Chem. Commun.* 8 (1984).
230. **Cuvigny, T., Julia, M., Jullien, L. and Rolando, C.** A simple way to some ketone homoenolates, *Tetrahedron Lett.* **23**, 2587 (1987).
231. **Majetich, G., Song, J-S., Ringold, C., Nemeth, G.A. and Newton, M.G.** Intramolecular additions of allylsilanes to conjugated dienones. A direct stereoselective synthesis of (+/-)-14-deoxyisoamijiol, *J. Org. Chem.* **56**, 3973 (1991).
232. **Olson, G.L., Cheung, H-C., Morgan, K.D., Neukon, C. and Saucy, G.** Vitamin A synthesis by sulfone alkylation-elimination. C_{15} halide, C_5 hydroxy sulfone approach, *J. Org. Chem.* **41**, 3287 (1976).
233. **McDonald, T.L. and Narayanan, B.A.** Pyrrolizidine alkaloid synthesis. (+/-)-supinidine, *J. Org. Chem.* **48**, 1131 (1983).
234. **Sturtz, G., Corbel, B. and Paugam, J-P.** Nouveau synthons phosphores: bianions d'hydroxy-1-propene-2 yl phosphonamides, carbanions en β d'acides carboxyliques potentiels, *Tetrahedron Lett.* 47 (1976).
235. **Corey, E.J. and Widiger, G.N.** The O,2-dilithio derivative of allyl alcohol, a useful synthetic reagent, *J. Org. Chem.* **40**, 2975 (1975).
236. **Schlosser, M. and Hammer, E.** Selective syntheses with organometallics III. (Z)-Crotylalcoholates with metal-carrying olefinic carbon atoms, *Helv. Chim. Acta* **57**, 2547 (1974).
237. **Barluenga, J., Fernandez, J.R. and Yus, M.** sp^2-Hybridized β-substituted organo-lithium, sodium, and -potassium dianions, preparation, stability and reactivity, *J. Chem. Soc. Perkin Trans. I.* 447 (1985).
238. **Barrett, A.G.M., Barta, T.E. and Flygare, J.A.** 4-[(tert-butyldiphenylsilyl)oxy]-2-(tributylstannyl)-(E)-2-buten-1-ol: a useful precursor for tetrahydrofuran synthesis, *J. Org. Chem.* **54**, 4246 (1989).
239. **Olsson, L-I. and Claesson, A.** Addition reactions of butyllithium with propargylic alcohols, *Tetrahedron Lett.* 2161 (1974).

240. **Saimoto, H., Hiyama, T. and Nozaki, H.** A new synthesis of dl-muscopyridine based on the regioselective cyclopentenone annulation, *Tetrahedron Lett.* **21**, 3897 (1980).
241. **Bulman Page, P.C., Rayner, C.M. and Sutherland, I.O.** Enantioselective synthesis of R-(+)-lipoic acid, *J. Chem. Soc. Chem. Commun.* 1408 (1986).
242. **Julia, M., Uguen, D. and Callipolitis, A.** Syntheses using sulfones X. Alkylation of hydroxysulfones; masked β-hydroxy carbanion. Application to the synthesis of terpenic alcohols, *Bull. Soc. Chim. Fr.* 519 (1976).
243. **Kim, K.S., Sohng, J-K, Ha, S.B., Cheong, C.S., Jung, D.I. and Hahn, C.S.** Synthesis of higher-carbon sugars; preparation and reaction of β-hydroxy sulfonyl sugars, *Tetrahedron Lett.* **29**, 2847 (1988).
244. **Tanaka, K., Ootake, K., Imai, N., Tanaka, N. and Kaji, A.** Dianions of β-hydroxy sulfones: new and general approach to selective synthesis of 2(5H)-furanones, *Chem. Lett.* 633 (1983).
245. **Zidani, A. and Vaultier, M.** Chelation assisted transmetalation of tetraalkyl tin derivatives: C-metalated lithiocarbamates as d^3 reagents for the synthesis of δ-hydroxycarbamates, *Tetrahedron Lett.* **27**, 857 (1986).
246. **Newman-Evans, R.B. and Carpenter, B.K.** Highly stereoselective synthesis of bicyclo[2.2.1]hexenes-5-d by transmetalation of a tetra-alkyl tin, *Tetrahedron Lett.* **26**, 1141 (1985).
247. **Poppe, L., Novak, L. and Szantay, C.** A convenient synthesis of (E)-β-farnesene, *Synth. Commun.* **17**, 173 (1987).
248. **Coppola, G.M.** A convenient preparation of 1,4-dihydro-1-methyl-4-oxo-2-quinolinecarboxaldehyde from N-methylisatoic anhydride, *J. Heterocyclic Chem.* **23**, 1717 (1986).
249. **Villieras, J., Bacquet, C. and Normant, J.F.** Nouvelle transposition de carbenoides lithiens α-alcolates: syntheses regiospecifique de α-chloroacetonees, *J. Organomet. Chem.* **40**, C1 (1972).
250. **Crandall, J.K. and Lin, L-H.C.** Base-promoted reactions of epoxides. III. *J. Am. Chem. Soc.* **89**, 4526 (1976).
251. **Crandall, J.K. and Lin, L-H.C.** Base-promoted reactions of epoxides. IV. *J. Am. Chem. Soc.* **89**, 4527 (1976).
252. **Saddler, J.C., Conrad, P.C. and Fuchs, P.L.** Conjugate addition reactions of organometallic reagents to γ-oxygenated-α,β-unsaturated sulfones: control of stereochemistry at the β-position, *Tetrahedron Lett.* 5079 (1978).
253. **Conrad, P.C. and Fuchs, P.L.** A versatile and expedient synthesis of α,β-unsaturated ketone. Utilization of the β-epoxy sulfone functional group as the enone progenitor, *J. Am. Chem. Soc.* **100**, 346 (1978).
254. **Najera, C. and Yus, M.** Substituted lithium (E)-3-lithio-3-tosyl-2-propenolates: useful intermediates in organic synthesis, *J. Org. Chem.* **54**, 1491 (1989).
255. **Dimmel, D.R. and Gharpure, S.B.** The isomerization of allyl phenyl ether to propiophenone by way of a dianion intermediate, *J. Am. Chem. Soc.* **93**, 3991 (1971).
256. **Degl'Innocenti, A., Dembech, P., Mordini, A., Ricci, A. and Seconi, G.** [3+n] Annulation reactions by means of 3-trimethylstannyl-2-[(trimethylstannyl)methyl]propene, an isobutene dianion synthetic equivalent, *Synthesis* 267 (1991).
257. **Comins. D.L. and Brown, J.D.** Ortho metalation directed by α-amino alkoxides, *J. Org. Chem.* **49**, 1078 (1984).
258. **Mudryk, B. and Cohen, T.** Synthetically useful dianions via reductive lithiation of tetrahydrofurans by aromatic radical anions, *J. Am. Chem. Soc.* **113**, 1866 (1991).
259. **Ramon, D.J. and Yus, M.** Arene-catalyzed reductive lithiation of tetrahydrofuran: improved synthesis of 1,5-diols, *Tetrahedron* **48**, 3585 (1992).
260. **Guijarro, D., Mancheno, B. and Yus, M.** C,O-dilithiated diarylmethanols: easy and improved preparation by naphthalene-catalyzed lithiation and reactivity toward electrophiles, *Tetrahedron* **49**, 1327 (1993).
261. **Gilman, H., Arntzen, C.E. and Webb, F.J.** Organometallic compounds in the Kolbe and Reimer-Tiemann reactions, *J. Org. Chem.* **10**, 374 (1945).
262. **Hlasta, D.J. and Bell, M.R.** Dianion metalation reactions of N,N-dimethylvanillylamine and N,N-dimethylisovanillylamine, *Tetrahedron Lett.* **26**, 2151 (1985).

263. **Gray, M. and Parsons, P.J.** Directed alkylation of phenols: a convenient synthesis of highly substituted aromatic rings, *SynLett.* 729 (1991).

264. **Bates, R.B. and Siahaan, T.J.** Preparation and reactions of dianions from the cresols, *J. Org. Chem.* **51**, 1432 (1985).

265. **Kende, A.S., Ebetino, F.H. and Ohta, T.** Synthesis of the spirocyclic center of Fredericamycin A by phenoxy-enoxy radical coupling, *Tetrahedron Lett.* **26**, 3063 (1985).

266. **Sinsheimer, J.E., Van den Eeckhout, E., Hewitt, L.E., Kido, Y., Wade, D.R., Hansen, D.W., Drach, J.C. and Burckhalter, J.H.** Metabolic formation and synthesis of 1-(3-chlorophenyl)-2-(4-hydroxyphenyl)-1-methyl-2-(2-pyridine)ethanol. A potential hypocholesteremic agent, *J. Med. Chem.* **19**, 647 (1976).

267. **Kendall, P.M., Johnson, J.V. and Cook, C.E.** Synthetic route to an aromatic analogue of strigol, *J. Org. Chem.* **44**, 1421 (1979).

268. **Cushman, M. and Matthew, J.** Nitration of the lithium potassium dianions of phenolic alkyl aryl ketones with propyl nitrate: synthesis of 1-nitroalkyl hydroxyphenyl ketones, *Synthesis* 397 (1982).

269. **Bates, R.B., Siahaan, T.J. and Suvannachut, K.** A new rearrangement of alkoxybenzyl anion, *J. Org. Chem.* **55**, 1328 (1990).

270. **Henoch, F.E., Hampton, K.G. and Hauser, C.R.** C-Alkylations of ketone or aldehyde phenylhydrazones, oximes, and azines having α-hydrogen through 1,4- or 1,6-dianions, *J. Am. Chem. Soc.* **91**, 676 (1969).

271. **Kofron, W.G. and Yeh, M-K.** Reaction of dilithium derivatives of oximes with electrophiles. Regiospecific substitution of ketones, *J. Org. Chem.* **41**, 439 (1976).

272. **Jung, M.E., Blair, P.A. and Lowe, J.A.** Reactions of oxime dianions: stereospecificity in alkylation, *Tetrahedron Lett.* 1439 (1976).

273. **Hassner, A. and Naumann, F.** Alkylation and aldol reactions of aldoxime anions, *Chem. Ber.* **121**, 1823 (1988).

274. **Glaser, R. and Streitweiser, A.** Ab initio study of the regiochemistry of dimetalated oximes. The importance of triple ions in isomeric lithium and sodium ion pairs of the acetaldoxime dianion, *J. Org. Chem.* **54**, 5491 (1989).

275. **Gawley, R.E. and Nagy, T.** Alkylation of aldehyde oxime dianions, *Tetrahedron Lett.* **25**, 263 (1984).

276. **Hiroi, K., Otsuka, M. and Sato, S.** Regioselective sulfenylation of oxime dianions. A novel and useful method for the regioselective ring cleavage, *Chem. Lett.* 1907 (1985).

277. **Lyle, R.E., Saavedra, J.E., Lyle, G.G., Fribush, H.M., Marshall, J.L., Lijinsky, W. and Singer, G.M.** Conformational stereospecificity in electrophilic reactions with cyclic anions, *Tetrahedron Lett.* 4431 (1976).

278. **Lyle, R.E., Fribush, H.M., Lyle, G.G., and Saavedra, J.E.** Regiospecificity and conformational specificity in oxime alkylation of a geometrical enantiomeric isomer, *J. Org. Chem.* **43**, 1275 (1978).

279. **Beam, C.F., Dyer, M.C.D., Schwarz, R.A. and Hauser, C.R.** A new synthesis of isoxazoles from 1,4,-dianions having an α-hydrogen. Mass spectroscopy, *J. Org. Chem.* **35**, 1806 (1970).

280. **Beam, C.F., Foote, R.S. and Hauser, C.R.** Reactions of dilithiooximes, dilithiophenylhydrazones, and trilithiohydrazones with nitriles. Acid-cyclization to isoxazoles or pyrazoles, *J. Heterocyclic Chem.* **9**, 183 (1972).

281. **Park, C.A., Beam, C.F., Kaiser, E.M., Kaufman, R.J., Henoch, F.E. and Hauser, C.R.** Preparation of 2-isoxazolines from C(α),O-dilithiooximes and aldehydes and ketones, *J. Heterocyclic Chem.* **13**, 449 (1976).

282. **Livingston, M.L., Chick, M.F., Shealy, E.O. and Beam, C.F.** The preparation of 5-phenacylisoxazoles and 5-hydroxyphenylisoxazoles- and pyrazoles by the condensation of C(α)-dianions with ethyl benzoylacetate and methylsalicylate, *J. Heterocyclic Chem.* **19**, 215 (1982).

283. **Huff, A.M., Hall, H.L., Smith, M.J., O'Grady, S.A., Waters, F.C., Fengl, R.W., Welsh, J.A. and Beam, C.F.** The preparation of phenacylpyrazole, imidazoylisoxazoles, and imidazolylpyrazoles from C(α)-dianions of oximes, phenylhydrazones, and acylhydrazones, *J. Heterocyclic Chem.* **22**, 501 (1985).

284. **Mack, H.M., Davis, E.A., Kadkhodayan, B., Taylor, R.A., Duncan, D.C. and Beam, C.F.** The preparation of quinolines and related fused-ring heterocycles from the dianion of benzoylacetone, certain cyclic ketone oximes, or certain substituted hydrazones, *J. Heterocyclic Chem.* **24**, 1733 (1987).
285. **Bellassoued, M., Chtara, R., Dardoize, F. and Gaudemar, M.** Reaction of oxime dianions with aldimines, a useful route to 2-hydroxyaminoazetidines, *Synthesis* 951 (1983).
286. **Yamamoto, Y., Morita, Y., Kikuchi, R., Yokoo, E., Ohtsuka, K. and Katoh, M.** Reaction of 6-methyl-2-phenyl-4H-1,3-oxazin-4-one with oxime dianions, *Heterocycles* **29**, 1443 (1989).
287. **Crowther, G., Johnston, M.Y. and Bradsher, C.K.** New acridizinium salts via alkylation of a dianion, *J. Heterocyclic Chem.* **8**, 157 (1971).
288. **Ager, D.J.** The synthesis of 2,5-disubstituted furans, *Tetrahedron Lett.* **24**, 5441 (1983).

Chapter 4

DIANIONS FROM N,C-BIS DEPROTONATIONS

4.1 INTRODUCTION

N,C-Dianions are intermediates formed by initial deprotonation of a N-H group. A wide array of nitrogen-containing compounds serve as useful starting materials for this process. The various types of N,C-dianions are assembled in Sections 4.2 through Section 4.8 according to the nature of the functional group(s) flanking the N-H deprotonation site. For example, Section 4.2 covers the class of amide-based, N,C-bis deprotonations where the N-H deprotonation site is stabilized by one juxtaposed π-moiety. Section 4.4 contains dianions derived from carbamates that differ from amide-type dianions by virtue of the fact that the α'-site is occupied by an oxygen rather than an alkyl or C-H containing group. The remaining categories include hydrazones, enaminones, pyrrole/indoles, amines, and imides, that were separated according to differences in their relative N-H acidity,[1] structure, and ease of reaction classification. Some examples of N,C-dianions are given below (Figure 4.1). Certain dianion reactants, for example the sulfonyl hydrazones, could find classification under two categories, sulfonamide- or hydrazone-type. Since sulfonylhydrazone dianions react with electrophiles to give products that more resemble those of the hydrazones than amides, these materials are categorized with the former. Other dianion types that bridge categories will be denoted as such and cross-referenced to related sections in the text where necessary.

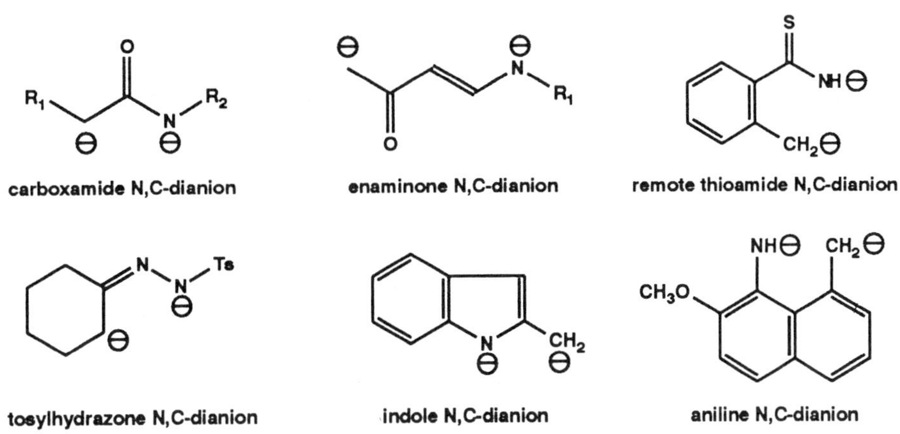

Figure 4.1. Select N,C-dianions

Although N,C-dianion coverage has appeared in general reviews of dianion chemistry,[2-4] N,C-dianions have not been the subject of a specific review. Yet, N,C-dianion intermediates continue to be used extensively, and have been creatively installed into synthetic pathways. A summary of the utility of N,C-dianions is conducted in this chapter with the generation and reaction utility of each dianion type discussed under individual categories. Representative experimental procedures are collected in Section 4.9.

In general, the N-H acidity is most affected by the nature of neighboring functional groups. Therefore a wide range of deprotonating bases have been employed to form the primary anion. For example, whereas the imide N-H (Section 4.7) can be deprotonated using weak bases, enaminones (Section 4.5) generally require alkyl lithium bases for satisfactory conversion to the nitrogen anion. The choice of base for the subsequent C-H deprotonation is usually governed by the relative reactivity of other functional groups present, and contributions by neighboring carbanion stabilizing groups.

Individual classifications of N,C-dianion types are assigned the designations identified in Chapter 1.

4.2 CARBOXAMIDE-, THIOAMIDE-, SULFINAMIDE- AND SULFONAMIDE-BASED DIANIONS

4.2.1 Generation of carboxamide, thioamide, sulfinamide, and sulfonamide N,C-dianions.

Carboxamide dianions may be divided into α,α′-, vicinal-, and remote-dianion types (Figure 4.2) similar to the carboxylate O,C-dianions (Chapter 3).

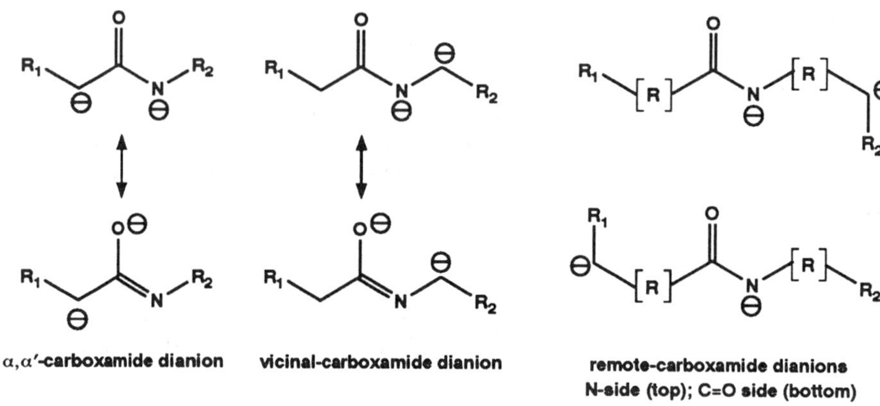

Figure 4.2. Representative carboxamide dianions

Certain carboxamide N,C-dianions undergo novel side reactions, which are important to note prior to a discussion of their utility. The formation of a nitrile through competitive formation of the trianion[5] of phenylacetamide (Eq. 4.1), and decarboxamidation from the diphenylacetamide monoanion[6] (Eq. 4.2) illustrate some concerns.

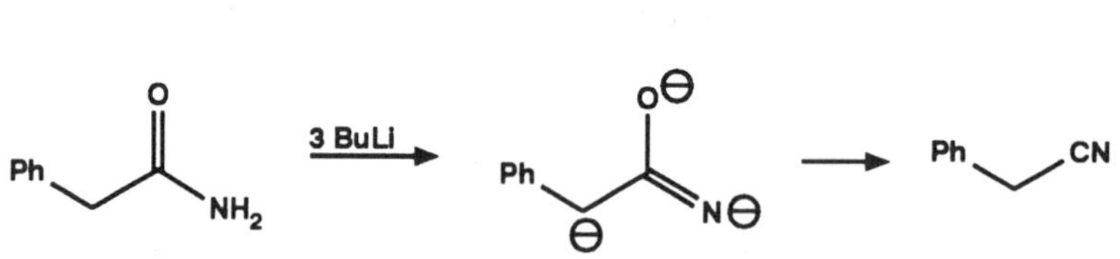

(4.1)

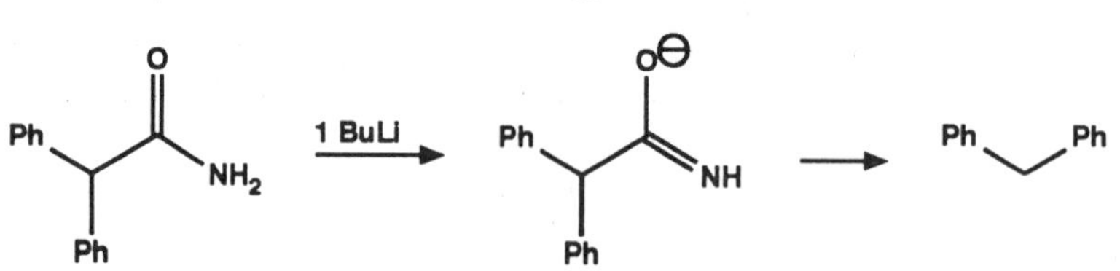

(4.2)

As mentioned in Section 4.1, bases of modest strength may be used to deprotonate N-H groups when flanked by a hetero-π bond (e.g., C=O, C=S, etc.). Alkoxides, metal hydrides, and metal amides (e.g., $NaNH_2$) all have been used frequently to effect N-H deprotonations of carboxamides, although substrates bearing nonreactive functional groups elsewhere in the molecule have been deprotonated with alkyl lithium and lithium amide (e.g., LDA) bases. Where compatible, alkyl lithium bases (BuLi) also are capable of generating the carbanion site offering the advantage of using a single base to achieve both deprotonations. Some base choices for the formation of carboxamide N,C-dianions are collected in Table 4.1.

In general, the formation of vicinal N,C-dianions requires stronger bases than α,α'-N,C-dianions owing to charge repulsion in the former. When the carbanion site of a vicinal N,C-dianion is also adjacent to an ester or related electron stabilizing group, LDA (2 equiv.) has been used most frequently as base. Formation of remote N,C-dianions usually require strong bases (LDA, BuLi) for deprotonation at the remote C-H site. Metal-exchange and metal-halogen exchange reactions also have been employed successfully to prepare remote N,C-dianions.

4.2.2 Reactions of carboxamide-based N,C-dianions

Three types of dianions can be formed from primary and secondary carboxamides; a) α,α'-, b) vicinal-, and c) remote-dianions. The α,α'-carboxamide dianions are structurally similar to α,α'-carboxylate (Chapter 3) and α,α'-carbonyl (Chapter 2) dianions. The vicinal carboxamide N,C-dianions differ from the α,α'-type because either a second electron stabilizing group (alpha to the carbanion site) must be present in order to overcome competing α'-carbonyl proton abstraction, or no α'-protons are present (e.g., benzoyl- or pivaloyl-amides). Dianions other than α,α'- and vicinal are categorized as remote since the second deprotonation site is insulated by at least one neutral atom or group. These carboxamide dianions include lateral and homoenolate types.

α,α'-Carboxamide dianions

α,α'-Dianions may be generated from primary or secondary carboxamides that carry a variety of substituents on either the carbonyl (R_1) or nitrogen (R_2) side. These dianions undergo reaction with alkyl halides, aldehydes, ketones, and esters to provide α-carbonyl substituted carboxamides (Eq 4.3) (Table 4.1).[2,4,7-19]

The reaction of α,α'-carboxamide dianions with epoxides (Entries 2 and 9) furnishes γ-hydroxy amides. The yields and rates of these reactions were markedly improved by the addition of BF_3-Et_2O.[15] The γ-hydroxy carboxamide products can be converted to lactams following reaction with base and Ts-Cl (Eq. 4.4).

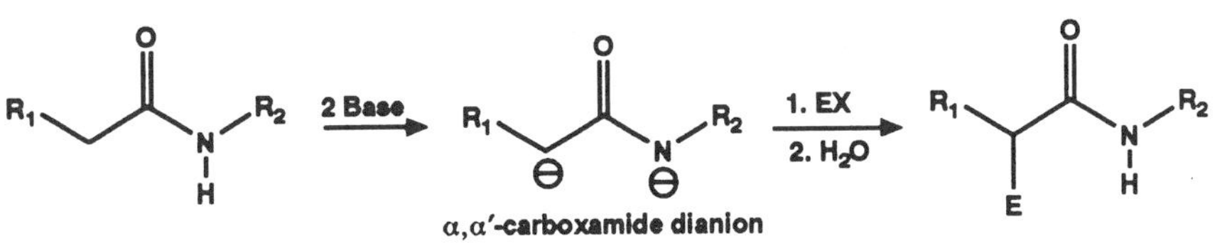

(4.3)

The α,α'-carboxamide dianions undergo Michael addition to β-nitrostyrenes,[12] although little exploration in 1,2- versus 1,4-addition to α,β-unsaturated carbonyls has been conducted.

Table 4.1.
Reaction of Carboxamide α,α'-Dianions with Electrophiles (Equation 4.3)

Entry	R_1	R_2	Base	E-X	E	Yield(%)	Ref.
1	H	H	NaH (xs)	PhC(O)OMe	PhC(O)-	100	7
2	H	H	2 LDA	epoxides	$R(CHOH)CH_2$-	65	8
3	Ph	H	2.2 $NaNH_2$	$Ph_2C{=}O$	$Ph_2C(OH)$-	83	9
4	Ph	H	2 BuLi	n-BuBr	Bu-	86	10
5	Ar	H	2 $NaNH_2$	R-Cl	R	46	11
6	H	Me	2 BuLi	$RCH{=}CHNO_2$	$R(CH_2NO_2)CH$-	75	12
7	H	TMS	2 BuLi	R-X	R	41-82	13
8	H	Ph	2 BuLi	$Ph_2C{=}O$	$Ph_2C(OH)$-	91	14
9	H	Ph	2 BuLi	epoxides	$R(CHOH)CH_2$-	38-81	15
10	H	Ar_1	2 BuLi	R-X	R	43-56	16,17
11	H	Ar_2	2 BuLi	ketones	alcohols	19-100	17
12	Ph	Ph	2 KNH_2	BnCl	Bn-	88	18
13	PhS	R	2 NaH	CH_2I_2	β-lactams	27-54	19

Ar = bicyclic, Ar_1 = 2-pyridyl, Ar_2 = 2-thiazole

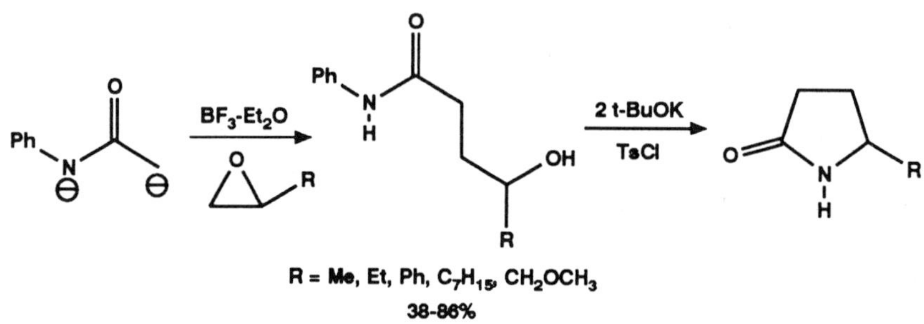

(4.4)

The dianion of N-(trimethylsilyl)acetamide undergoes alkylation, aldol condensation and acylation reactions (Eq. 4.5).[13] The silyl group both aids formation of the dianion owing to delocalization of charge into the d-orbitals, and also confers increased solubility of the dianion in THF. Easy removal of the TMS-group permits reentry to the unsubstituted amide.

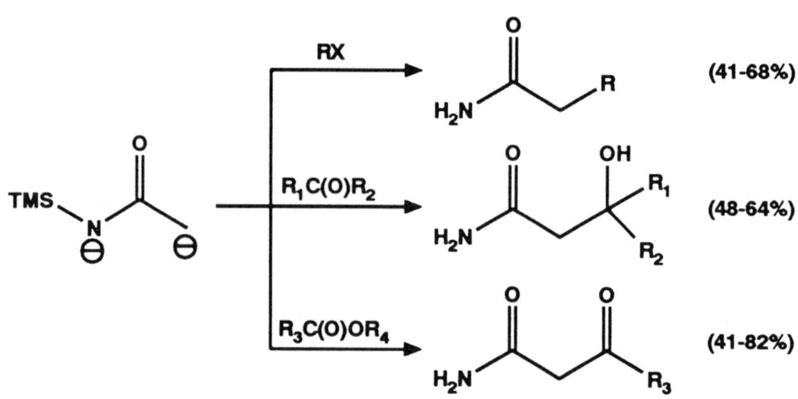

(4.5)

The α-phenylthioacetamide dianion is formed with two equivalents of NaH in DMF. Intramolecular N,C-bis-alkylation with CH_2I_2 gave a new preparation of substituted β-lactams (Eq. 4.6).[19]

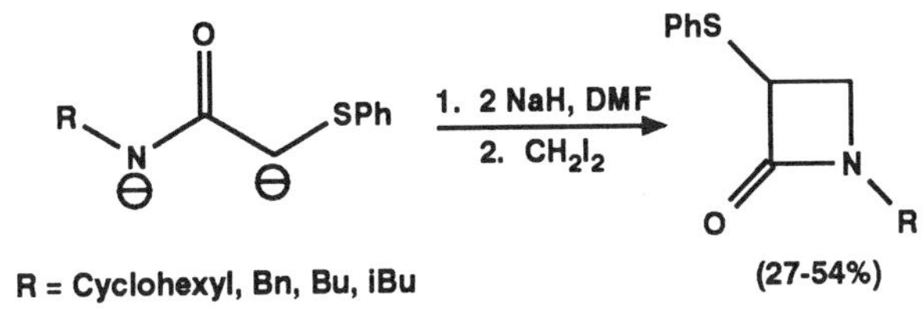

(4.6)

Carboxamide dianions of the α,α′-type also can be prepared from the heterocyclic analogues (lactams). The α,α′-dianions of β-lactams (Eq.4.7),[20] 2-piperazinones (Eq.4.8),[21] and oxindoles (Eq. 4.9)[22] all react regiospecifically with electrophiles to yield the corresponding α′-substituted heterocycles.

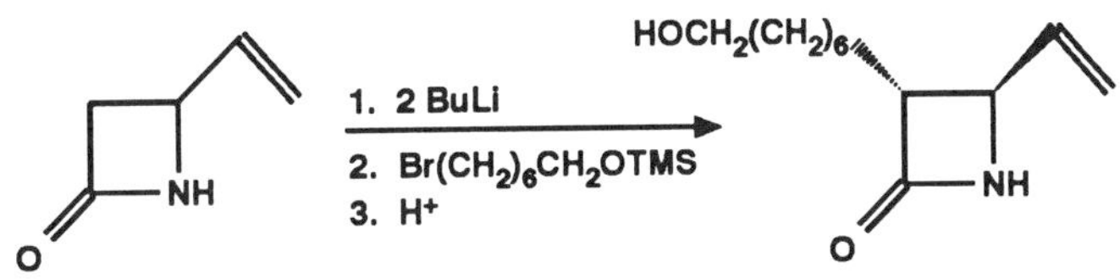

(4.7)

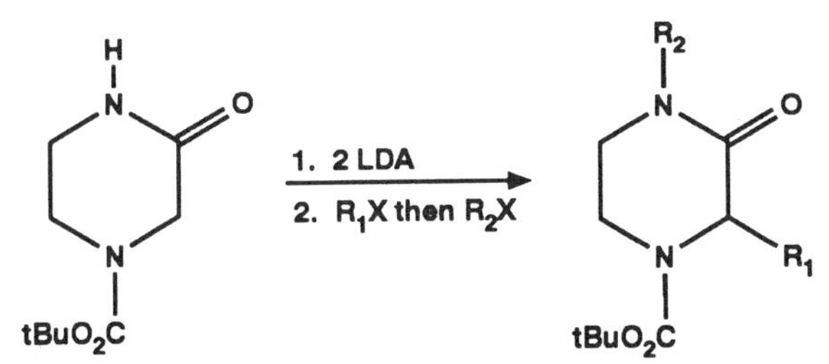

(4.8)

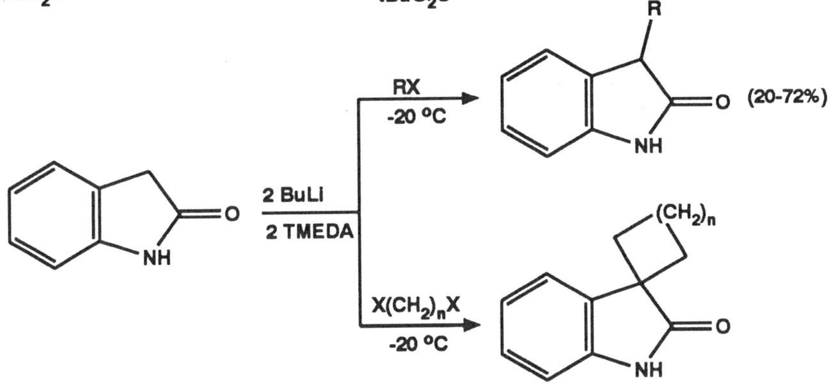

(4.9)

Vicinal carboxamide dianions

When a carboxamide is deprotonated at the C-H group adjacent to the amide nitrogen anion, a vicinal dianion is formed (Fig. 4.2). Two carboxamide substituent arrangements allow for this dianion to form. The first pattern requires a strong anion-stabilizing group alpha to the vicinal C-H site, which lowers the acidity of this position making it more amenable to proton abstraction than the competing α′-C-H position. The second arrangement occurs when the competing α′-position does not bear a C-H proton, as in the hippurates (Scheme 4.1).

The vicinal carboxamide dianion of a penicillin derivative was generated to epimerize an α′-center to the desired stereochemistry (Eq. 4.10).[23]

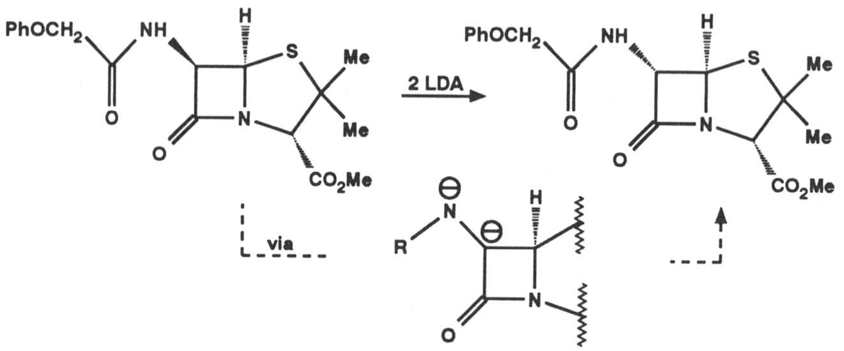

(4.10)

N-Acylamino acid ester dianions were reacted with aldehydes to give diastereomeric β-substituted serine derivatives in 36-72% yield (Eq. 4.11).[24]

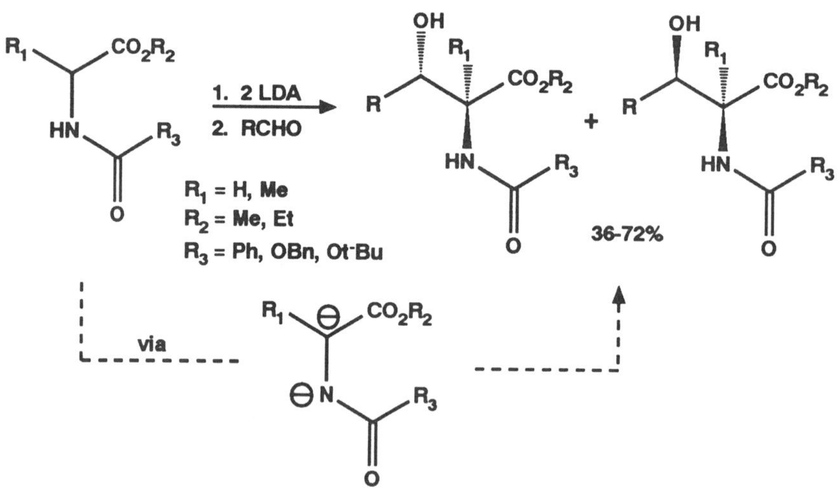

(4.11)

The methyl or ethyl hippurate vicinal dianion undergoes reaction with electrophiles to give an interesting array of highly functionalized products.[25-27] The addition of this dianion to succinic or glutaric anhydrides gives high yields of the corresponding keto acids.

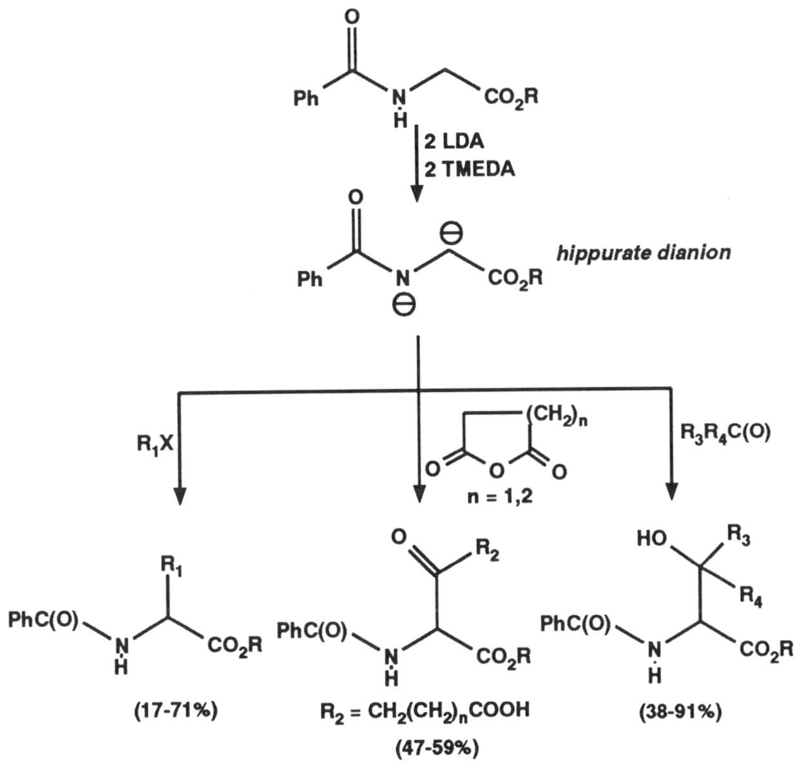

Scheme 4.1. Reactions of the hippurate dianion

Ethyl-(E)-2-formylamino-4-methoxy-3-butenoate forms the resonance stabilized dianion upon treatment with 2 LDA and undergoes alkylation predominantly at the vicinal position in 65-70% yield.[28]

Remote carboxamide dianions

There has been a significant amount of activity in the usage of remote carboxamide dianions. As examples, carboxamide nitrogen anions have been further deprotonated at a remote C-H site using sulfones, or resonance delocalization at an allylic or benzylic position for added anion stabilization.

The phenylsulfone moiety offers a unique anchor for α-(CH)-deprotonation. As revealed in Chapters 2 and 3, the sulfone group normally imparts greater acidity to its α-protons than do the corresponding carbonyl functional groups, especially when the carbonyl is already part of a deprotonated amide or carboxylic acid. Tanaka and coworkers explored the reactions of dianions derived from 3-(phenylsulfonyl)-propanamide and 3-(phenylsulfonyl)-propenamide.[29,31-34]

The propanamide congener,[29] formed by treatment with 2 BuLi or 2 LDA, reacts with aldehydes and ketones to give γ-hydroxy carboxamides in good yield (Eq. 4.12). Good yields of 5-substituted furanones can be obtained following acid-promoted cyclization. The corresponding N-benzyl analog also was used in the preparation of δ-lactams.[30]

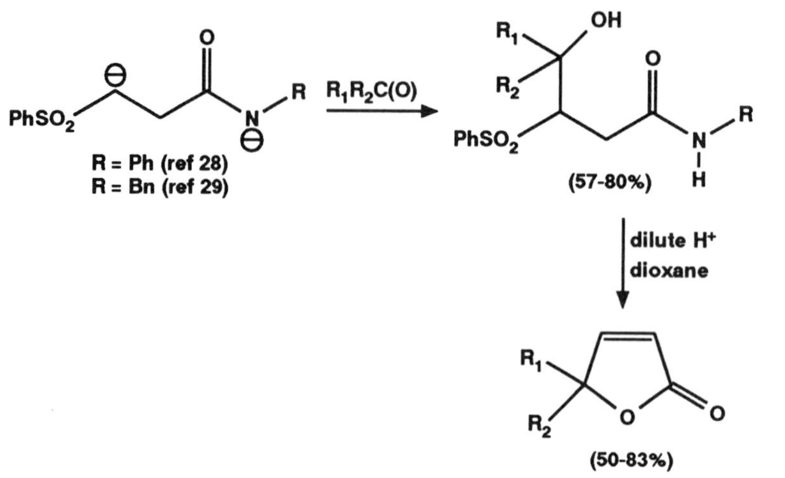

(4.12)

Use of an enantioenriched auxiliary in an analogous sequence gave furanones in high optical purity following separation of the 4-hydroxybutanamide diastereomers (approx. 1:1) (Eq. 4.13).[29]

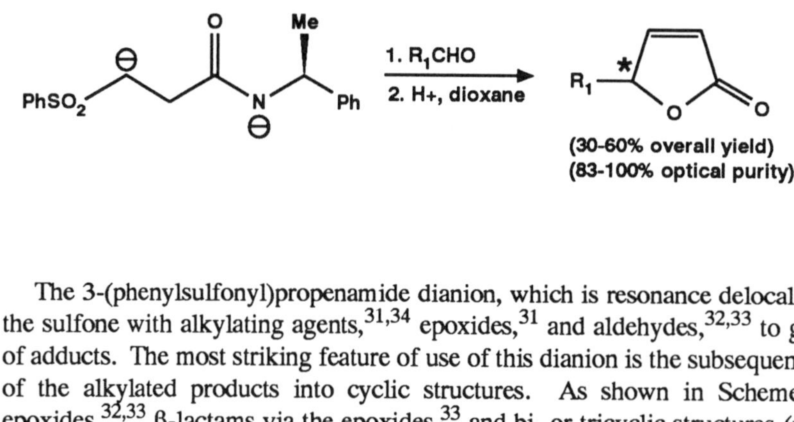

(4.13)

The 3-(phenylsulfonyl)propenamide dianion, which is resonance delocalized, reacts α to the sulfone with alkylating agents,[31,34] epoxides,[31] and aldehydes,[32,33] to give good yields of adducts. The most striking feature of use of this dianion is the subsequent transformation of the alkylated products into cyclic structures. As shown in Scheme 4.2, pyrans,[31] epoxides,[32,33] β-lactams via the epoxides,[33] and bi- or tricyclic structures (following Lewis acid mediated cyclization; not shown) may be accessed via this dianion.

The dianions of substituted allyl- and acrylamides have been studied.[35-37] These dianions give symmetric resonance stabilization leading to unambiguous regiochemical outcomes (Scheme 4.3). Specifically, acrylamide dianions bearing N-tert-butyl and N-(o-methoxyphenyl) groups were generated with 2 BuLi/TMEDA or tBuOK/BuLi base combinations, and reacted with alkyl halides, carbonyl compounds, and epoxides. Formation of the exocyclic methylene lactones via this reagent is a noteworthy transformation.

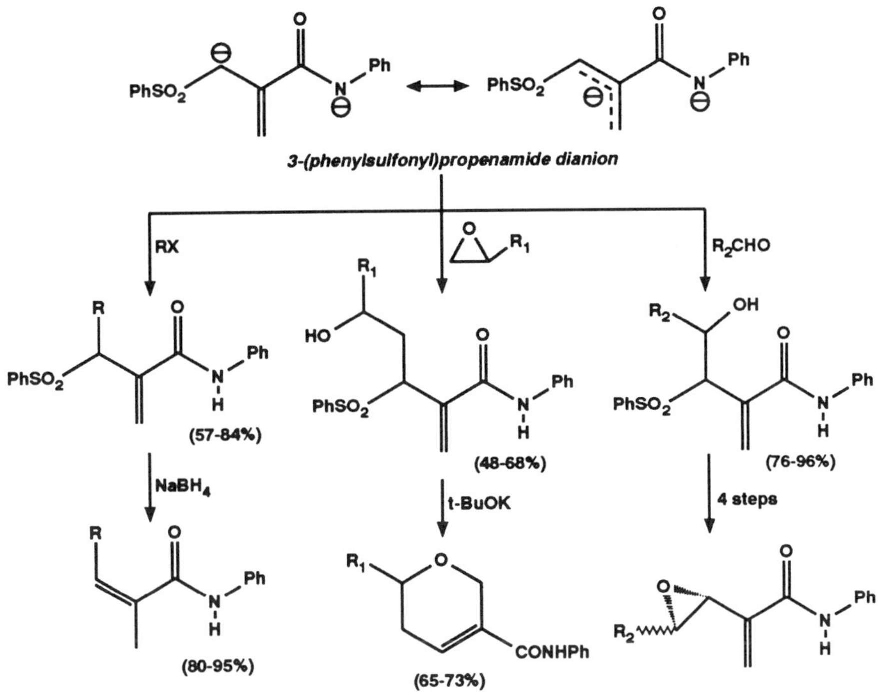

Scheme 4.2. Reactions of the 3-(phenylsulfonyl)propenamide dianion

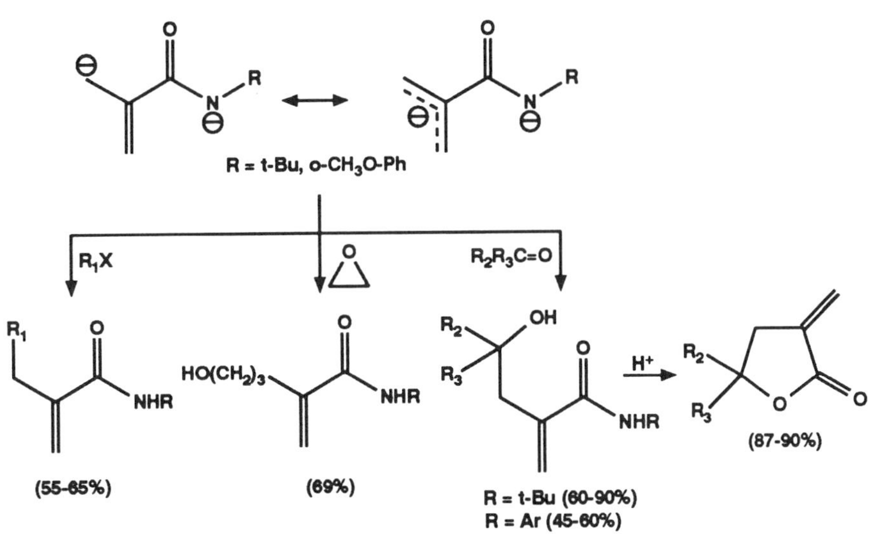

Scheme 4.3. Reactions of resonance-stabilized carboxamide dianions

The N-(p-methoxyphenyl)-butanamide dianion undergoes an interesting addition to certain benzynes to provide, after rearrangement of the primary addition adduct, indans and 3-arylpropenamides in a ratio of 10:1 in an overall yield of 60% (Eq. 4.14).[37]

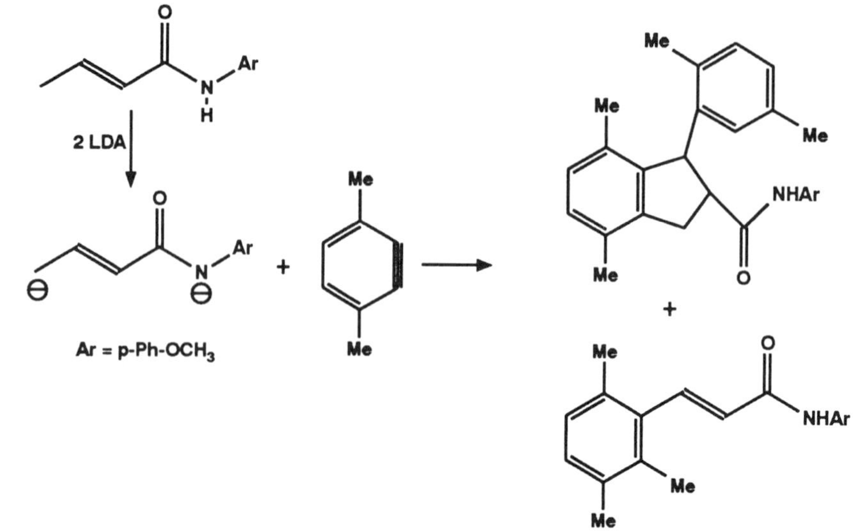

(4.14)

Remote carboxamide dianions also have taken advantage of the benzylic C-H acidity (Eqs. 4.15 and 4.16).[38,39] Alkylation of a carboxamide dibenzylic dianion followed by hydrolysis of a ketal led to a desired α-diketone precursor (Eq. 4.15). Goswami and Corcoran showed that formation of a homoenolate carboxamide dianion rearranged to the more stable benzylic dianion, which reacts with electrophiles (Eq. 4.16).[39] The rearrangement was slow at -40°C and presumably proceeded through a cyclopropane intermediate.

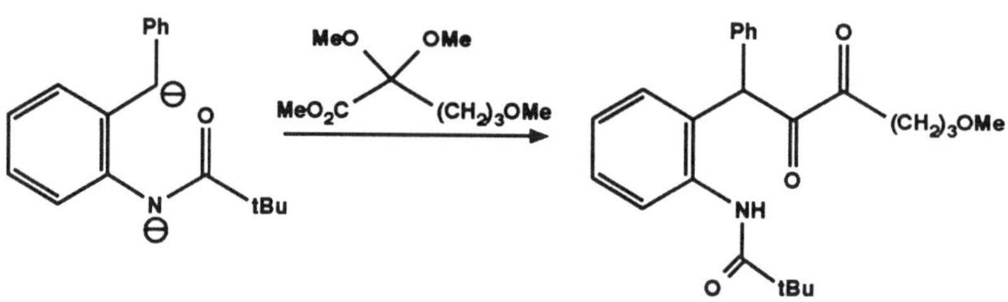

(4.15)

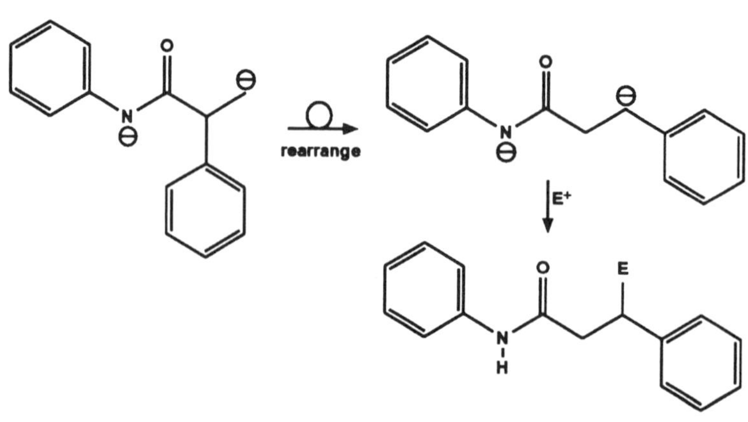

(4.16)

Ring metalation may follow carboxamide N-H deprotonation to provide remote dianion intermediates.[40-42] For example, N-pivaloyl anilides may be metalated directly[40] or brought to form the N,C-dianion via metal-halogen exchange,[42] and reacted with electrophiles (Eq. 4.17). Benzanilide undergoes metalation at the ortho position, and interestingly adds elemental sulfur (81%), selenium (76%), or tellurium (70%) (Eq. 4.18).[41] The main difference in the ortho-lithiation procedures resides in the formation of a metalated aniline ring (Eq. 4.17) versus a metalated benzoyl ring (Eq. 4.18). Preferential formation of the benzoyl-based dianion appearing in Eq. 4.18 versus possible metalated aniline ring may arise from the electron-withdrawing nature of the carbonyl group and/or possible chelation to the amide enolate oxygen.

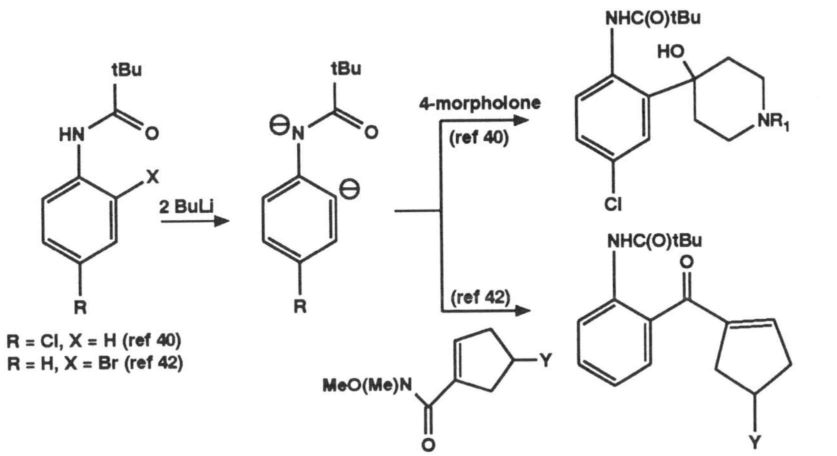

(4.17)

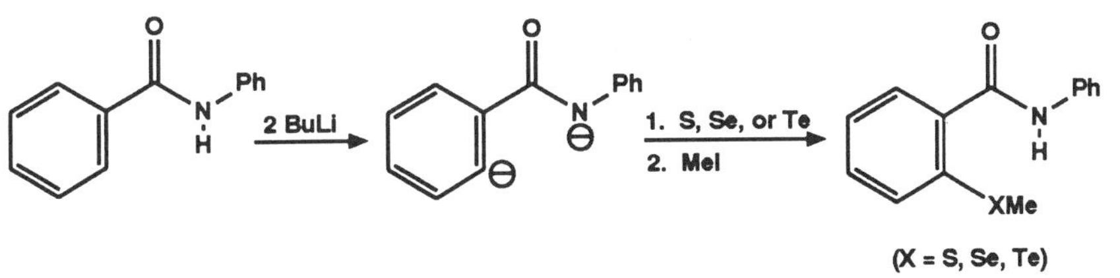

(4.18)

Artz cleverly employed iterative ring metalation of carboxamide dianions to access 2,6-disubstituted benzamides (Eq. 4.19).[43]

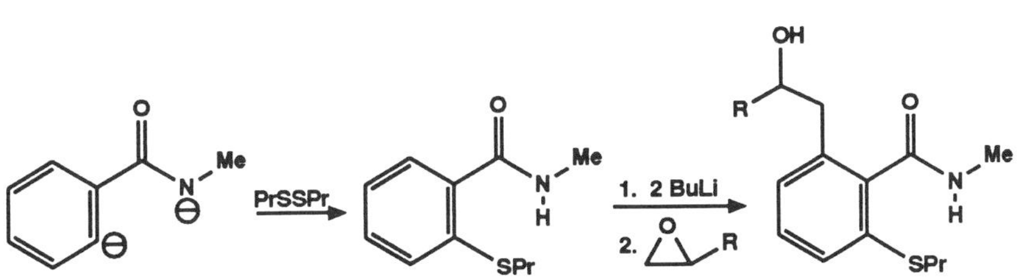

(4.19)

Remote carboxamide dianions featuring metalation also have been explored with heterocycles. N-tert-Butyl-2-carboxamide-furan (X = O), -thiophene (X = S), and -1-methylpyrrole (X = NCH_3) bis-deprotonations were examined (Eq. 4.20), and control of the dianion regiochemistry [N,C(3)-dianion versus N,C(5)-dianion] was found to be dependent upon the reaction time, solvent, temperature and base choice.[44] For example, identical experimental conditions (2.2 BuLi, -10°C, 0.5 h, DME as solvent) gave exclusive N,C(3)-dianion formation for the thiophene carboxamide dianions, yet exclusive N,C(5)-dianion formation for the furan carboxamide dianion. When sec-butyllithium was used (-78°C) exclusive formation of the N,C(3)-furanyl dianion was observed, whereas the thiophene analog favors N,C(5)-dianion formation after reaction with 2 LDA. 1-Methylpyrrole analogs afford ratios of 4:1 to 10:1 of the N,C(3)- to N,C(5)-dianions, respectively, depending upon the conditions.

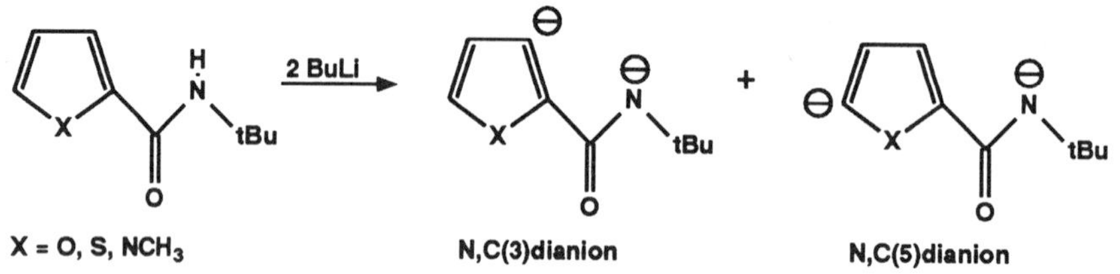

(4.20)

Heterocyclic congeners of remote carboxamide dianions may be formed. Domagala and coworkers showed that 6-methyl-2-pyridones or 2-methyl-4-pyridones form benzylic-type dianions that condense readily with carbonyls (Scheme 4.4).[45,46] Metal-halogen exchange also was performed on 5-bromo-2-pyridone to access a ring metalated remote dianion that reacted with electrophiles to afford 5-substituted 2-pyridones.[46]

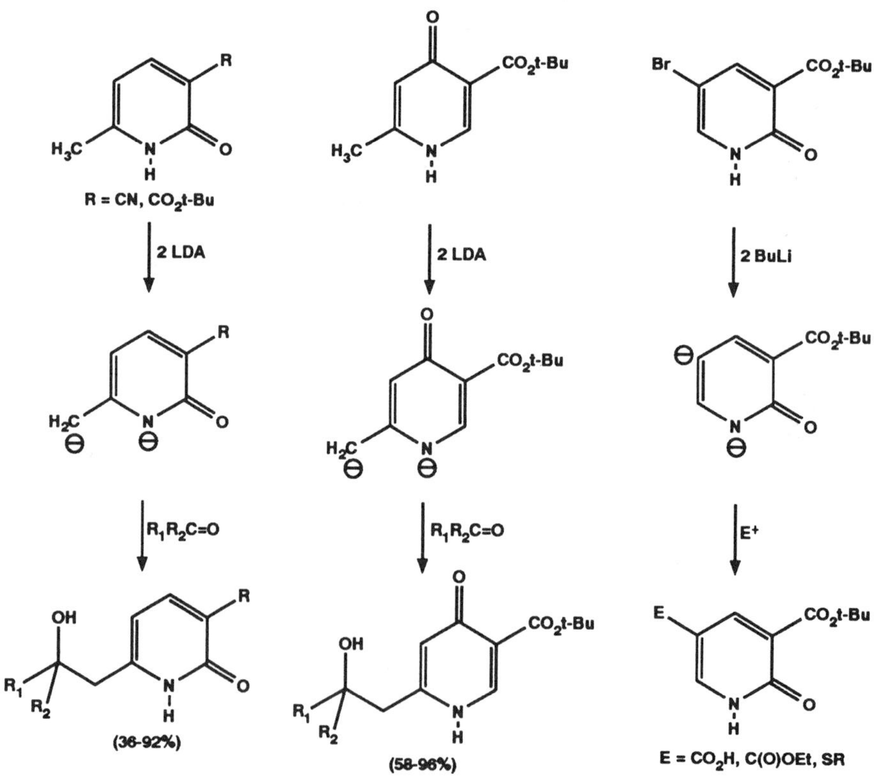

Scheme 4.4. Formation and reactions of pyridone dianions

Homoenolate carboxamide dianions have been produced by tin-metal exchange reactions and reacted with electrophiles (Eq. 4.21).[47]

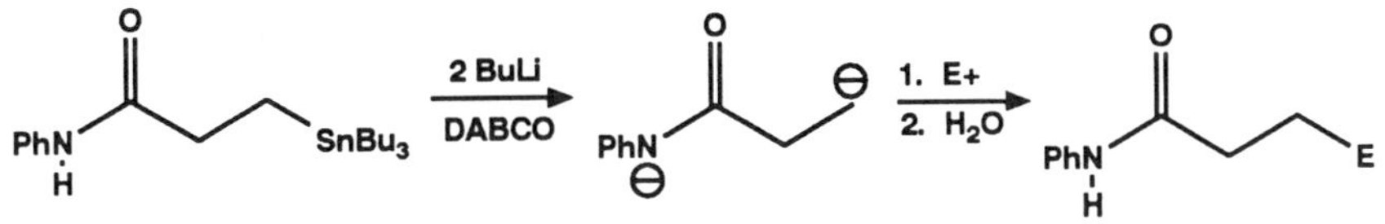

(4.21)

4.2.3 Reactions of thioamide-based N,C-dianions

Thioamide-based dianions have not received the level of attention of carboxamide dianions, but some recent detailed explorations of their use indicate distinct advantages. Yoshida[48-50] and Takahata[51,52] studied the generation of certain thioamide dianions, and incorporated the reaction into stereoselective syntheses. The N-phenyl thiopropanamide dianion, generated by treatment with 2.2 equiv. of BuLi undergoes regioselective reactions with electrophiles (57-92%) according to Equation 4.22.[48] Alkyl chain extended thioamide dianion analogs also were prepared by Michael addition of organometallics to the thiopropenamides.[48]

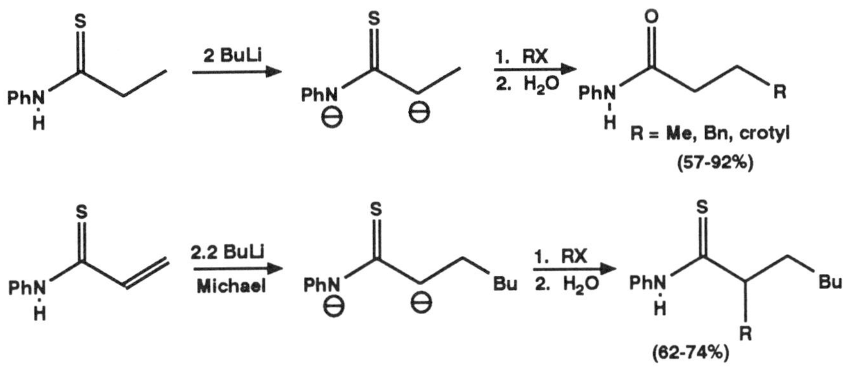

(4.22)

Simple N-phenyl thioamide dianions undergo aldol condensation reactions to give an excess of the *threo* stereoisomer (approx. 95:5) (Scheme 4.5).[49] This result is in contrast to stereochemical outcomes with similar conditions used with tertiary amide monoanion condensations that give *erythro* products. The authors linked this dianion-based diastereoselectivity to a transition state geometry that assumed the Z-configuration in route to the enolate (Scheme 4.5).[49,50]

Through a very detailed subsequent investigation, Yoshida and coworkers explored α-allylation of the thioamide dianion, and found that a thio-Claisen rearrangement supported the enolate Z-configuration in route to *threo* products.[50]

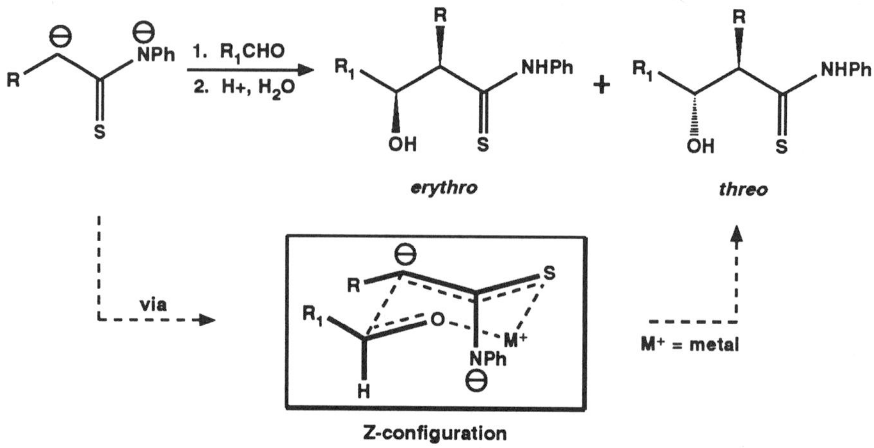

R	R_1	threo:erythro	yield (%)
Me	Me	69:31	95
Me	iPr	88:12	95
Me	Ph	62:38	97
iPr	iPr	93:7	92
iPr	Ph	69:31	99

Scheme 4.5. Stereoselective aldol condensations using thioamide dianions

The N-benzylthiopropanamide homoenolate dianion was generated by tin-metal exchange and reacted with electrophiles.[51] The study was extended by substitution of the amide benzyl substituent for the chiral R-2-phenylethylamine auxiliary, and reacted with acrolein.[52] Unfortunately, low stereoselectivities were obtained (3:2) (Eq. 4.23). The stereoisomers were separated to give a useful precursor that was used in the synthesis of N-protected statines.[52]

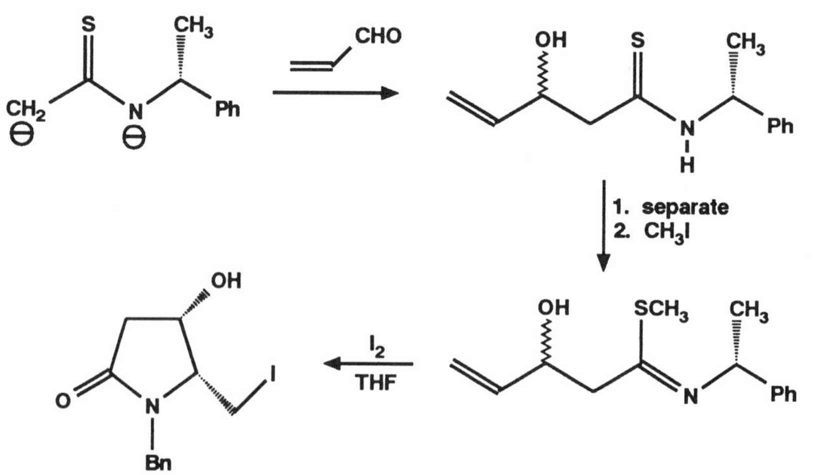

(4.23)

4.2.4 Reactions of sulfinamide and sulfonamide-based N,C-dianions

Sulfinamides

Few reports have used sulfinamides as precursors for dianion reagents.[53-55] However, recent advances in the preparation of enantioenriched sulfinate esters and sulfinamides should increase future interest in this dianion group. Corey and Durst generated sulfinamide α,α'-dianions, which reacted readily with carbonyl compounds (Eq. 4.24).[53] Thermolysis led to alkene formation. However, attempts to C-alkylate a cyclic sulfinamide α,α'-dianion with benzyl bromide failed despite variation in the base selection.[54] In this case, N-benzylation was the major product observed.

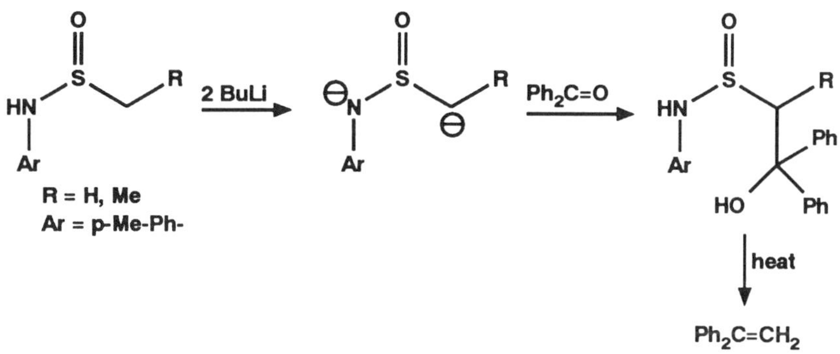

(4.24)

Sulfonamides

Although sulfonamide dianions have not been explored to the extent of the corresponding carboxamides, several reports suggest that sulfonamide dianions are equally versatile. The most common sulfonamide dianions reported have been the α,α'- and remote types.

N-Phenyl alkylsulfonamide dianions react with carbonyls to furnish the β-hydroxysulfonamides.[55] Like sulfinimides, these intermediates undergo thermal elimination to form alkenes (Eq. 4.25). The corresponding N-alkyl sulfonamide dianions react with aldehydes, alkyl halides, and nitriles to give excellent (62-97%) yields of adducts (Eq. 4.26).[56] When the N-alkyl group is tert-butyl, subsequent hydrolysis with trifluoroacetic acid

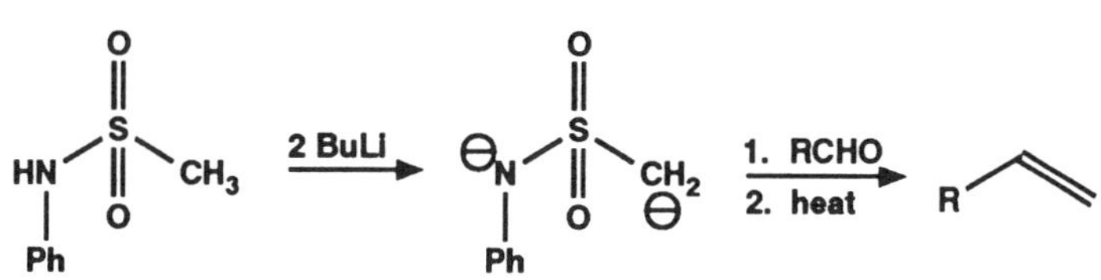

(4.25)

provides facile entry to several α-substituted, primary sulfonamides (Eq. 4.26). In an interesting concomitant study, the intermediates formed from condensation of the sulfonamide dianion with nitriles and aldehydes were trapped with phosgene or phenylchloroformate to yield 1,2,4-thiadiazinone 1,1-dioxides and 1,4,3-oxathiazinone 4,4-dioxides, respectively (Eq. 4.26).[57]

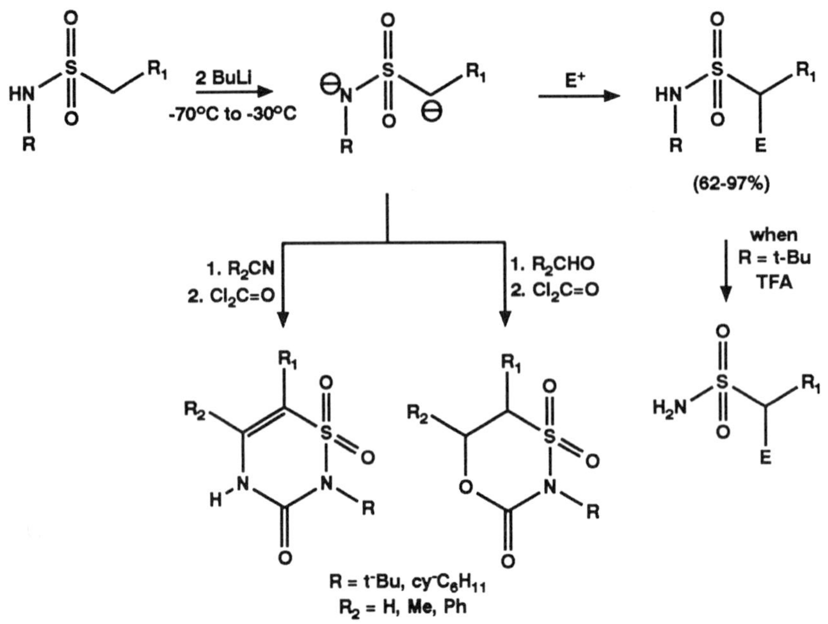

(4.26)

The α,α′-dianion of N-benzyl benzylsulfonamide is formed with 2.1 equivalents of BuLi and undergoes condensation with excess CH_2O, trioxane, and isobutyraldehyde. Excess CH_2O adds to both carbanion and the nitrogen anion to give a new route to 1,4,3-oxathiazine 4,4-dioxides (Eq. 4.27).[58]

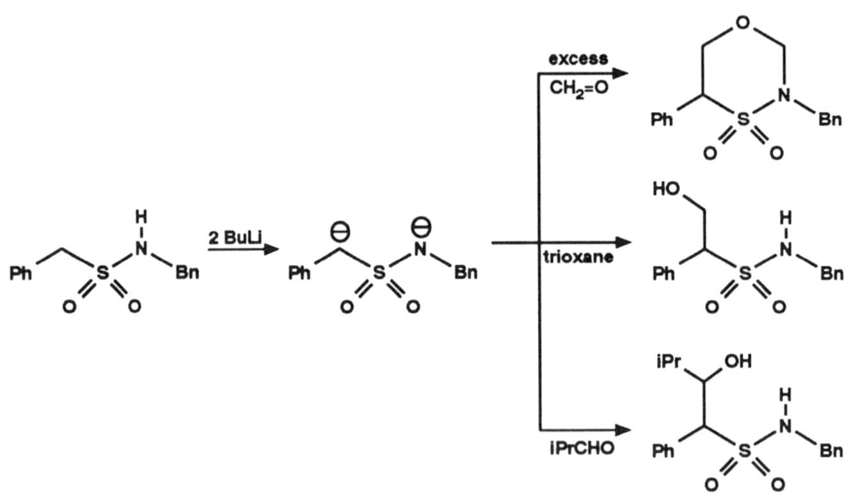

(4.27)

Cyclic variations of the sulfonamide α,α′-dianion have been explored. Sultams give the α,α′-dianion following treatment with 2 BuLi and undergo silylation, alkylation and condensation with aldehydes in 41-70% yield (Eq. 4.28).[59] An interesting side reaction occurs when N-substituted sultams are reacted with 1 equivalent of BuLi to give the monoanion, which dimerizes to the dithiazine tetraoxides (Eq. 4.28).

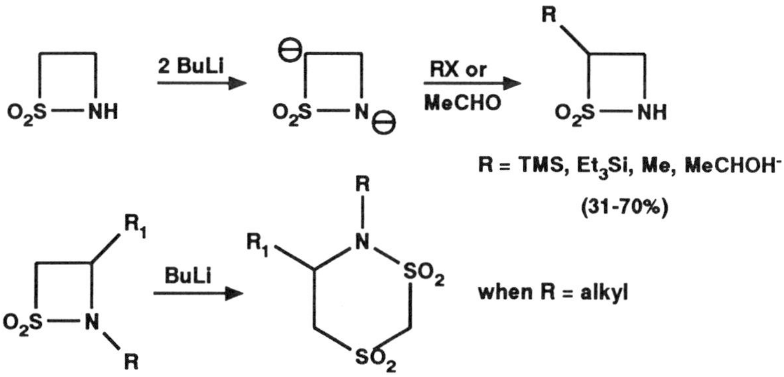

(4.28)

Davis and co-workers used the dianion formed from enantioenriched (camphorsulfonyl)imine as entry toward new oxidizing agents capable of transferring optical purity (Eq. 4.29).[60] Alkylation of this resonance delocalized α,α′-dianion with benzyl

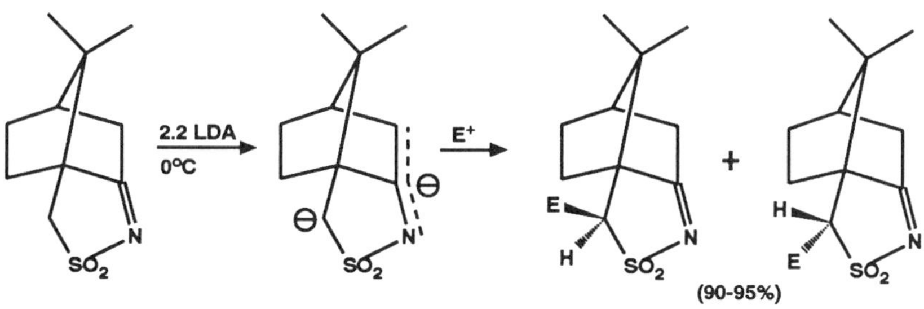

(4.29)

E	*exo*	*endo*
BnBr	50	50
ethylene oxide	30	70

bromide led to equal amounts of exo and endo products, whereas condensation with ethylene oxide gave a 3:7 mixture of exo-endo products.

Several investigations of remote sulfonamide dianions indicate that useful regioselective ortho-metalations can be achieved. Simple N-alkyl benzenesulfonamides undergo high conversion to the N,C(ortho)-dianion, and condense readily with electrophiles including carbonyls, carbon dioxide, nitriles, isocyanates, dimethylformamide, and epoxides (Eq. 4.30) (Table 4.2).[43,61-65] The presence of the 2-chloro group (entries 9/10) did not adversely affect formation of the 6-positional [ortho] anion.[43]

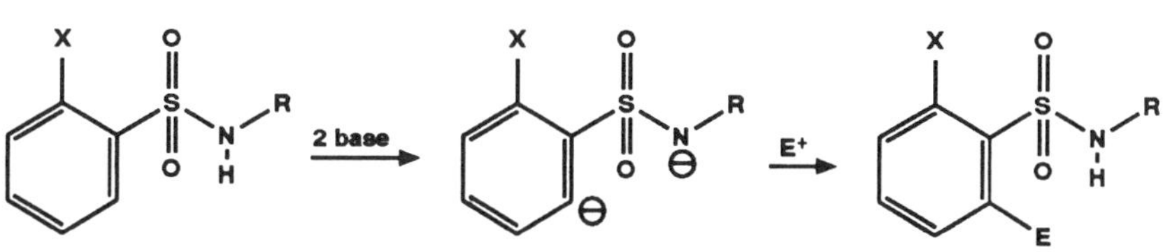

(4.30)

Table 4.2
Reactions of ring-metalated benzenesulfonamides (Eq. 4.30)

Entry	R	X	E	ref.
1	Me	H	$RR_1C(OH)$-	61
2	Me	H	CO_2H	64
3	Me	H	PhNHC(O)-	62
4	Me	H	PhC(=NH)-	63
5	Ph	H	$RR_1C(OH)$-	61
6	Ph	H	PhC(=NH)-	63
7	Ph	H	PhNHC(O)-	62
8	t-Bu	H	-CHO	65
9	R	Cl	CH_2CH_2OH	43
10	R	Cl	-CHO	43

When a benzylic C-H is available on the benzenesulfonamide, the second proton abstraction generally occurs at this site (Figure 4.3).[43,63,65,66] An ester-stabilized, benzylic benzenesulfonamide remote dianion was condensed with ethylene oxide, which undergoes immediate cyclization to the γ-lactone.[65]

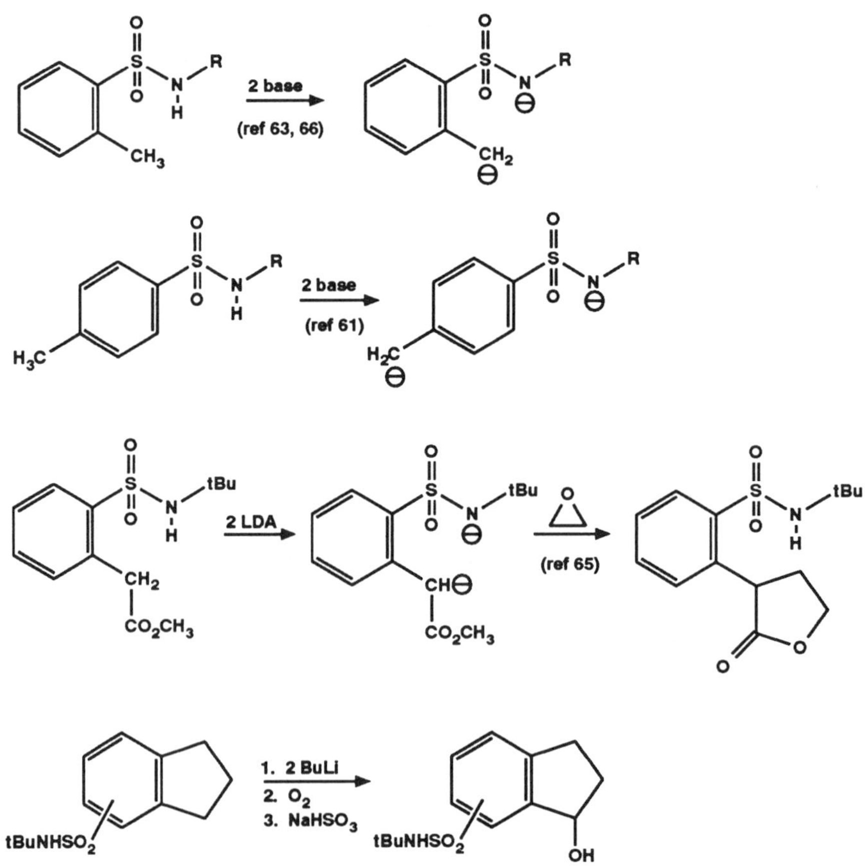

Figure 4.3. Reactions of benzylic-type benzenesulfonamide dianions

(N-tert-butyl)-4-Substituted-thiophene-3-sulfonamides give ring metalation at the 2-position after reaction with two equivalents of LDA (Eq 4.31).[67]

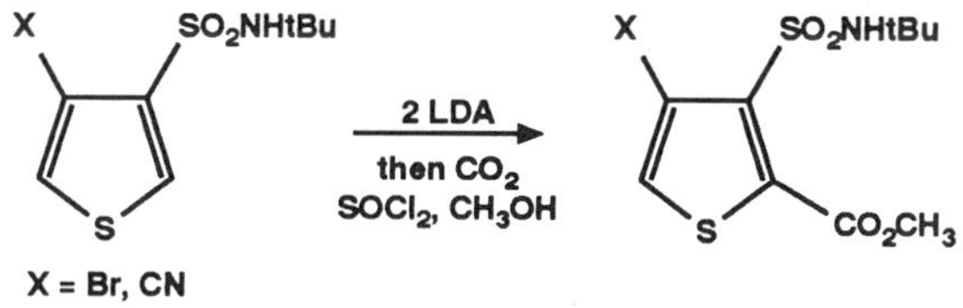

(4.31)

Graham and Scholtz[68] closely examined the formation of the ring metalated dianion of N-tert-butyl-2-thiophenesulfonamide, and found that a 2:1 ratio of the 3- to 5-metalated species formed following treatment with 2 BuLi (Eq. 4.32) (Table 4.3). Increasing the time prior to quenching the reaction showed that formation of the C-5 anion increased. The increase in regioisomer composition was attributed to precipitation of the 5-carbanion salt, which drives the equilibrium. This observation led the authors to use 1.95 equivalents of BuLi for formation of the 5-lithio species following equilibration. Treatment of the starting sulfonamide with 2 LDA rather than 2 BuLi led to direct formation of the N,C-5 dianion, which was condensed with various methoxyamides to yield 2-sulfonamide-5-acylated thiophenes. These dianion equilibria are in contrast to that of the corresponding thiophene-2-carboxamide (Eq. 4.20), which under identical deprotonation conditions, give exclusive 3-position metallation.[44]

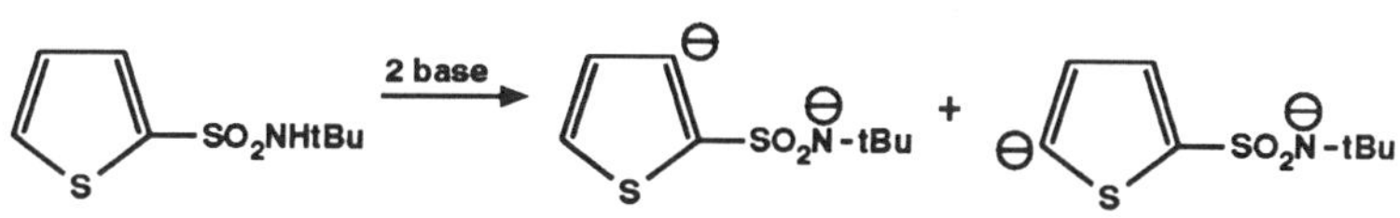

(4.32)

Table 4.3
Conditions for the formation of 3-lithio and 5-lithio thiophenes

Base	temp (°C).	time	% 3-lithio	% 5-lithio
2 BuLi	-70	60 min	55	45
2 BuLi	0	15 min	68	32
2 BuLi	-10	120 min	20	80
1.95 BuLi	-10	120 min	4	86
2.2 LDA	-78	30 min	0	100

4.3 HYDRAZONE- AND SULFONYLHYDRAZONE-BASED DIANIONS

Two prior reviews covered the major usages of hydrazone- and sulfonylhydrazone-based dianions.[2,3] A third review updated hydrazone-based dianion advances.[4] In this section, a summary from those reports, and some more recent additions of hydrazone dianions will be covered. Certain general aspects of hydrazone-dianions are worthy of note. First, with the exception of N-carboalkoxy derivatives, many aldehyde hydrazone dianions are difficult to form.[3] Following deprotonation of the N-H moiety, alkyl lithium reagents may add to these compounds to give, after hydrolysis, substituted alkanes (Eq. 4.33).[69]

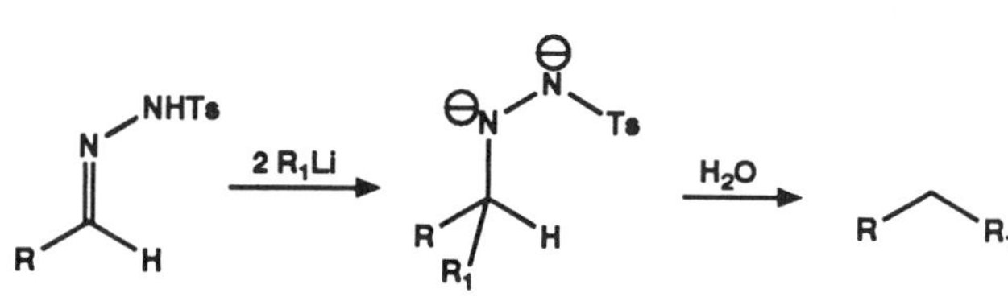

(4.33)

Second, unsubstituted hydrazones (e.g., $R_1R_2C{=}NNH_2$) react with 2 equivalents of base to form the terminal N,N-dianion, to give the N,N-dialkyl hydrazones after reaction with excess alkyl halide. Since these intermediates have not been used to construct a new C-C bond, these dianions are not covered. Last, owing to the atom arrangement, sulfonylhydrazones could have been categorized with the sulfonylamides (Section 4.2.4) as did a prior review,[2] but the unique reactivity and regiochemical outcome following reaction with electrophiles warranted inclusion in this section. For example, sulfonylhydrazone dianions are particularly useful in alkene synthesis via the Shapiro reaction, heterocycle synthesis, carbonyl transposition reactions, and remote metalation. Carboalkoxyhydrazone dianions are covered here rather than in Section 4.4 for similar reasons. This organization differs from our prior report.[4]

4.3.1 Generation of hydrazone- and sulfonylhydrazone-based N,C-dianions

Most alkyl- and arylhydrazone dianions were generated using metal amide or alkyllithium bases.[2] Representative experimental conditions also indicate that alkyllithium and LDA base combinations are used most often to generate sulfonylhydrazone dianions. In the latter case, additional equivalents of alkyllithium base are added to form an alkene via elimination of toluenesulfinate and nitrogen (Shapiro reaction, Section 4.33). TMEDA is frequently used as additive. When a specific alkene regiochemistry is desired, judicious choice of solvent and base is essential.[3]

4.3.2 Reactions of hydrazone-based N, C-dianions

Alkyl and aryl hydrazone-based dianions have not been thoroughly investigated owing largely to the more frequent use and expanded versatility of the sulfonylhydrazone-based dianions. However, in addition to straightforward alkylations,[70] acylations,[71-73,77] and condensations[74] at the α-carbonion, hydrazone dianions also offer entry to 1,2-azoles via cyclodehydration (Scheme 4.6).[71,73,74,77]

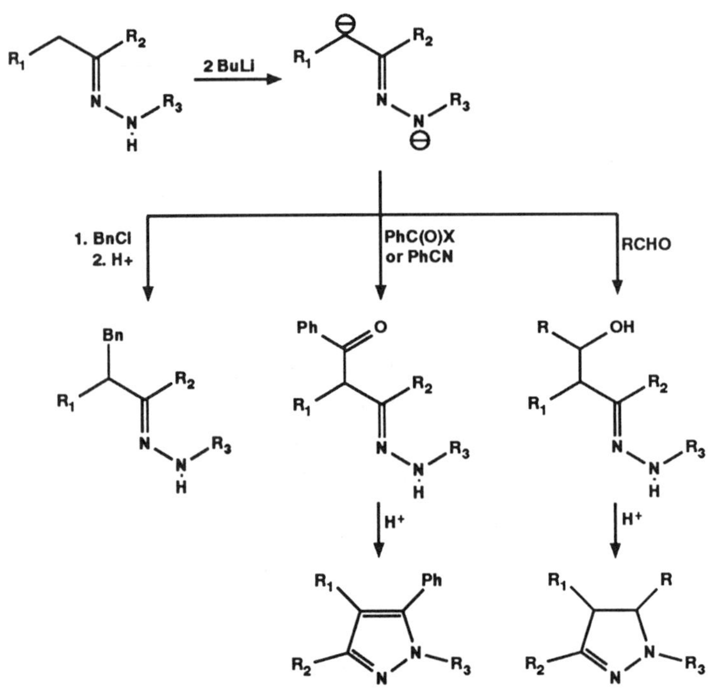

Scheme 4.6. Representative reactions of hydrazone dianions

Hydrazones bearing a carboalkoxy moiety at the N-terminus may be bis-deprotonated and enter into reactions with electrophiles.[75-77] Acetophenone N-ethoxycarbonylhydrazone forms the 1,4-N,C-dianion upon treatment with 2 BuLi at -78°C and reacts with; 1. aldehydes and ketones to give β-hydroxyketones and/or α,β-unsaturated ketones (60-90%),[75] 2. acylating agents to give N-carboalkoxy pyrazoles (25-77%),[75] 3. α-haloketones to give pyrazoline derivatives (39-62%) (Scheme 4.7).[76] Carboalkoxyhydrazone N,C-dianions also may be prepared by reaction with excess LDA and reacted with aromatic or heteroaromatic methyl esters to furnish N-carboalkoxypyrazoles in 11-85% yield.[77]

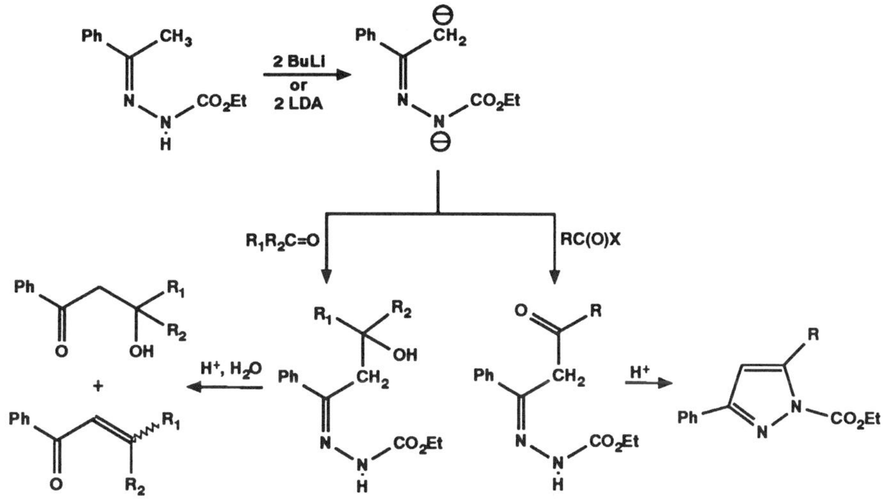

Scheme 4.7. Reactions of carboalkoxyhydrazone dianions

4.3.3 Reactions of sulfonylhydrazone-based N,C-dianions

In contrast to the N-alkylhydrazones, sulfonylhydrazone dianions have been well explored.[3] The most popular use of these dianions has been as intermediates for alkene synthesis. This process, known as the Shapiro reaction has been the subject of several reviews.[3,78,79] This particularly important transformation and other uses of sulfonylhydrazone-based dianions are individually categorized in the following discussions.

Sulfonylhydrazone dianions and the Shapiro reaction

The Shapiro reaction,[78-80] which centers upon formation of a sulfonylhydrazone dianion, may be brought through two possible pathways (Scheme 4.8; path a or b). Treatment of a sulfonylhydrazone with two equivalents of base leads to the N,C-dianion, which upon warming expels the organic sulfinate and nitrogen gas to give the vinyl lithium (or alkene anion; path a). The vinyl lithium undergoes reaction with a variety of electrophiles to give substitution at the hydrazo-carbon center. A range of alkenes may be prepared from hydrazones using this method as summarized in Table 4.4, including the synthesis of deuterated alkene and hydrazone products.[81,82] More extensive listings and its scope and limitations are available from the aforementioned review articles. Alternatively, the N,C-sulfonylhydrazone dianion solution can be maintained at -78°C and reacted directly with electrophiles to give substitution α to the hydrazo carbon (path b).[82] The newly derivatized sulfonylhydrazones may be isolated, or re-subjected to the Shapiro conditions to give, for example, homoallylic alcohols (Scheme 4.8).[89] It is important to note that pathways a and b offer complementary regiochemical outcomes. Trisylhydrazone (triisopropylphenyl-hydrazones) dianions have been shown to be superior to tosyl- and benzenesulfonylhydrazone dianions in the Shapiro reaction sequence because they do not contain an acidic ortho

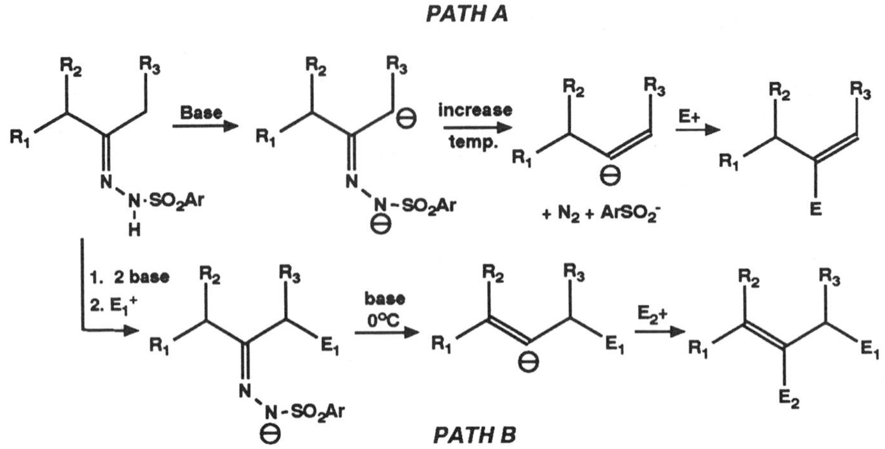

Scheme 4.8. Sulfonylhydrazone dianions; reaction pathways

Table 4.4
Reaction of Sulfonylhydrazone-based Dianions with Electrophiles (Scheme 4.8)

R_1	R_2	R_3	Base	Ar	E	E_1	E_2	Yield(%)	Ref.
					PATH A: vinyl lithium formation				
Me	H	H	s	2	TMSCH=CHCH(OH)-	-	-	90	79
Me	H	Me	n	2	OHC-	-	-	70	87
Me	Me	H	n	2	C_5H_{11}-	-	-	45	79
Et	H	Me	n	3	TMS-	-	-	97	79
Et	H	Me	n	1	TMS	-	-	38	85
C_5H_{11}	H	H	n	1	PhCH(OH)-	-	-	78	83
C_5H_{11}	H	H	n	2	PhCH(OH)-	-	-	62	87
C_5H_{11}	H	H	n	1	HOOC-	-	-	52	83
C_5H_{11}	H	H	n	1	Bu-	-	-	65	83
Ph	H	H	n	1	Me-	-	-	55	82
H	-$(CH_2)_2$-		n	2	Bu-, pentyl-, hexyl- via R_3B, I_2	-	-	90-98	79
H	-$(CH_2)_3$-		n	1	Pr-	-	-	67	82
H	-$(CH_2)_3$-		n	1	$(Me)_3Sn$-	-	-	47	85
H	-$(CH_2)_3$-		n	1	OHC-	-	-	54	84
H	-$(CH_2)_4$-		n	1	Me_3M-, M = Sn,Ge,Si	-	-	56-64	85
Me	-$(CH_2)_3$-		s	2	Bu-	-	-	59	87
Me	-$(CH_2)_3$-		n	2	D-	-	-	88	87
Ph	-$(CH_2)_3$-		n	3	TMS-	-	-	48	79
					PATH B: tandem dianion functionalization-vinyl lithium formation				
H	H	H	-	2	-	C_5H_{11}-	$HOCH_2$, Et-, OHC-	47-56	88
Et	H	Et	-	2	-	Bu-	OHC-	86 (4:1, Z/E)	91
Me	-$(CH_2)_3$-		-	1	-	Bu-	OHC-	47	91

1 = tosyl, **2** = trisyl, **3** = phenyl; n = n-BuLi, s = s-BuLi

hydrogen atom thereby saving the use of one equivalent of base needed to decompose the hydrazone anion to the vinyl lithium.[87]

α-Ketoester, α-ketocarboxamide, and α-ketoacids also were converted to the corresponding trisylhydrazones, reacted with BuLi to give polymetalated intermediates, and condensed with carbonyls to give acrylates.[90] A tandem electrophilic substitution process can be conducted with sulfonylhydrazone dianions to give high regiochemical control and good yields of alkenes (Scheme 4.8, path b) (Table 4.4).[79,91] The reader is referred to the review by Chamberlin and Bloom for extensive listings and experiments for the Shapiro-type reactions.[79]

α-Methylene lactones via sulfonylhydrazone dianions

Adlington and Barrett reacted acetone trisylhydrazone dianions with carbonyl compounds to afford the β-hydroxy N,O-dilithio hydrazones, which are subjected to Shapiro conditions to form the vinyl carbanion. The resultant O,C-dianion is trapped with carbon dioxide, and cyclizes following quench with trifluoroacetic acid to give α-methylene lactones in 40-74% yield (Eq. 4.34).[92] Clever usage of [2.2.1]hept-2-en-5-one as the carbonyl substrate gives a spirocyclic α-methylene compound following reaction with the dianion. The spirocyclic lactone is then warmed to induce a retro Diels-Alder reaction to give 3,5-dimethylene lactones (Eq. 4.34).[93]

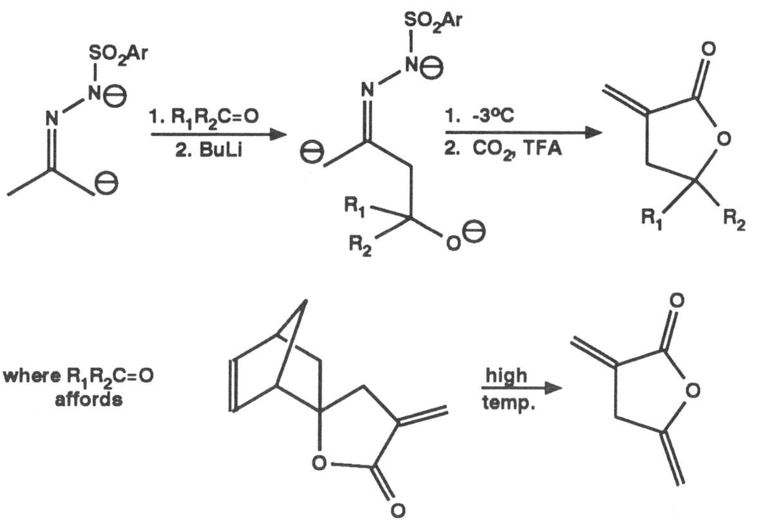

(4.34)

1,2-Carbonyl transposition using sulfonylhydrazone dianions

A procedure for a 1,2-carbonyl transposition has been developed that uses tosyl- or trisylhydrazone dianions as intermediates.[94,95] As shown in Equation 4.35, sulfonylhydrazones (e.g., hydrazones derived from 4-methylcyclohexanone) were converted to the dianion and trapped as the methylthioether with dimethyl disulfide. Addition of BuLi leads to the α-thioether carbanion, which undergoes the Shapiro reaction to form an enol thioether. Mercuric chloride-promoted hydrolysis forms the transposed ketone. A similar reaction sequence strategy can access phenylalkynes.[96]

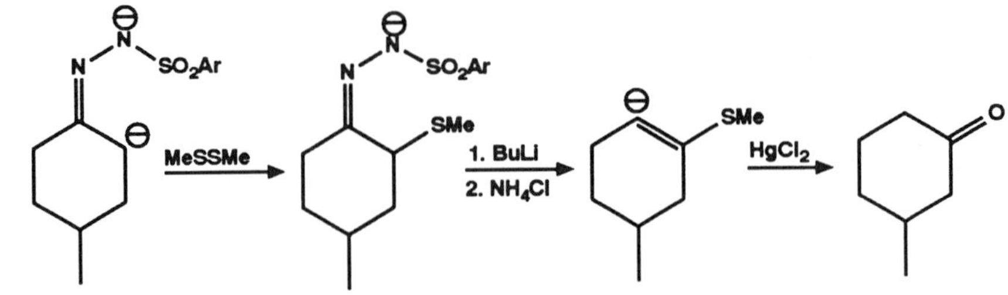

(4.35)

Remote metalation via sulfonylhydrazone dianions

The benzophenone tosylhydrazone forms the N,C(ortho)-dianion following deprotonation and metal-halogen exchange using CH_3Li and BuLi in sequence. Attempts to use 2 BuLi as the base led to competing reactions. The resulting remote sulfonyl-hydrazone dianions react with alkylating agents to furnish the ortho-substituted hydrazones in 81% yield, or with acylating agents to give phthalazines following cyclization (Eq. 4.36).[97]

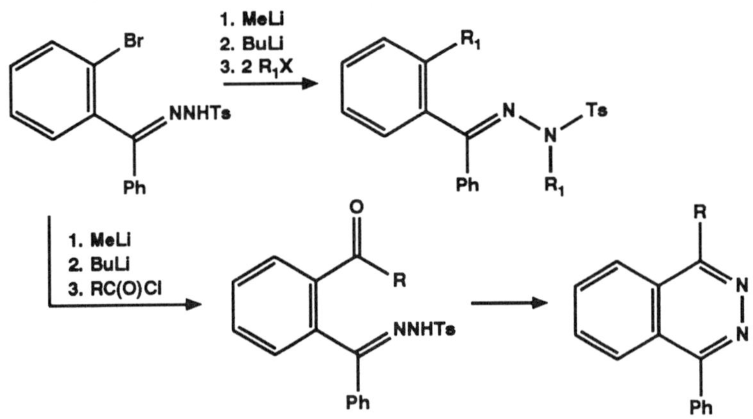

(4.36)

Other

Allylic dithiocarbonate tosylhydrazone dianions undergo regio- and stereoselective alkylations (Eq. 4.37).[98]

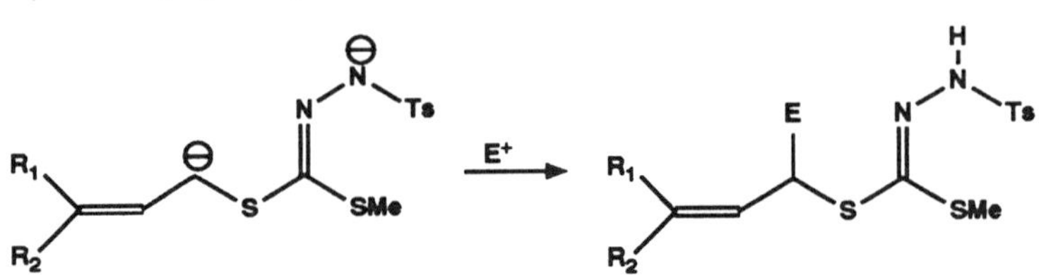

(4.37)

4.4 CARBAMATE-BASED DIANIONS

N,C-Dianions that result from primary deprotonation at a carbamate N-H moiety can be generated and reacted with electrophiles. Two predominant types of carbamate dianions have appeared, vicinal and remote, the latter inclusive of ortho-lithiation procedures. Another form of carbamate-derived dianions, the carboalkoxyhydrazones, were covered in Section 4.3.2. Therefore, specific benefits to the usage of the carbamate group is best found through examination of both sections. A frequent justification for the use of a carbamate-based dianion is its eventual ease of deprotection relative to the carboxamide moiety.

4.4.1 Generation of carbamate-based dianions

Vicinal carbamate dianions have been generated primarily with LDA (2-3 equivalents), whereas remote dianions are best prepared by reaction with alkyl lithiums, Grignard bases, and LiHMDS, depending upon the substituents present. A thorough investigation of the optimum conditions needed for ortho-lithiation of N-boc-aniline showed that 2.2 equiv. of BuLi in diethyl ether at -10°C for 3 hours gave 90% formation to the desired dianion with high conversion to products.[99]

4.4.2 Reactions of carbamate-based dianions

Vicinal carbamate dianions

Most reported vicinal carbamate dianions are propargylic, the alkyne serving for added stabilization of the carbanion. In these reports, the vicinal carbamate dianions were prepared with greater than 2 equivalents of LDA, and reacted with alkyl halides, Michael acceptors (CH_2=$CHCO_2Me$), and epoxides to give good yields of products (Eq. 4.38) (Table 4.5).[100-102] Following removal of the carbamate, new substituted propargylic amines can be accessed.

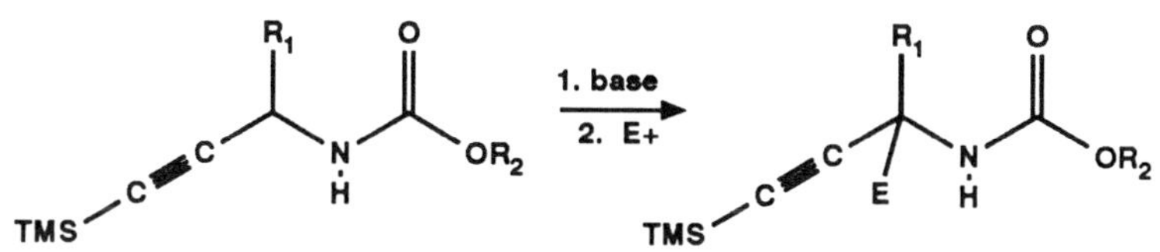

(4.38)

Table 4.5
Reaction of Vicinal Carbamate Dianions with Electrophiles (Eq. 4.38)

R_1	R_2	Base	E	Yield(%)	Ref.
MeO_2C	Et	3 LDA HMPA	Bn-, allyl-, Bu-, $-CH_2CH_2CO_2Me$	60-75	100
H	tBu	2 LDA	Bn-, allyl-, Bu-, $-CH_2CH_2CO_2Me$	65-95	101
H	tBu	2 LDA	$-CH_2CH_2(CHCH_2O)$	60	102

Remote carbamate dianions

In general, remote carbamate dianions have been little explored. Recent improvements in the experimental conditions for the ortho-lithiation of N-boc-anilines may increase their acceptance.[99]

In the same report utilizing vicinal carbamate dianions, Metcalf and Casara functionalized the terminal alkyne position using a remote carbamate dianion to obtain the starting material used in Eq. 4.38 (Eq. 4.39).[101] The authors also showed that the carbamate moiety can be easily removed with $SiHCl_3$ offering an alternative to acid hydrolysis that led to degradation.

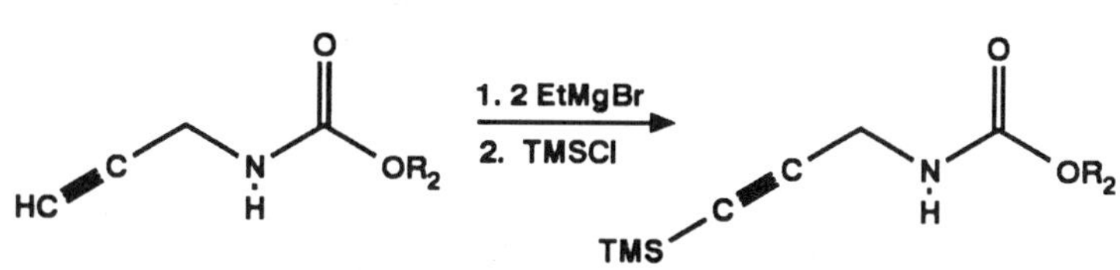

(4.39)

The metalated ortho dianion of N-boc-aniline may be formed, which undergoes reaction with the complete battery of electrophiles including subsequent cyclization to oxaza- and diazaheterocycles (Scheme 4.9).[99,103]

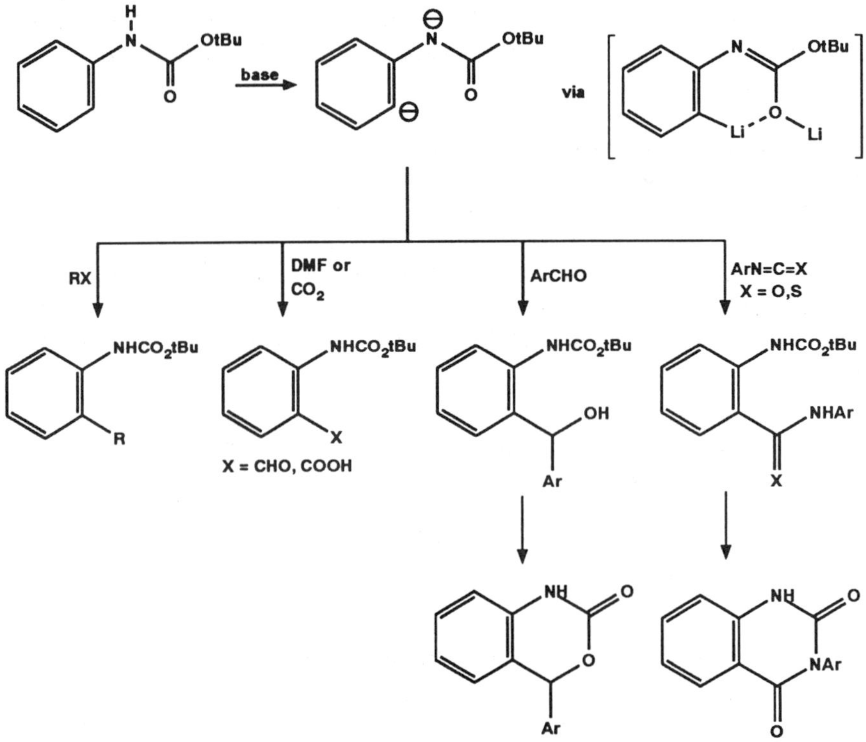

Scheme 4.9. Reactions of ortho-metalated carbamate dianions with electrophiles

Two additional examples of remote carbamate dianions are illustrated below. Suitably protected aspartate esters form the corresponding dianion with lithium hexamethyldisilazide and react at -78°C with methyl iodide to give 96% *threo* isomer, but give a 43:57 *threo*:*erythro* ratio when the temperature is raised to -7°C (Eq. 4.40).[104]

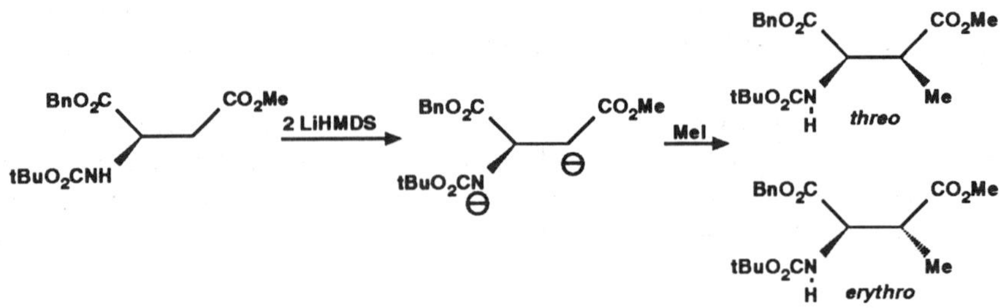

(4.40)

Dianions of N-tert-butyl benzylic carbamates react with alkyl halides or aldehydes.[105] The process may be repeated for the eventual preparation of tertiary carbinols or vicinal diols following reductive removal of the carbamate (Eq. 4.41).

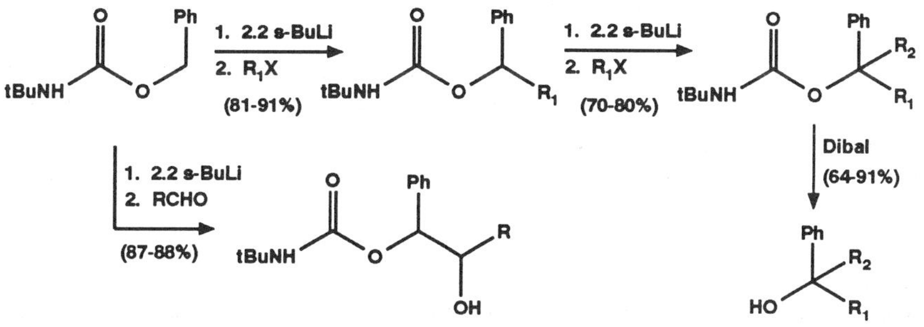

(4.41)

Zimmerman and coworkers chose a carbamate protection route when they found that 2-amino-6-methylpyrimidine analogues were insoluble upon formation of the polymetalated intermediate. Following protection of the amino moiety as the methylcarbamate, reaction of the resultant N,C-dianion with electrophiles at the 6-methyl carbanion was achieved with 2 LDA (Scheme 4.10).[106]

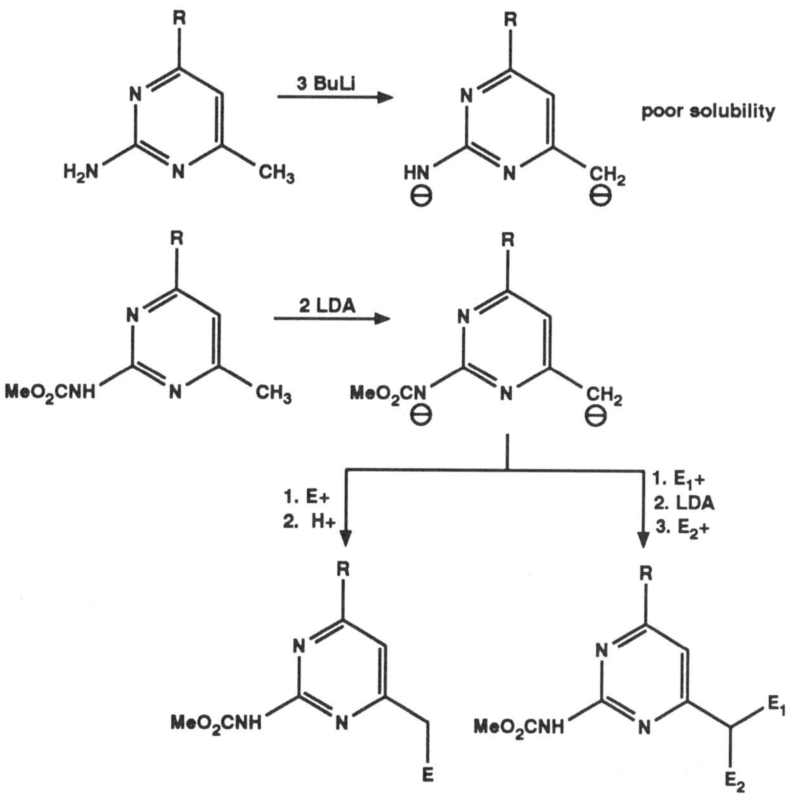

Scheme 4.10. Pyrimidine derivatization using N,C carbamate dianions

4.5 ENAMINONE-BASED DIANIONS

Enaminones [RC(O)CH=CHR-NHR] are a class of enamines that are formed from the reaction between amines and 1,3-dicarbonyl compounds. This section describes the formation and reaction of N,C-dianions derived from this highly functionalized species. One purpose of this section is to establish differences between enaminone dianions and amine dianions (Section 4.7), the former leading to unique regiochemical outcomes owing to the extended π-system. For this reason, an example of an enaminone hydrazone[107] is included in this section. Other hydrazone dianions are covered in Section 4.3.

4.5.1 Generation of enaminone-based N,C-dianions

By virtue of extended conjugation, certain enaminones form resonance delocalized dianions. As a result, secondary enaminones derived from acetylacetone can form the α,α′- and/or α,γ-dianion types (Figure 4.4), leading to regioisomeric products.

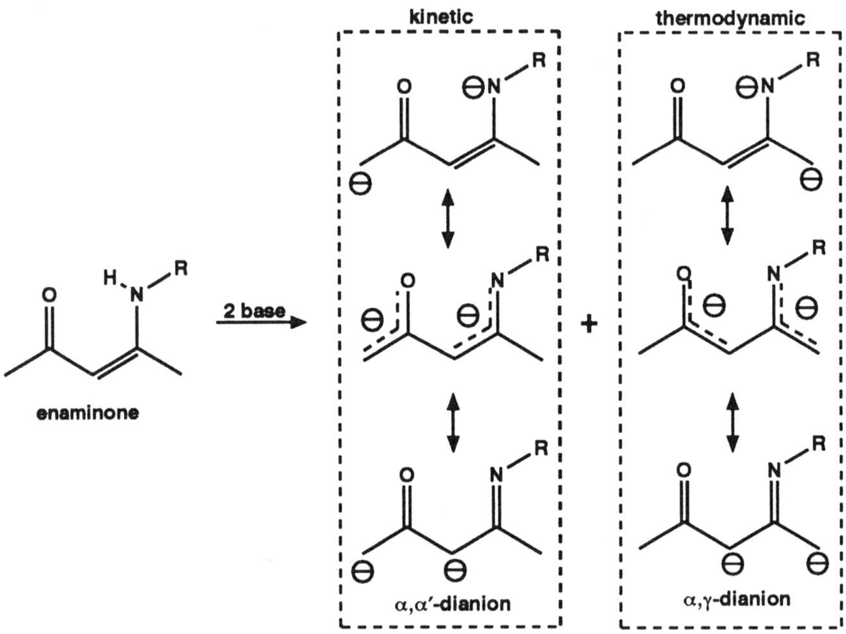

Figure 4.4. Isomeric enaminone dianions

This problem may be circumvented using enaminones that contain terminal (α′ or γ) groups that do not bear enolizable hydrogens (e.g., Ph, t-Bu). However, Bartoli and coworkers[108-112] found that exclusive formation of the γ-dianion could be achieved using CH_3Li-TMEDA (0°C-RT) as the deprotonating conditions, while highly favorable ratios of the kinetic α′-dianion could be obtained using lithium tetramethylpiperidine (LiTMP) in THF (0°C).

Two features contribute to selection and eventual usage of enaminone α′-dianions. First, the extent and composition of α′-dianion formation depends greatly upon the nitrogen substituent, and second, the α′-dianion slowly converts to the γ-dianion. Conversely, the γ-dianion rate of equilibration is slower than that of its formation giving an exclusive regiochemical outcome. Under identical dianion forming conditions, the α′:γ ratio ranged from 2:1 to 6:1 to >100:1 when the nitrogen substituents were Me-, iPr-, and t-Bu, respectively. However, the size of the substituent does not greatly alter the γ-selectivity.

4.5.2 Reactions of enaminone-based N,C-dianions

Early studies of acyclic enaminone dianions showed that these dianions offered distinct advantages over the β-dicarbonyl dianions, particularly the avoidance of O-alkylation side products. Specifically, 1-aryl-3-isopropylamino-2-butene-1-one reacts with LiTMP/THF to afford the unambiguous γ-dianion intermediate. The γ-dianion forms products following reaction with alkyl halides (54-97%), aldehydes (42-47%), or acetone (28%) (Eq. 4.42).[108] An earlier report indicated that cyclic enaminone dianions gave significant N-alkylation,[113] although virtually no N-alkylation was observed with this open chain analogue.

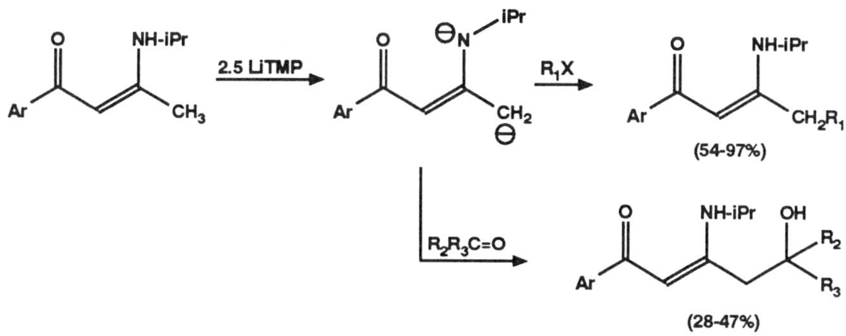

(4.42)

Interestingly, reaction of the same enaminone (Eq. 4.42) with excess base and excess trimethylsilyl chloride gave elimination to the alkyne (Eq. 4.43).[109] Control over the pH of the acid quench leads to the conjugated imino-alkyne (43%) or ynone (30-83%).

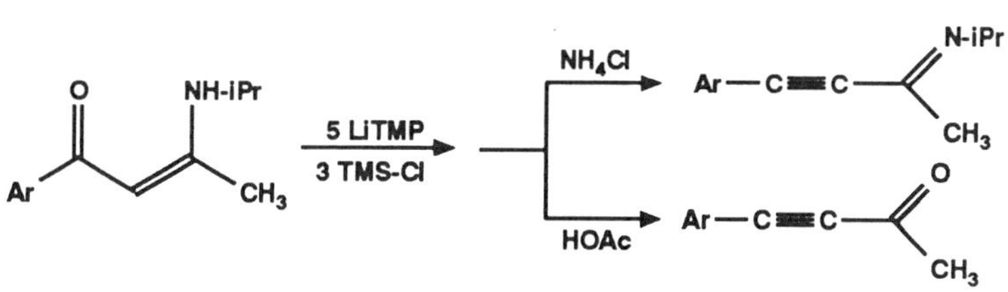

(4.43)

Continued investigations by Bartoli and coworkers[110-112] have posited clear experimental conditions to form either α′- or γ-enaminone dianion types. As determined by deuterium incorporation and alkylation reactions (75-94%), α′-dianions form in > 99% conversion when an N-isopropyl enaminone is reacted with 2.5 LiTMP in THF at 0°C followed by warming to 20°C and stirred for 2 h. The N-isopropyl enaminone γ-dianion is similarly prepared in > 99% conversion, but requires 2.5 MeLI and 2.5 TMEDA as additive. Early studies suggested that the complexing agent appeared to alter the interconversion rate between kinetic and thermodynamic species. This hypothesis was given less importance since the α′-dianion interconverts to the γ-form slowly if TMEDA is added. *Ab initio* calculations support the formation of different geometries under the α′- and γ-dianion conditions,[110] one of which is more capable of a 1,5-hydrogen shift.

Without governance by a steric contributor (N-methyl enaminone), excellent regiochemical control and yield were still accomplished in the reaction with epoxides (Eq. 4.44).[111] The resultant 2-carbon chain extended alcohols could be cyclized to furylidenes with HOAc at 70°C (66-85%).[111]

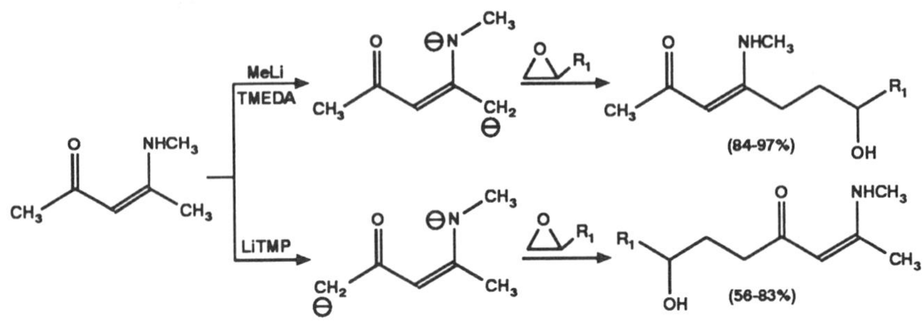

(4.44)

Reaction of the enaminone γ-dianion with RCN yields 4-alkylaminopyridines, whereas use of 2 RCN leads to pyrimidines (Eq. 4.45).[112]

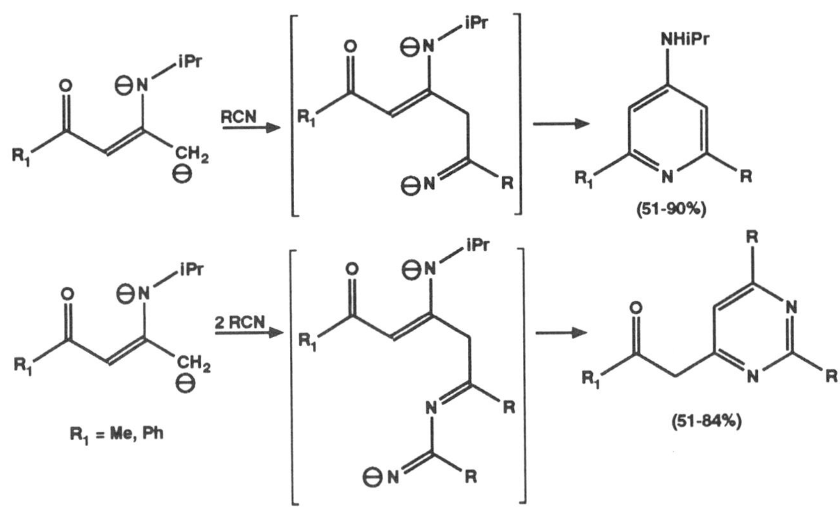

(4.45)

Cyclic enaminone dianions, the first enaminone dianions to be studied,[113] can be easily generated with 2 BuLi or 2 LDA. However, the isolated 5:8 ratio of α':γ methylation with CH_3I along with significant N-alkylation were limiting (Eq. 4.46). A tricyclic system, however, was converted to the desired N,γ-C dimethyl product in 73% yield.

Use of a chiral auxiliary in combination with an exocyclic, nonenolizable acyl moiety permitted an examination of diastereofacial selectivity in the alkylation of enaminone dianions (Eq. 4.46).[114] The γ-alkylations occurred in excellent yields (75-93%) and noteworthy si-facial diastereomeric excesses ranged from 27-98% depending upon the steric constitution of the alkylating agent.

Hydrazoenaminone dianions, formed from bis-deprotonation with BuLi or KDA in the presence of TMEDA, give high yields of γ-alkylation (Eq. 4.47).[107] The usage of TMEDA/alkyllithium conditions to give exclusive γ-alkylation is consistent with those reported by Bartoli, and may indicate that these complexing agents play significant roles in the stabilization of the dianions.

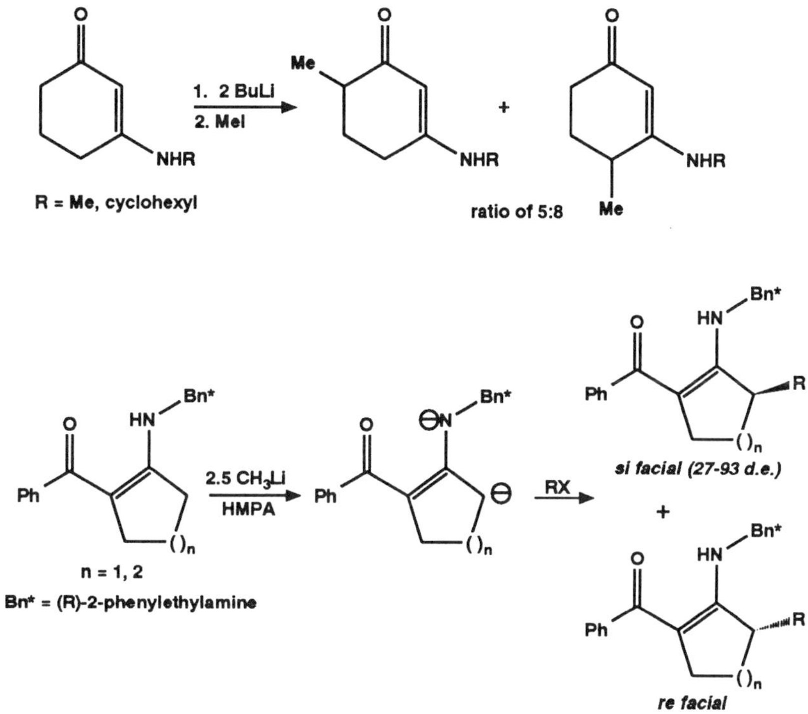

(4.46)

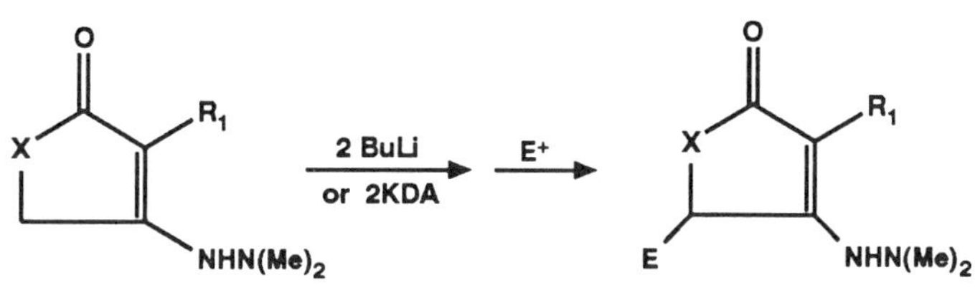

(4.47)

4.6 PYRROLE- AND BENZAZOLE-BASED DIANIONS

4.6.1 Generation of indole- and pyrrole-based dianions

Few studies have reported N-H deprotonations of heteroaromatics in route to dianions, although the monometalated species are commonplace. Among the N,C-heteroaromatic dianions, pyrroles, indoles, indazoles, and benzimidazoles have been studied. Given the large difference in the N-H pKa's, no uniform experimental condition exists. Further, countercation and temperature play an important role in both the alkylation regiochemistry, and the amount of N,C-dialkylated product. For example, reaction of 2-methylindole with 3 BuLi followed by 2 tBuOK gave 77% of the desired C-alkylated product, whereas reversal of the base order gave only 11%. Permitting the reaction to warm to room temperature led to equal amounts of C-alkylated and N-alkylated products.[115]

Such reversals in reaction site also have been linked to the countercation for the monoanions in this compound class. The N-sodio or N-potassio pyrrole reacts with electrophiles to give mainly N-substituted products, but pyrryl-magnesium halides react predominantly at C-2. Indolyl magnesium bromide reacts with alkylating agents at C-3, but the sodio derivative alkylates mostly at nitrogen. N-Methylindole metalates with BuLi at C-2 and further reacts at this site. Therefore, before meaningful and extended synthetic use is made of these heteroaromatic N,C-dianions, a comprehensive study of the effect of countercations on regiochemistry should be undertaken.

4.6.2 Reactions of pyrrole-based dianions

In studies aimed at the total synthesis of CC-1065 (a potent anti-tumor agent), substituted pyrrole dianions were generated with 2 LDA with HMPA and reacted at a remote carbanion site (Eq. 4.48).[116] Subsequent intramolecular alkylation gave cyclopropanes.

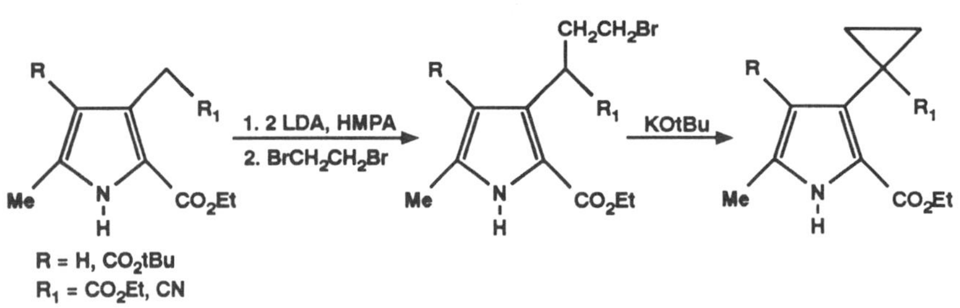

(4.48)

4.6.3 Reactions of indole- and indazole-based dianions

Several indole and ring-fused indole systems have been converted to N,C-dianions and reacted with electrophiles.[115-119] These reactions are illustrated in Eq. 4.49 through Eq. 4.52. Regioselective C-2 methyl (methylene) metalation is evidenced in Equations 4.49 and 4.51. In the latter reaction, the authors found that the dianion could not be formed when the indole nitrogen was lithiated, whereas the sodium and potassium salt proceeded to dianion readily. The work by Levy[119] (Eq. 4.52) further illustrates that highly regiocontrolled carbanion generation can be achieved following indole N-H deprotonation.

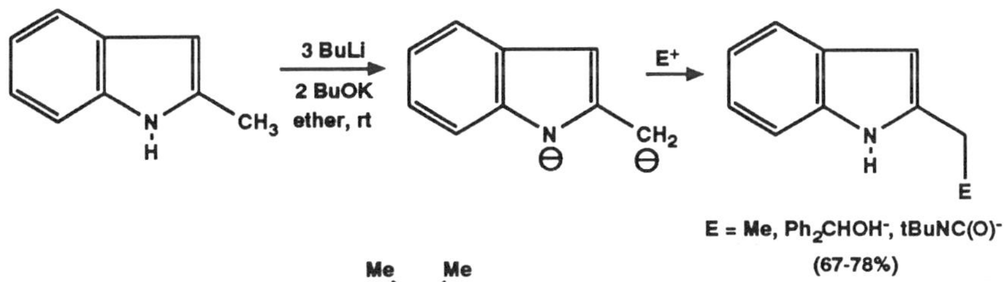

(4.49)

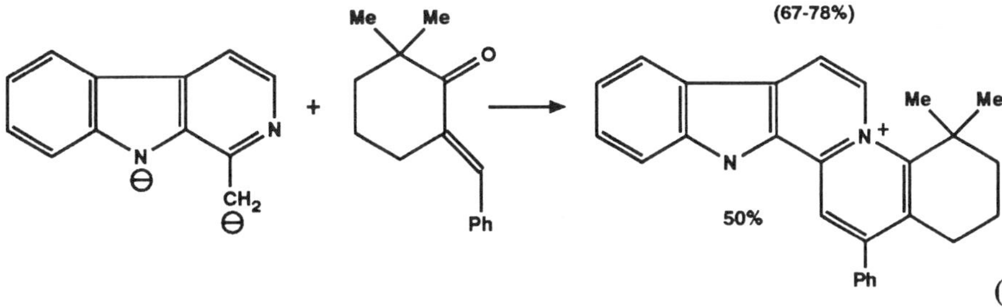

(4.50)

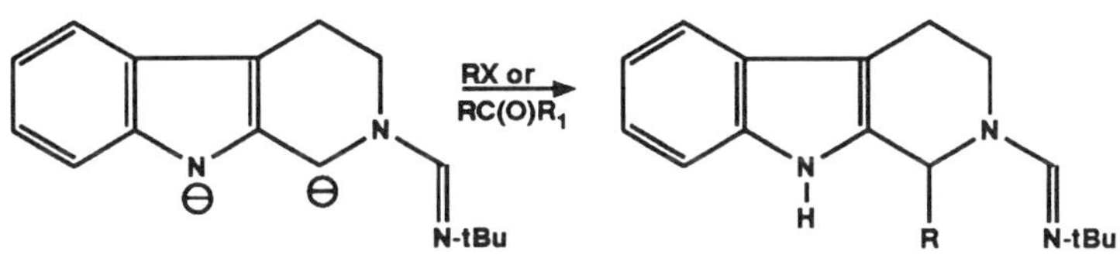

(4.51)

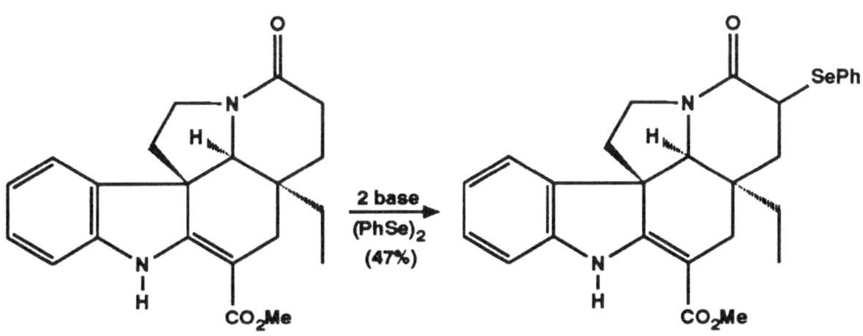

(4.52)

Benzimidazoles undergo N,C-dianion formation when a C-2 alkyl group is present (Eq. 4.53). These dianions react with alkyl halides or carbonyls to give good yields of adducts.[120]

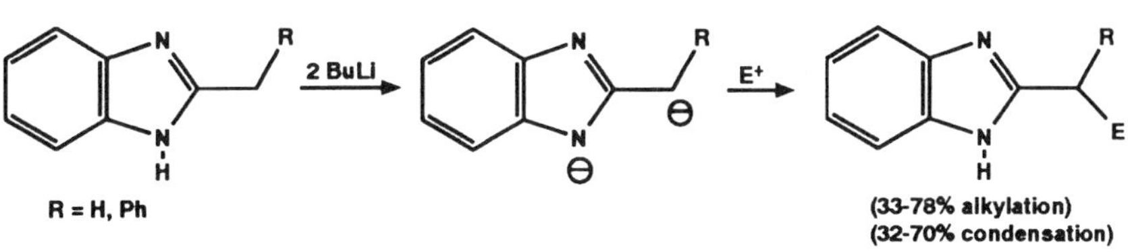

(4.53)

Welch and coworkers recently showed that indazoles give N,C-dianion formation via metal-halogen exchange.[121] These intermediates give average to low yields of condensation products. However, direct dianion formation leads to ring opening of the azole.

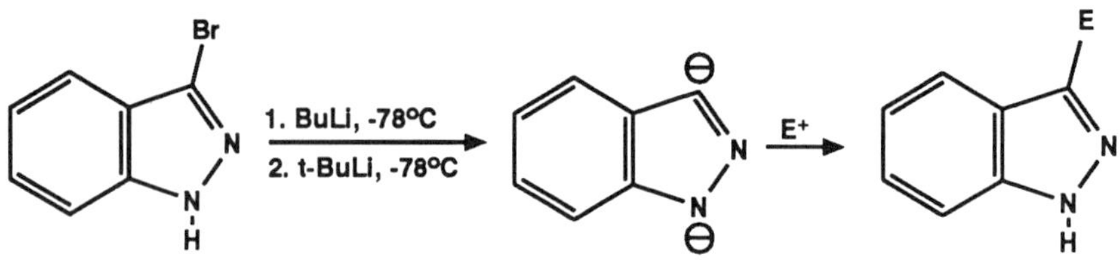

(4.54)

4.7 IMIDE- AND N-ACYLSULFONAMIDE-BASED DIANIONS

4.7.1 Generation of imide, and N-acylsulfonamide-based N,C-dianions

Dianions derived from imides and N-acylsulfonamides may be viewed as aza analogues of the β-dicarbonyl dianions (Chapter 2). Bis-deprotonation of imides and N-acylsulfonamides largely give rise to dianions of the α,α′-type (Fig. 4.5). There has been very little exploration of the remote forms. Acyclic and cyclic versions of these dianions have been prepared and, despite high N-H acidity owing to two adjacent π-moieties, relatively strong amide bases ($NaNH_2$ and KNH_2) in liquid ammonia have been used to access these dianions.[2] In part, this is due to limited availability of alkyllithium reagents at the time these intermediates were originally explored.

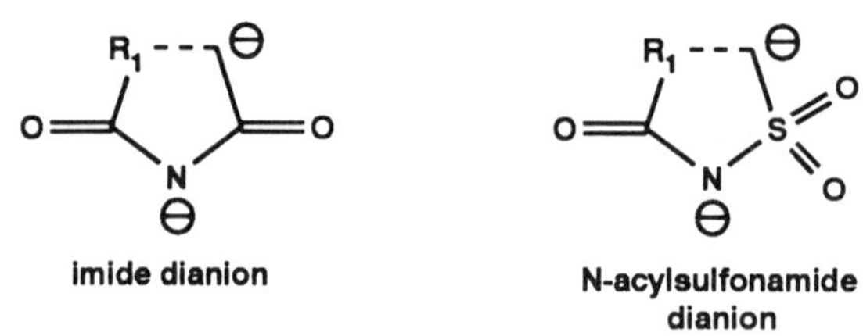

Figure 4.5. Imide and N-acylsulfonamide α,α′-dianions

4.7.2 Reactions of imide-based dianions

Linear imides furnish α,α′-type N,C-dianions and react exclusively at the carbanionic center with electrophilic reagents (Eq. 4.55) (Table 4.6).[122-127] Yields of alkylated and acylated products were modest, but interesting structural alterations, for example, polyanions, of N-acetylureas (entries 10, 12) and salicylates (entries 4-6) indicate marked flexibility.

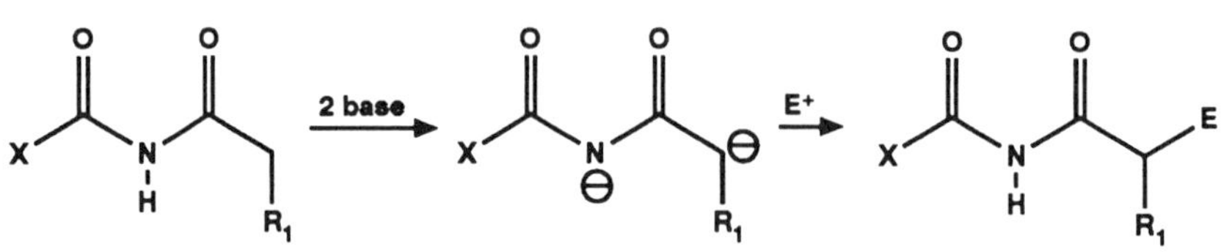

(4.55)

Table 4.6
Reaction of Imide Dianions with Electrophiles (Eq. 4.55)

Entry	X	R_1	Base	E	Yield(%)	Ref.
1	Me	H	2 KNH_2	Bn-	63	122
2	Me	H	2 KNH_2	PhC(O)-	13	122
3	Me	H	2 KNH_2	$Ph_2C(OH)$-	32	122
4	o-PhOH	H	3 $NaNH_2$	Bn-	52	123
5	o-PhOH	H	3 $NaNH_2$	$Ph_2C(OH)$-	72	123
6	o-PhOH	H	3 $NaNH_2$	PhC(O)-	43	123
7	Ph	H	2 KNH_2	Bn-	64	122
8	Ph	H	2 KNH_2	PhC(O)-	28-85	122, 124
9	Ph	H	2 KNH_2	$Ph_2C(OH)$-	75	122
10	NH_2	Ph	KNH_2*	Bn	66	125
11	$PhC(O)CH_2$	H	3 KNH_2	Bn	69	126
12	PhNH	Ph	KNH_2*	Bn	73	125
13	$PhCH_2$	Ph	2 $LiNH_2$	Bn	58	127

* *used in excess*

The condensation of N-acetylbenzamide dianion with nitriles led to pyrimidones (Eq. 4.56).[128]

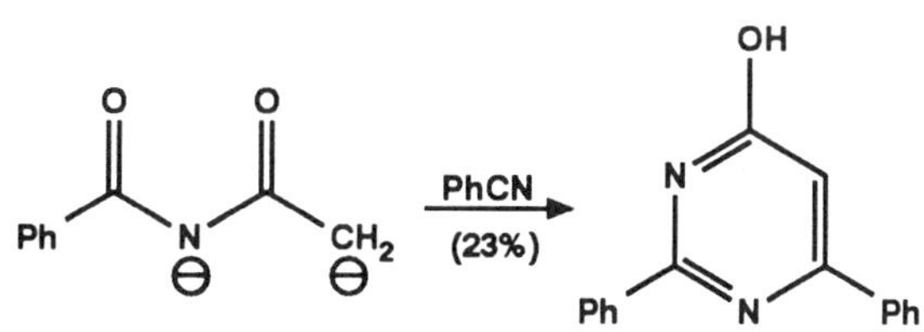

(4.56)

Cyclic versions of imide dianions, namely succinimide and glutarimide, morpholinedione, and heteroatom-containing succinimides have been reacted with electrophiles according to the general illustration below (Eq. 4.57).[124,129-132] Again, metal amides have been most often used to generate the dianion, but more recent reports[130,133,134] indicate that BuLi may be used to form the corresponding dianion with equal success. Modest to good yields of the derivatized heterocycles were obtained (Table 4.7).

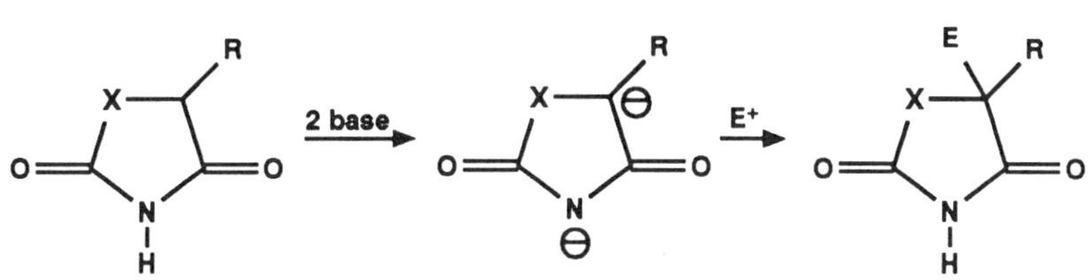

(4.57)

Table 4.7
Reaction of Cyclic Imide Dianions with Electrophiles (Eq. 4.57)

Entry	X	R	Base	E	Yield(%)	Ref.
1	-S-	H	2 $LiNH_2$	Bn-	57	129
2	-S-	Me	2 KNH_2	Bn-	76	129
3	-S-	H	2 BuLi	$ArSO_2$-	12-63	130
4	-CH_2O-	H	2 $NaNH_2$	Bn-	46	131
5	-CH_2O-	H	2 $NaNH_2$	Ph_2C(OH)-	45	131
6	-CH_2O-	H	2 $NaNH_2$	PhC(O)-	23	131
7	-CH_2-	H	2 KNH_2	PhC(O)-	44	132
8	-CH_2-	H	NaH*	PhC(O)-	56	124
9	-CH_2-	H	2 KNH_2	Bn-	29	132
10	-$(CH_2)_2$-	H	2 $NaNH_2$	Bn-	80	131
11	-$(CH_2)_2$-	H	2 $NaNH_2$	PhC(O)-	72	131

* *used in excess*

The oxa- and thia-containing cyclic imide dianions, 2,4-oxazolidinedione (X = O) and 2,4-thiazolidinedione (X = S), show good reactivity toward alkylation and sulfonation, and further utility in the construction of furanones (Scheme 4.11).[133,134] In the latter sequence, the heterocyclic dianions are reacted with α-haloketones to give transient formation of the halohydrin, which presumably undergoes immediate cyclization to the epoxide. Addition of aqueous HCl gives the rearranged allylic alcohols in excellent yield. Treatment of the allylic alcohols with 6 N HCl at 100°C gives 3-hydroxy- or 3-mercapto-2(5H)-furanones. Alternatively, oxidation of the allylic alcohol to the aldehyde, and subsequent cyclization gave 3,5-dihydroxy-2(5H)-furanones.

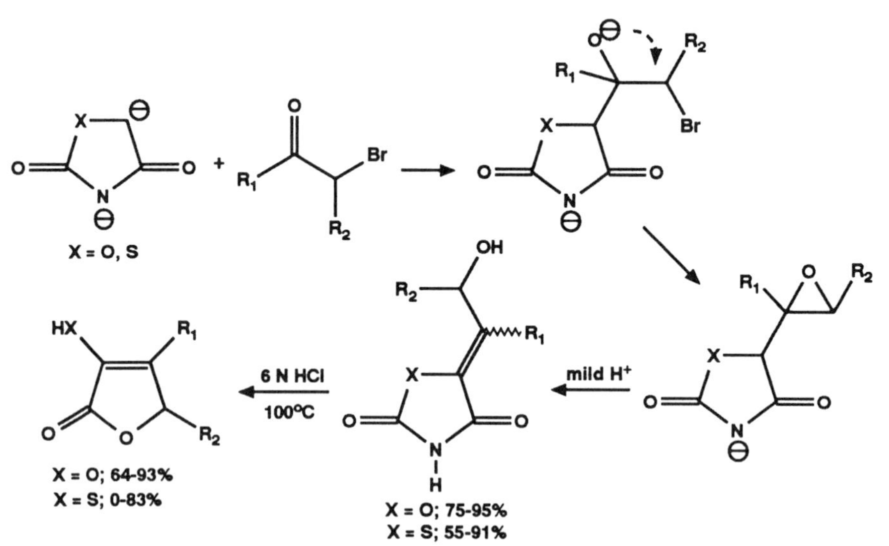

Scheme 4.11. Synthesis of furanones via oxa- and thia-containing cyclic imide dianions

The 2,4-oxazolidinedione was reluctant to form the dianion using alkyllithium reagents because of precluding precipitation of the monoanion. However, addition of LiCl (6 equiv.) greatly aided the solubility leading to high conversion to the dianion with 2.1 equiv. of t-BuLi.[134]

4.7.3 Reactions of N-acylsulfonamide dianions

Acyclic and cyclic N-acylsulfonamides have been investigated. In most instances, alkyl lithiums and LDA have been used to generate the dianion.

Belletire and Mahmoodi showed that the α,α′- (carbonyl side)-dianion of N-acylsulfonamides could be alkylated in good yield, and converted into butyrolactones (Eq. 4.58).[135]

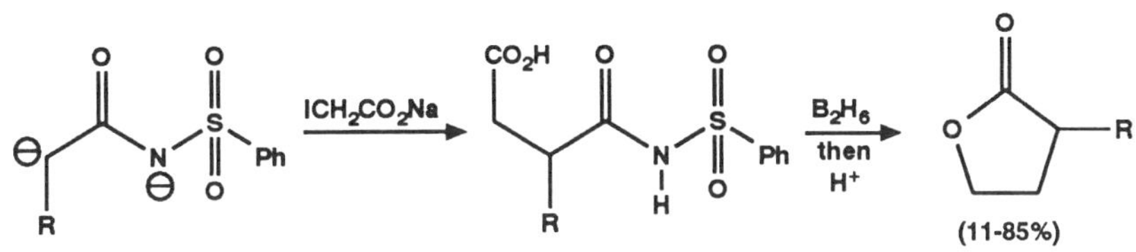

(4.58)

N-Acetylbenzenesulfonamide forms the α,α′-type dianion upon reaction with 2 equiv. of base whereas N-benzoylbenzenesulfonamide and N-isopropyl-2-chlorobenzenesulfonamide undergo ortho-directed and metal halogen exchange reactions, respectively.[136] In the latter instance, 1,2-benzoisothioazole 1,1-dioxides are formed (Eqs. 4.59 and 4.60).

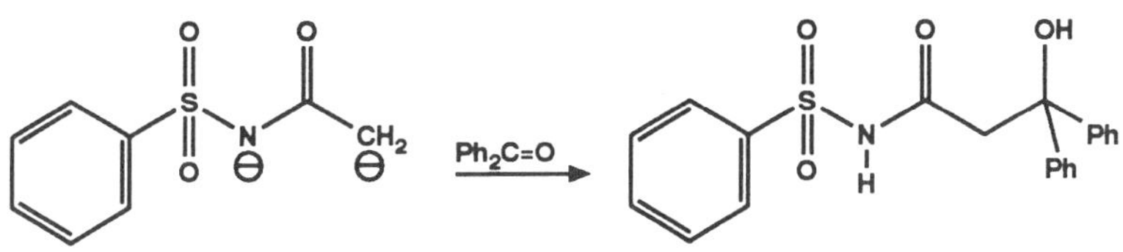

(4.59)

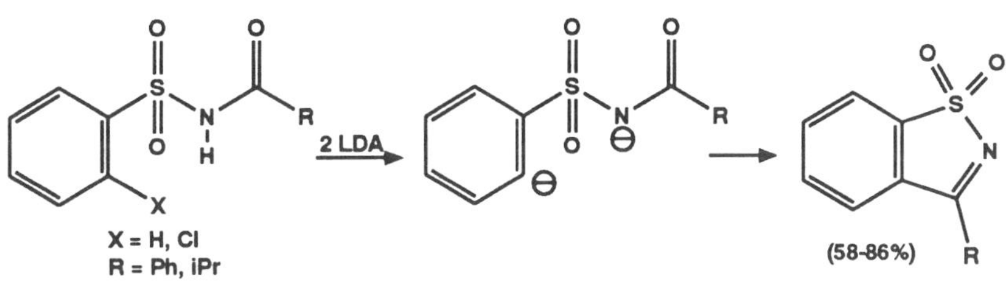

(4.60)

Alkylation of a cyclic N-acylsulfonamide dianion also has been reported (Eq. 4.61).[65]

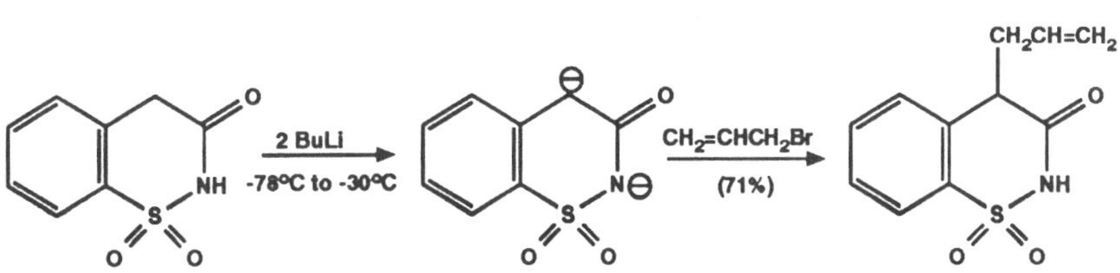

(4.61)

4.8 MISCELLANEOUS AMINE- AND ANILINE-BASED DIANIONS

Included in this section are N,C-dianions that have escaped categorization in Sections 4.2 through Section 4.7 yet warrant entry in this chapter.

4.8.1 Reactions of amine-based dianions

Little exploration has been devoted to simple amine and aniline type N,C-dianion formation and reactivity, although Ludt and Hauser showed over 20 years ago that N-substituted benzylamines undergo ortho-lithiation and reaction with electrophiles (Eq. 4.62).[137]

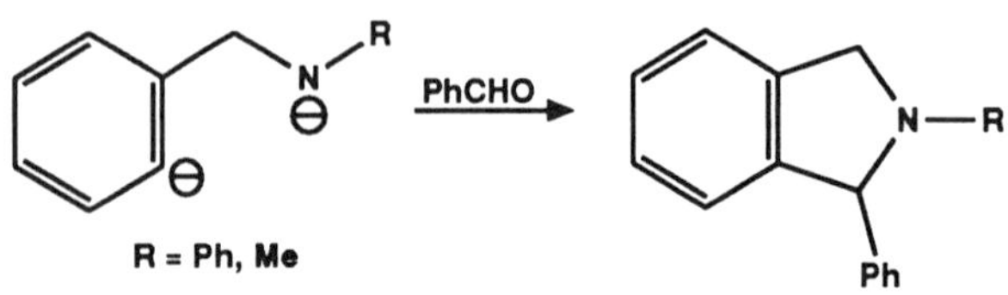

(4.62)

N,C-Dianions derived from initial deprotonation of anilines[138] and aminopyrimidines[139] lead directly to products that otherwise would have required tedious protection/deprotection sequences (Eqs. 4.63 and 4.64).

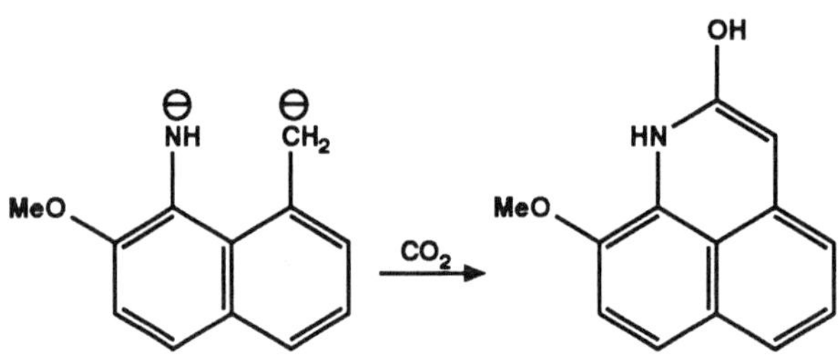

(4.63)

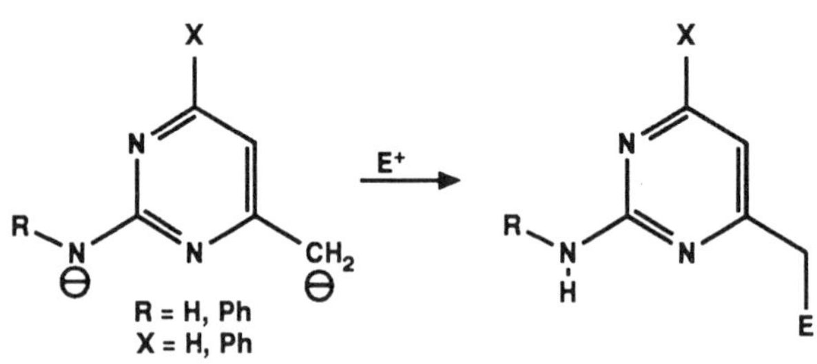

(4.64)

Vicinal N,C-amino dianions of benzophenone anils have been investigated by Smith and coworkers.[140-144] These dianions, however, are derived from alkali metal reduction, and not via N,C-deprotonation. The disodio intermediate undergoes alkylation at carbon,[140] but reaction with acid halides gives rearrangement from the C-acyl to N-acyl product upon warming (Scheme 4.12).[141-143]

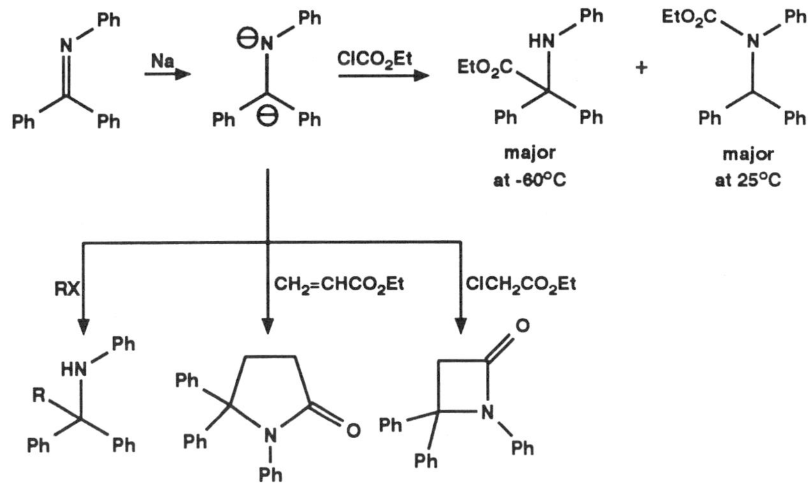

Scheme 4.12. Reaction of benzophenone anil dianions

In similar studies, Mehrotra and coworkers found that C-alkylation occurred with the dianion prepared from 2-imino-1,2-diphenylethanone (Eq. 4.65).[145] Oxazolines, oxazocines, and oxazonines also were prepared in this study via reaction at the N,O-anionic sites.

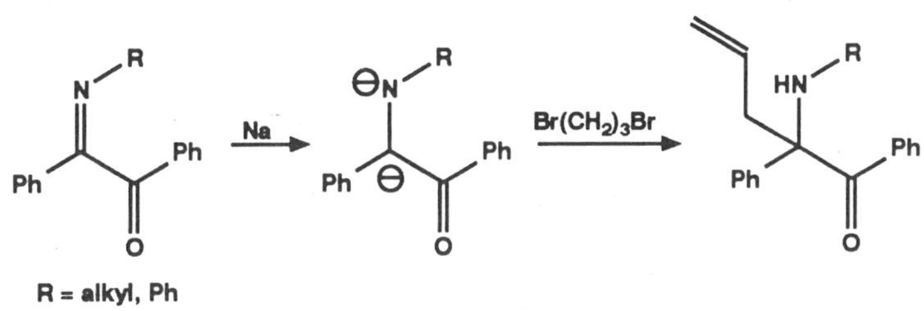

(4.65)

4.9 EXPERIMENTAL PROCEDURES

N-Phenyl-4(hydroxy)hexanamide (Eq. 4.4) - reaction of 1,2-butylene oxide with the dianion of acetanilide using BF_3-Et_2O and the subsequent conversion to δ-lactams (from Takahata, Wang, and Yamakazi).[15]

A solution of n-BuLi (22 mmol) was added to a solution of acetanilide (10 mmol) in THF (10 mL) and HMPA (3.8 mL) at -78°C. The mixture was stirred for 30 min, and BF_3-Et_2O (0.6 mL) and the oxirane were added stepwise at -78°C. The reaction temperature was raised to 0°C over a 2 h period, at which point the reaction was quenched by the addition of aqueous ammonium chloride. The reaction was diluted with ethyl acetate, and the organic layer extracted with brine, and dried over magnesium sulfate. Column chromatography using ethyl acetate/hexane gave the desired γ-hydroxyamide in 80% yield.

An HMPA solution of N-phenyl-4-(hydroxy)hexanamide (1 mmol/7 mL) was added to a suspension of t-BuOK (2.2 mmol) in THF at 0°C and the mixture stirred for 1 h. Toluenesulfonyl chloride (1 mmol) in THF (2 mL) was added and stirred for 1 h at 0°C, then heated at 50°C for 3 h. The reaction was cooled to room temperature and quenched with ammonium chloride. The same workup as above gave 25% of N-phenyl-5-ethyl-γ-lactam.

(S)-(+)- and (R)-(-)-5-octyl-2(5H)-furanone (Eq. 4.13) - reaction of remote carboxamide dianions bearing a chiral auxiliary with aldehydes, and subsequent cyclization to enantioenriched furanones (from Tanaka, Wakita, Yoda, and Kaji).[29]

A solution of N-(S)-α-methylbenzyl-3-(phenylsulfonyl)-propanamide in THF was reacted with 2 equivalents of n-BuLi at -78°C to give a pale yellow solution of the dianion. Nonanal (1.1 equiv.) was added as a solution in THF and the reaction stirred for 2 h. Standard workup gave a 1:1 diastereomeric mixture of N-(S)-α-methylbenzyl-4-hydroxy-3-(phenylsulfonyl)-dodecanamides in a combined yield of 66%. The more chromatographically mobile stereoisomer was isolated as an oil. Treatment of this stereoisomer with dilute hydrochloric acid in refluxing dioxane for 1 h gave the phenylsulfonyl furanone, which without purification, was reacted with triethylamine in chloroform at room temperature for 18 h to afford (S)-(+)-5-octyl-2(5H)-furanone in 84% yield and 98% optical purity.

The less mobile stereoisomer also was carried through an identical procedure to give (R)-(-)-5-octyl-2(5H)-furanone in 84% yield and 100% optical purity.

4-Hydroxy-6-methyl-2-methylene-N-phenyl-3-(phenylsulfonyl)-heptanamide (Scheme 4.2) - reaction between the dianion of N-phenyl-2-[(phenylsulfonyl)methyl]propenamide and aldehydes (from Tanaka, Horiuchi, and Yoda).[33]

To a solution of N-phenyl-2-[(phenylsulfonyl)methyl]-propenamide (1.5 g; 5.0 mmol) in dry THF (40 mL) and HMPA (1.8 mL; 11 mmol) at -78°C was added n-BuLi (11 mmol). After stirring for 1 h, a solution of isovaleraldehyde (5.5 mmol) in THF (5 mL) was added, and the mixture stirred for 2 h at -78°C. The reaction was quenched with 10 mL aqueous ammonium chloride, and diluted with 100 mL water. The product was extracted with ethyl acetate (3 x 50 mL), the extracts combined, washed with brine, and dried over sodium sulfate. The crude product was purified using silica gel chromatography (ethyl acetate: hexane, 1:5) to give a 56% of the product.

1-Benzyl-4-hydroxy-4-(2-pivaloylamino)-5-chlorophenyl piperidine (Eq. 4.17) - reaction between ortho-metalated benzamides and ketones (from Takai et. al.).[40]

An n-BuLi solution (374.4 mmol) was added dropwise to a solution of 1-pivaloylamino-4-chlorobenzene (31.8 g; 150.2 mmol) in THF (750 mL) over 0.5 h while maintaining the temperature at -5 to 0°C. The reaction was held at 0°C for 2 h, whereupon a solution of 1-benzyl-4-piperidone (30 g; 158.5 mmol) in dry THF (45 mL) was added in portions over 1 h. Stirring was continued an additional 3 h at 0°C, and then stirred overnight at room temperature. The solution was partitioned between THF and water, and the organic layer was washed with brine, dried over sodium sulfate, and concentrated in vacuo. The product may be isolated using chromatography (ethyl acetate: hexane, 1:4) or crystallized from the crude material to give 38.5% or 29.1% yield, respectively. mp = 202 -203°C.

2-(N-tert-butylcarbamoyl)thiophene-3-carboxylic acid (Eq. 4.20) - reaction between a lateral N,C(3)-metalated thiophene-2-carboxamide and CO_2 (from Carpenter and Chadwick).[44]

To 2-(N-tert-butylcarbamoyl)thiophene (5 mmol) in THF (50 mL) was added a solution of n-BuLi (11 mmol in hexane) at -78°C for 0.5 h under argon. The solution was then poured with stirring onto a slurry of solid CO_2 and ether, and the mixture was left until the solvents had evaporated. The white solids were taken up in 0.01 M KOH (50 mL), and the solution washed with ethyl acetate (2 x 10 mL). The aqueous layer was acidified to pH 1 with concentrated HCl, saturated with NaCl, and extracted with methylene chloride (5 x 20 mL). The organic extracts were combined, dried over magnesium sulfate, and evaporated to yield the crude acid. Recrystallization from cyclohexane/CH_2Cl_2 gave the pure carboxylic acid (74%) as white needles. mp = 185 - 186°C.

***erythro*-N-phenyl-2,3-dimethylthiopent-4-enamide and *threo*-N-phenyl-2,3-dimethylthiopent-4-enamide** (Scheme 4.5) - diastereoselective α-allylation of thioamides via a thio-Claisen rearrangement, and conditions for stereospecific control of erythro and threo isomers (from Tamaru, Furukawa, Mizutani, Kitao, and Yoshida).[50]

Into a solution of N-phenyl-thiopropanamide (1 equiv.) in THF was added n-BuLi (2.2 equiv. in hexanes) at 0°C. After stirring for 1 h, the mixture was cooled to -78°C and 1.1 equiv. of TMS-Cl was added. The mixture was stirred at -78°C for 0.5 h, and warmed to 0°C. After stirring for 1 h, the solution was brought to -78°C and a THF solution of the alkylating agent (1.1 equiv.; prepared in situ by equimolar reaction of 2-butene-1-ol with n-BuLi in THF, followed by toluenesulfonyl chloride at -78°C) was added. After stirring for 0.5 h, the reaction was warmed to room temperature and subjected to individual thio-Claisen rearrangement conditions (below). After an extractive workup, the products were purified by column or preparative thin-layer chromatography.

Conditions for thio-Claisen rearrangement:

Use of the *trans*-2-butene-1-tosylate to alkylate N-phenyl-thiopropanamide as above leads to a 51% yield of *erythro*-N-phenyl-2,3-dimethylthiopent-4-enamide (less than 1% threo) after stirring at room temperature for 40 h. mp = 66.5 - 67°C.

Use of the *cis*-2-butene-1-tosylate to alkylate N-phenyl-thiopropanamide leads to a 50% yield of *threo*-N-phenyl-2,3-dimethylthiopent-4-enamide (less than 1% erythro) following reflux of the reaction mixture for 32 h. mp = 95 - 95.5°C.

N-tert-butylethanesulfonamide and ethanesulfonamide (Eq. 4.26) - reaction of the α,α'-dianion of an alkanesulfonamide with an alkyl halide and subsequent amide deprotection (from Thompson).[56]

Lithium diisopropylamide was generated by the addition of 17.3 mL (27.7 mmol) of n-BuLi in hexane to a solution of diisopropylamine (30.4 mmol) in THF (20 mL) at -78°C. After stirring briefly at 0°C, the solution was returned to -78°C, and reacted dropwise with N-tert-butylmethanesulfonamide (2.0 g; 13 mmol) in THF (24 mL) over 10 min. The reaction was warmed to -30°C over 45 min to form the dianion, then brought to -78°C, and treated rapidly with methyl iodide (17 mmol). The solution was allowed to come to room temperature, stirred for 12 h, recooled to 0°C, and quenched with 5% HCl. The layers were separated, and the aqueous layer was extracted with ether (3 x 20 mL). The combined organic layers were washed with water, and dried over magnesium sulfate. Removal of the solvent deposited the crystalline product in 73% yield (1.6 g). mp = 53-54.5°C.

The alkylated product, N-tert-butylethanesulfonamide (1.2 mmol) was added to 4 mL of trifluoroacetic acid and stirred at room temperature for 24 h using a drying tube. The trifluoroacetic acid was removed in vacuo to afford a yellow oil that crystallized upon the addition of a mixture of ether-hexane (1:1), v/v). Filtration and drying affords ethanesulfonamide as a white powder (92%). mp = 57-59°C.

***erythro*-(2R,3S)-1-benzyl-4-methyl-N-tert-butoxycarbonyl-3-methylaspartate** (Eq. 4.40) - reaction between a remote carbamate dianion bearing an asymmetric center and an alkyl halide (from Beatty, Jennings-White, and Avery).[104]

A stirred solution of 1,1,1,3,3,3-hexamethyldisilazane (29.4 mmol) in THF (40 mL) was cooled to 0°C and treated with a solution of BuLi in hexane (30.9 mmol). The resulting mixture was stirred for 10 min at 0°C, and then cooled to -78°C. A solution of (2R)-1-benzyl-4-methyl-N-tert-butoxycarbonyl-aspartate (4.5 g; 13.4 mmol) in THF (10 mL) was added dropwise over 20 min and the mixture stirred at -78°C for 1 h. The reaction was warmed to -7°C, methyl iodide (16.4 mmol) was added, the reaction stirred for 1 h, and quenched by pouring the reaction mixture onto saturated aqueous ammonium chloride-ice. The solution was extracted with ether, the organic layer extracted with brine, and dried over magnesium sulfate. Filtration and removal of the solvent left an orange oil that was flash chromatographed using ethyl acetate-hexane (3:1) to give the *erythro* isomer (37%) as a colorless oil. This material was accompanied by 46% of the *threo* isomer. NOTE: if the alkylation of the dianion was conducted at -78°C, 96% of the *threo* isomer can be obtained.

7-hydroxy-4-(N-methylamino)hept-3-en-2-one (Eq. 4.44) - reaction of the γ-dianion of 4-(N-methylamino)pent-3-en-2-one with ethylene oxide (from Bartoli, Bosco, Cimarelli, Dalpozzo, and Palmieri).[111]

A solution of methyl lithium (12.5 mmol) was dropped into a stirred solution of 4-(N-methylamino)pent-3-en-2-one (5 mmol) and TMEDA (12.5 mmol) at 0°C under nitrogen. The temperature was allowed to rise to 20°C, and the mixture was stirred at this temperature for 2 h. The dianion solution was rechilled to -50°C, and a THF solution of ethylene oxide (8 mmol) was added. The ice bath was removed allowing the reaction to warm to room temperature. After 30 min a gel appeared, and the reaction was continued for an additional 1-3 h, whereupon aqueous ammonium chloride was added to dissolve the gel. The reaction was extracted with hexane, dried over sodium sulfate, and purified by column chromatography using ethyl acetate-hexane as eluent (1:19) to give 80% of the desired γ-product. The α′-alkylation product was formed in about 7% yield.

7-hydroxy-7-phenyl-2-(N-phenylamino)hept-2-en-4-one (Eq. 4.44) - reaction of the α′-dianion of 4-(N-phenylamino)pent-3-en-2-one with styrene oxide (from Bartoli, Bosco, Cimarelli, Dalpozzo, and Palmieri).[111]

A solution of LiTMP (12.5 mmol) was added to a stirred THF solution of 4-(N-phenylamino)pent-3-en-2-one (5 mmol) at 0°C under nitrogen. The temperature was allowed to rise to 20°C and the mixture was maintained at these conditions for 2 h. The dianion solution was rechilled to -50°C before the addition of a THF solution of styrene oxide (8 mmol). The procedure then followed that for γ-alkylation as above. The γ-alkylation product was isolated as an oil in 77% yield with no detectable α′-alkylation product.

5-(1-(hydroxymethyl)-2,2-dimethylpropylidene)-2,4-oxazolidinedione (Scheme 4.11) - reaction between the dianion of 2,4-oxazolidinedione and α-haloketones to give rearranged allylic alcohols (from Zask).[134]

To a rapidly stirred solution of lithium chloride (10.1 g; 238 mmol) and 2,4-oxazolidinedione (10.8 g; 39.6 mmol) in THF (400 mL) under nitrogen at -78°C was added t-BuLi (83 mmol) dropwise over 15 min. The resultant yellow reaction mixture was stirred for 15-30 min, warmed briefly to 0°C, and rechilled to -78°C whereupon a solution of 1-bromo-3,3-dimethyl-2-butanone (7.45 g; 41.6 mmol) was added in THF (50 mL) in one portion. After 10 min at -78°C, the mixture was warmed to room temperature, stirred for 5 h, then quenched by the addition of aqueous HCl (100 mL, 1 N). The organic phase was removed in vacuo, and the aqueous phase extracted with ethyl acetate (3 x 100 mL). The organic layers were combined, dried over magnesium sulfate, and concentrated to a solid to give 86% of the allylic alcohol as a powder following trituration and recrystallization from ethyl acetate. mp = 148-150°C.

REFERENCES

1. **Bordwell, F.G.** Equilibrium acidities in dimethyl sulfoxide solution, *Accts. Chem. Res.* **21**, 456 (1988).
2. **Kaiser, E.M., Petty, J.D. and Knutson, P.L.A.** Di- and polyalkali metal derivatives of heterofunctionally substituted molecules, *Synthesis* 509 (1979).
3. **Tanaka, K. and Kaji, A.** The synthetic utility of sulfur-containing dianions, *Sulfur Rep.* **1**, 97 (1980).
4. **Thompson, C.M. and Green, D.L.C.** Recent advances in dianion chemistry, *Tetrahedron* **47**, 4223 (1991).
5. **Kaiser, E.M., Vaulx, R.L. and Hauser, C.R.** Conversions of certain amides into nitriles by means of n-butyllithium, *J. Org. Chem.* **32**, 3640 (1967).
6. **Smith, H.A. and Hauser, C.R.** Novel decarboxamidation of certain primary carboxamides by means of n-butyllithium, *J. Am. Chem. Soc.* **91**, 7774 (1969).
7. **Wolfe, J.F. and Timitsis, G.B.** Twofold arylations of certain amides by means of sodium hydride, *J. Org. Chem.* **33**, 894 (1968).
8. **Creger, P.L.** Metalated carboxylic acids IV. Reactions of metalated carboxylic acids with epoxides. Substituted steroidal spiro-γ-lactones from spiro-β-epoxides, *J. Org. Chem.* **37**, 1907 (1972).
9. **Kaiser, E.M. and Hauser, C.R.** Kinetic versus thermodynamic control in carbonyl addition reactions of carbanions, *Chem. Ind. (London)* 1299 (1965).
10. **Meyer, R.B. and Hauser, C.R.** Alkylations at the α-carbon of phenylacetamide and phenylacetic acid through their disodio salts, *J. Org. Chem.* **26** 3696 (1961).
11. **Lassen, N. and Ammityboll, T.** Synthesis of thiophthalans, *Acta Chem. Scand.* **25**, 2879 (1971).
12. **Ehrig, V. and Seebach, D.** Reacktionen polymetallierter carbonylverbindungen mit nitroolefin. Eine neue ringschluβ- und anellierungsmethode, *Chem. Ber.* **108**, 1961 (1975).
13. **Kuzma, P.C., Brown, L.E. and Harris, T.E.** Generation of the dianion of N-(trimethylsilyl)acetamide and reaction of the dianion with electrophilic reagents, *J. Org. Chem.* **49**, 2015 (1984).
14. **Gay, R.L., Boatman, S. and Hauser, C.R.** Condensations at the methyl group of acetanilide and 6-methyl-2(1)-pyridone by means of n-butyllithium, *Chem. Ind. (London)* 1789 (1965).
15. **Takahata, H., Wang, E-C. and Yamazaki, K.** Reaction of oxiranes with dianions of secondary amides. New entry to 5-alkylsubstituted γ-lactams, *Synth. Commun.* **18**, 1159 (1988).
16. **Barnish, I.T., Hauser, C.R. and Wolfe, J.F.** Formation of 1,3-dianions of 2- and 3-acetamidopyridines by means of n-butyllithium. Condensation with carbonyl compounds and nitriles. *J. Org. Chem.* **33**, 2116 (1968).
17. **Clemence, F., Le Martret, O., Devevallee, F., Benzoni, J., Jouanen, A., Jouquey, S., Mouren, M. and Deraedt, R.** 4-Hydroxy-3-quinolinecarboxamides with antiarthritic and analgesic activities, *J. Med. Chem.* **31**, 1453 (1988).
18. **Work, S.D., Bryant, D.R. and Hauser, C.R.** Condensations at the α-carbon of phenylacetanilide, phenylacetamide, and phenylacetic acid through their dialkali salts in liquid ammonia, *J. Org. Chem.* **29**, 722 (1964).
19. **Hirai, K. and Iwano, Y.** A convenient synthesis of the β-lactam ring by 3 + 1 cyclization, *Tetrahedron Lett.* 2031 (1979).
20. **Depres, J-P., Greene, A.E. and Crabbe, P.** β-Lactam prostaglandins, *Tetrahedron Lett.* 2191 (1978).
21. **Kane, J.M. and Carr, A.A.** Alkylation of protected piperazinone dianions, *Tetrahedron Lett.* **21**, 3019 (1980).
22. **Kende, A.S. and Hodges, J.C.** Regioselective C-3 alkylations of oxindole dianion, *Synth. Commun.* **12**, 1 (1982).
23. **Koppel, G.A.** Direct C_6 epimerization of penicillin V methyl ester via the vicinal dianion, *Tetrahedron Lett.* 4233 (1973).
24. **Dikshit, D.K. and Singh, S.** 1,2-Addition of α-amino acid derivatives to conjugated aldehydes: synthesis of β-substituted serines, *Tetrahedron Lett.* **29**, 3109 (1988).
25. **Krapcho, A.P. and Dundulis, E.A.** Synthetic routes to α-amino acids. C-alkylations of the hippuric acid trianion and the ethyl hippurate dianion, *Tetrahedron Lett.* 2205 (1976).

26. **Evans, D.A. and Sidebottom, P.J.** A simple route to α-aminoketones and related derivatives by dianion acylation reactions; an improved preparation of δ-aminolevulinic acid, *J. Chem. Soc. Chem. Commun.* 753 (1978).
27. **Harding, K.E., Moreno, L.N. and Nace, V.M.** Hydroxyalkylation and acylation reactions of methyl hippurate, *J. Org. Chem.* **46**, 2809 (1981).
28. **Hoppe, I. and Schollkopf, U.** Synthesis of ethyl (E)-2-formylamino-4-methoxy-3-butenoate and its 2-alkyl derivative, *Synthesis* 129 (1982).
29. **Tanaka, K., Wakita, H., Yoda, H. and Kaji, A.** Dianions of N-monosubstituted-3-(phenylsulfonyl)propanamides. Convenient reagents for the synthesis of 5-alkyl-2(5H)-furanone, *Chem. Lett.* 1359 (1984).
30. **Takahata, H., Wang, E-C., Ikuro, K., Yamazaki, T. and Momose, T.** Electrophilic olefin heterocyclization in organic synthesis. Formation of δ-lactams by iodine induced lactamization of δ,ε-unsaturated thioimidates, *Heterocycles* **34**, 435 (1992).
31. **Tanaka, K., Yoda, H. and Kaji, A.** Dianion of N-phenyl-2-(phenylsulfonyl)propenamide: an efficient synthetic reagent for (E)-trisubstituted olefins and 5,6-dihydro-2-pyrans, *Tetrahedron Lett.* **26**, 4747 (1985).
32. **Tanaka, K., Yoda, H. and Kaji, A.** A general method for the synthesis of 3,4-epoxy-2-methylenealkanoic acid acid derivatives using dianion of N-phenyl-2-(phenylsulfonylmethyl)-propenamide, *Tetrahedron Lett.* **26**, 4751 (1985).
33. **Tanaka, K., Horiuchi, H. and Yoda, H.** Dianion of N-phenyl-2-[(phenylsulfonyl)methyl]propenamide as a versatile reagent for the preparation of α-methylene carbonyl compounds, *J. Org. Chem.* **54**, 63 (1989).
34. **Tanaka, K., Ushio, H. and Horiuchi, H.** A new synthetic route to α-methylenecarboxamides using dianion of N-phenyl-2-[(phenylsulfonyl)methyl]propenamide, *Bull. Inst. Chem. Res., Kyoto Univ.* **67**, 128 (1989).
35. **Tanaka, K., Nozaki, Y., Tamura, N., Tanikaga, R. and Kaji, A.** Dianion of N-(o-methoxyphenyl)-2-methylpropenamide: A new reagent for the convenient synthesis of α-methylene-γ-butyrolactones from carbonyl compounds, *Chem. Lett.* 1567 (1980).
36. **Fitt, J.J. and Gschwend, H.W.** Dimetalation of N-tert-butylmethacryamide: A new synthetic reagent, *J. Org. Chem.* **45**, 42, (1980).
37. **Deshmukh, A.R., Zhang, H., Tran, L.D. and Biehl, E.** Tandem-addition rearrangements in the reaction of 2-butenoic acid and N-(4-methoxyphenyl)-2-butenamide with methyl-substituted benzynes, *J. Org. Chem.* **57**, 2485 (1992).
38. **Yates, P. and Schwartz, D.A.** Synthetic approaches to haplophytine. 1. Synthesis of 1-methoxy- and 1-hydroxy-6-phenyl-6-(2-pivaloylamidophenyl)non-8-ene-4,5-dione and 1-methyl-3-phenyl-3-(2-pivaloylamidophenyl)-2-pyrrolidone, *Can. J. Chem.* **61**, 509 (1983).
39. **Goswami, R. and Corcoran, D.E.** Homoenolate dianions. Intramolecular rearrangements of 1,4-dilithio-N-phenyl amides, *J. Am. Chem. Soc.* **105**, 7182 (1983).
40. **Takai, H., Obase, H., Teranishi, M., Karasawa, A., Kubo, K., Shuto, K., Kasuya, Y., Hashikami, M., Karashima, N. and Shigenobu, K.** Spiropiperidines. 1. Synthesis of 1′-substituted spiro[4H-3,1-benzoxazine-4,4′-piperidin]-2-(1H)-one derivatives and evaluation of their antihypertensive activity, *Chem. Pharm. Bull. (Japan)* **33**, 1129 (1985).
41. **Engmann, L. and Hallberg, A.** Expedient synthesis of ebselen and related compounds, *J. Org. Chem.* **54**, 2964 (1989).
42. **Fevig, J.M., Marquis, R.W. Jr. and Overman, L.E.** A new approach to *Strychnos* alkaloids. Stereocontrolled total synthesis of (+/-)-dehydrotubifoline, *J. Am. Chem. Soc.* **113**, 5085 (1991).
43. **Artz, S.P., Finkelstein, B.L., Hanagan, M.A., Moon, M.P., Pasteris, R.J. and Rorer, M.P.** Synthesis and herbicidal activity of bicyclic sulfonylureas, in *Synthesis and Chemistry of Agrochemicals II* (Baker, D.R., Fenyes, J.G. and Moberg, W.K., Eds.), ACS Symposium Series 443, ACS, Washington, DC, 1991, pp. 51-61.
44. **Carpenter, A.J. and Chadwick, D.J.** The scope and limitations of carboxamide induced β-metalation of 2-substituted furan, thiophene, and 1-methylpyrrole derivatives. Application of the method to the synthesis of 2,3-disubstituted thiophenes and furans, *J. Org. Chem.* **50** 4362 (1985).

45. **DeJohn, D., Domagala, J.M., Kaltenbronn, J.S. and Krolls, U.** Functionalization of substituted 2(1H)- and 4((1H)-pyridones. III. The preparation of substituted 6-vinyl-1,2-dihydro-2-oxo- and 1,4-dihydro-4-oxo-3-pyridinecarboxylic acids through the chemistry of pyridone dianions, *J. Heterocyclic. Chem.* **20**, 1295 (1983).
46. **Domagala, J.M.** Functionalization of substituted 2(1H)-pyridones. V. The synthesis of 1,2-dihydro-2-oxo-3,5-pyridinedicarboxylic acid derivatives through a novel dianion route, *J. Heterocyclic Chem.* **21**, 1705 (1984).
47. **Goswami, R. and Corcoran, D.E.** Homoenolate dianions of secondary amides via tin/lithium exchange, *Tetrahedron Lett.* **23**, 1463 (1982).
48. **Tamaru, Y., Kagotani, M., Furukawa, Y., Amino, Y. and Yoshida, Z.** Generation of sec-thioamide dianions and their regioselective reaction with electrophiles, *Tetrahedron Lett.* **22**, 3413 (1981).
49. **Tamaru, Y., Amino, Y., Furukawa, Y., Kagotani, M. and Yoshida, Z.** Stereoselective generation of Z secondary thioamide dianion: application to diastereoselective aldol condensations and thio Claisen rearrangements, *J. Am. Chem. Soc.* **104**, 4018 (1982).
50. **Tamaru, Y., Furukawa, Y., Mizutani, M., Kitao, O. and Yoshida, Z.** Diastereoselective α-allylation of secondary and tertiary thioamides via thio-Claisen rearrangement. A structural proof of Z secondary thioamide dianions and Z tertiary thioamide anions, *J. Org. Chem.* **48**, 3631 (1983).
51. **Takahata, H., Yamazaki, K., Takamatsu, T., Yamazaki, T. and Momose, T.** A facile synthesis of N-protected statine and its analogue via stereoselective iodolactamization, *J. Org. Chem.* **55**, 3947 (1990).
52. **Takahata, H., Ohkura, E., Ikuro, K. and Yamazaki, K.** N-Benzyl-3-(tri-n-butylstannyl)thiopropionamides as a thiohomoenolate synthon. Application to a synthesis of some heterocycles, *Synth. Commun.* **20**, 285 (1990).
53. **Corey, E.J. and Durst, T.** The synthesis of olefins and ketones from carbonyl compounds and sulfinamides, *J. Am. Chem. Soc.* **90**, 5548 (1968).
54. **Bell, S.I., Parvez, M. and Weinreb, S.M.** Metalation and alkylation of 3,6-dihydrothiazine 1-oxides prepared via Diels-Alder cycloadditions of N-sulfinyl dienophiles, *J. Org. Chem.* **56**, 373 (1991).
55. **Corey, E.J. and Chaykovsky, M.** Methylsulfinyl carbanion (CH_3SOCH_2-). Formation and application to organic synthesis, *J. Am. Chem. Soc.* **87**, 1345 (1965).
56. **Thompson, M.E.** N-Alkanesulfonamide dianions: formation and chemoselective C-alkylations, *J. Org. Chem.* **49**, 1700 (1984).
57. **Thompson, M.E.** A one-pot synthesis of 2H-1,2,4-thiadiazin-3(4H)-one 1,1-dioxides and 5,6-dihydro-1,4,3-oxathiazin-2(3H)-one 4,4-dioxides, *Synthesis* 733 (1988).
58. **Grunder-Klotz, E. and Ehrhardt, J-D.** Unusual *C,N*-dialkylation of α,N-alkanesulfonamide dianions: A new route to tetrahydro-1,4,3-oxazathiazine 4,4-dioxides, *Synlett.* 800 (1991).
59. **Szymonifka, M.J. and Heck, J.V.** The preparation and alkylation of β-sultam mono- and dianions, *Tetrahedron Lett.* **30**, 2873 (1989).
60. **Davis, F.A., Weismiller, M.C., Lal, G.S., Chen, B.C. and Przeslawski, R.M.** (Camphorsulfonyl)imine dianion in the synthesis of new optically pure (camphorsulfonyl)oxaziridine derivatives, *Tetrahedron Lett.* **30**, 1613 (1989).
61. **Watanabe, H., Gay, R.L. and Hauser, C.R.** Ortho metalation of N-substituted benzenesulfonamides by excess butyllithium. Condensation with carbonyl compounds, *J. Org. Chem.* **33**, 900 (1968).
62. **Watanabe, H., Mao, C.-L. and Hauser, C.R.** Beckman rearrangment of some benzophenone oximes having an ortho-N-substituted carboxamide or sulfonamide group leading to cyclization, *J. Org. Chem.* **34**, 1786 (1969).
63. **Watanabe, H., Mao, C.-L., Barnish, I.T. and Hauser, C.R.** Ring-chain tautomerism of imine-amine or ketone-carbinol systems obtained by condensation of dilithio-N-substituted-carboxamides or sulfonamides with benzonitrile. Synthesis of cyclic products, *J. Org. Chem.* **34**, 919 (1969).
64. **Lombardino, J.G.** Preparation of substituted 1,2-benzothiazolin-3-one 1,1-dioxides (o-benzoic sulfimides), *J. Org. Chem.* **36**, 1843 (1971).

65. **Thompson, M.E. and Liang, P.H.** Synthesis and herbicidal activity of conformationally restricted butyrolactone sulfonylureas, in *Synthesis and Chemistry of Agrochemicals II* (Baker, D.R., Fenyes, J.G. and Moberg, W.K., Eds), ACS Symposium Series 443, ACS, Washington, DC, 1991, pp. 87-97.
66. **Watanabe, H. and Hauser, C.R.** Metalation at methyl groups of N-substituted o-toluenesulfonamides by excess butyllithium. Condensation with benzophenone, *J. Org. Chem.* **33**, 4278 (1968).
67. **Cuomo, J., Gee, S.K. and Hartzell, S.L.** Synthesis of heterocyclic sulfonamides, in *Synthesis and Chemistry of Agrochemicals II* (Baker, D.R., Fenyes, J.G. and Moberg, W.K., Eds), ACS Symposium Series 443, ACS, Washington, DC, 1991, pp. 62-73.
68. **Graham, S.L. and Scholz, T.H.** Preparation and utility of dianions from N-tert-butylthiophene-2-sulfonamide, *J. Org. Chem.* **56**, 4260 (1991).
69. **Vedejs, E. and Stolle, W.T.** Reductive alkylation of aldehyde tosylhydrazones with organolithium reagents, *Tetrahedron Lett.* 135 (1977).
70. **Henoch, F.E., Hampton, K.G. and Hauser, C.R.** C-alkylations of ketone or aldehyde phenylhydrazones, oximes, and azines having α-hydrogens through 1,4- or 1,6-dianions, *J. Am. Chem. Soc.* **91**, 676 (1969).
71. **Beam, C.F., Foote, R.S. and Hauser, C.R.** Reactions of dilithiooximes, dilithiophenylhydrazones, and trilithiohydrazones with nitriles. Acid cyclization to isoxazoles or pyrazoles, *J. Heterocyclic Chem.* **9**, 183 (1972).
72. **Foote, R.S., Beam, C.F. and Hauser, C.R.** New synthesis of substituted pyrazoles from 1,4-dianions of phenylhydrazones having an α-hydrogen, *J. Heterocyclic Chem.* **7**, 589 (1970).
73. **Beam, C.F., Reames, D.C., Harris, C.E., Dasher, L.W., Holinger, W.M., Shealy, N.L., Sandifer, R.M., Perkins, M. and Hauser, C.R.** New method for the preparation of 4-acylpyrazoles. Reaction of C(α),N-dianions of phenylhydrazones with acid chlorides, *J. Org. Chem.* **40**, 514 (1975).
74. **Reames, D.C., Harris,, C.E., Dasher, L.W., Sandifer, R.M., Hollinger, W.M. and Beam, C.F.** Reactions of C(α),N-dilithiophenylhydrazones with aldehydes. Acid cyclization to 2-pyrazolines, *J. Heterocyclic Chem.* **12**, 779 (1975).
75. **Matsumura, N., Kunugihara, A. and Yoneda, S.** 1,4-Dianion of acetophenone N-ethoxycarbonylhydrazone as a synthetic intermediate, *Tetrahedron Lett.* **24**, 3239 (1983).
76. **Matsumura, N., Kunugihara, A. and Yoneda, S.** Reaction of α-chloroketones with 1,4-dianion of acetophenone N-ethoxycarbonylhydrazone, *Tetrahedron Lett.* **25**, 4529 (1984).
77. **Duncan, D.C., Trumbo, T.A., Almquist, C.D., Lentz, T.A. and Beam, C.F.** The preparation of N-carboalkoxypyrazoles and N-phenylpyrazoles from C(α)-dianions of carboalkoxyhydrazones and phenylhydrazones, *J. Heterocyclic Chem.* **24**, 555 (1987).
78. **Shapiro, R.H.** Alkenes from tosylhydrazones, *Org. React.* **23**, 405 (1976).
79. **Chamberlin, A.R. and Bloom, S.H.** Lithioalkenes from arenesulfonylhydrazones, *Org. React.* **39**, 1 (1990).
80. **Kolonko, K.J. and Shapiro, R.H.** Elimination of tertiary α-hydrogens from tosylhydrazones with lithium diisopropyl amide: preparation of trisubstituted alkenes, *J. Org. Chem.* **43**, 1404 (1978).
81. **Stemke, J.E. and Bond, F.T.** Efficient synthesis of deuterated olefins from arenesulfonyl hydrazones, *Tetrahedron Lett.* 1815 (1975).
82. **Shapiro, R.H., Lipton, M.F. and Kolonko, K.J., Buswell, R.L. and Capuano, L.A.** Tolylsulfonyl hydrazones and alkyllithium reagents. Regiospecifity of the reaction and the trapping of three intermediates, *Tetrahedron Lett.* 1811 (1975).
83. **Stemke, J.E., Chamberlin, A.R. and Bond, F.T.** A convenient route to vinyllithium reagents, *Tetrahedron Lett.* 2947 (1976).
84. **Traas, P.C., Boelens, H. and Takken, H.J.** A versatile synthesis of α,β-unsaturated aldehydes from tosylhydrazones, *Tetrahedron Lett.* 2287 (1976).
85. **Taylor, R.T., Degenhardt, C.R., Melega, W.P. and Paquette, L.A.** Direct conversion of ketones to vinylsilanes, germanes, and -stannanes, *Tetrahedron Lett.* 159 (1977).
86. **Bunnell, C.A. and Fuchs, P.L.** Synthesis of β,γ-unsaturated esters. Direct generation of ester dienolates from β-keto ester tosylhydrazone trianions, *J. Am. Chem. Soc.* **99**, 5184 (1977).

87. **Chamberlin, A.R., Stemke, J.E. and Bond, F.T.** Vinyllithium reagents from arenesulfonylhydrazones, *J. Org. Chem.* **43**, 147 (1978).
88. **Chamberlin, A.R. and Bond, F.T.** Regiochemical control in alkenyllithium generation from arenesulfonylhydrazones, *Synthesis* 44 (1979).
89. **Lipton, M.F. and Shapiro, R.H.** Regiospecific synthesis of homoallylic alcohols from tosylhydrazones, *J. Org. Chem.* **43**, 1409 (1978).
90. **Adlington, R.M. and Barrett, A.G.M.** The preparation and reactions of 1-lithio-alkylamino-1-lithio-oxy-allene derivatives, *Tetrahedron* **37**, 3935 (1981).
91. **Bond, F.T. and DiPietro, R.A.** Alkylation and subsequent decomposition of arenesulfonylhydrazone dianions, *J. Org. Chem.* **46**, 1315 (1981).
92. **Adlington, R.M. and Barrett, A.G.M.** Novel synthesis of α-methylene-γ-lactones, *J. Chem. Soc. Chem. Commun.* 1071 (1978).
93. **Adlington, R.M. and Barrett, A.G.M.** Novel synthesis of 3-methylene- and 3,6-dimethylene-tetrahydropyran-2-one and 3,5-dimethylenetetrahydrofuran-2-one derivatives, *J. Chem. Soc. Chem. Commun.* 1122 (1979).
94. **Nakai, T. and Mimura, T.** A facile procedure for regioselective 1,2-carbonyl transposition. One-pot conversion of ketone tosylhydrazones to enol thioethers of transposed ketones, *Tetrahedron Lett.* 531 (1979).
95. **Mimura, T. and Nakai, T.** Regiochemical control in the 1,2-carbonyl transposition of α,α'-dimethylene systems through regioselective sulfenylation of tosylhydrazone derivatives, *Chem. Lett.* 931 (1980).
96. **Kano, S., Yokomatsu, T., Ono, T., Hibino, S. and Shibuya, S.** A new synthesis of phenylalkynes from the p-toluenesulfonylhydrazone of α-methylthioacetophenone, *Synthesis* 305 (1978).
97. **Sharp, J.T. and Skinner, C.E.D.** The generation and reactions of C,N-dianions of aromatic tosylhydrazones: ortho-N-dilithiated benzophenone tosylhydrazone, *Tetrahedron Lett.* **27**, 869 (1986).
98. **Nakai, T. and Mikami, K.** Regio- and stereoselectivity in alkylations of dianions derived from allylic dithiocarbonate p-toluenesulfonylhydrazones, *Chem. Lett.* 469 (1979).
99. **Stanetty, P., Koller, H. and Mihovilovic, M.** Directed ortho-lithiation of phenylcarbamic acid 1,1-dimethylethyl ester (N-Boc-aniline). Revision and improvements, *J. Org. Chem.* **57**, 6833 (1992).
100. **Casara, P. and Metcalf, B.W.** Trimethylsilylacetylene-N-carboethoxy glycinate dianion - a general synthon for α-acetylenic α-amino acids, *Tetrahedron Lett.* 1581 (1978).
101. **Metcalf, B.W. and Casara, P.** Synthetic access to a α-substituted prop-2-ynylamines and α-acetylenic amino acids via the t-butyl N-trimethylsilylprop-2-ynyl carbamate dianion, *J. Chem. Soc. Chem. Commun.* 119 (1979).
102. **Manfre, F., Kern, J-M. and Biellmann, J-F.** Synthesis of proline analogues as potential mechanism-based inhibitors of proline dehydrogenase: 4-methylene-*L*-, (E)- and (Z)-4-(fluoromethylene)-*L*-, cis- and trans-5-ethynyl-(+/-), and cis- and trans-5-vinyl-*L*-proline, *J. Org. Chem.* **57**, 2060 (1992).
103. **Muchowski, J.M. and Venuti, M.C.** Ortho functionalization of N-(tert-butoxycarbonyl)aniline, *J. Org. Chem.* **45**, 4798 (1980).
104. **Beatty, M.F., Jennings-White, C. and Avery, M.A.** Stereocontrolled synthesis of (2S,3S,8S,9S,4E,6E)-3-amino-9-methoxy-2,6,8-trimethyl-10-phenyldeca-4,6-dienoicacid(Adda), the amino acid characteristic of microcystins and nodularin, *J.C.S. Perkin Trans. 1*. 1637 (1992).
105. **Barner, B.A. and Mani, R.S.** Carbamate dianions; generation and alkylation of α-oxo carbanions, *Tetrahedron Lett.* **30**, 5413 (1989).
106. **Zimmerman, W.T., Hillemann, C.L., Selby, T.P., Shapiro, R., Tseng, C.P. and Wexler, B.A.** Heterocyclic amine precursors to sulfonylurea herbicides, in *Synthesis and Chemistry of Agrochemicals II* (Baker, D.R., Fenyes, J.G. and Moberg, W.K., Eds), ACS Symposium Series 443, ACS, Washington, DC, 1991, pp. 74-86.
107. **Demir, A.S. and Enders, D.** Regioselective alkylation of dianions of cyclic 1,3-diketones via their mono-dimethylhydrazones. Synthesis of C-4 substituted cyclic 1,3-diketones, *Tetrahedron Lett.* **30**, 1705 (1989).

108. **Bartoli, G., Bosco, M., Cimarelli, C., Dalpozzo, R. and Palmieri, G.** C-Alkylation of dianions of β-(isopropylamino)-α,β-enones, *SynLett.* 229 (1991).
109. **Bartoli, G., Cimarelli, C., Palmieri, G., Bosco, M. and Dalpozzo, R.** Unexpected elimination to α,β-alkynylketones in the reaction of dianions of 1-arylenaminones with trimethylchlorosilane, *Tetrahedron Lett.* **32**, 7091 (1991).
110. **Bartoli, G., Bosco, M., Cimarelli, C., Dalpozzo, R., Guerra, M. and Palmieri, G.** Selectivity in C-alkylation of dianions of acyclic β-enamino ketones, *J. Chem. Soc. Perkin Trans. 2.* 649 (1992).
111. **Bartoli, G., Bosco, M., Cimarelli, C., Dalpozzo, R. and Palmieri, G.** Reaction of dianions of acyclic β-enamino ketones with electrophiles. Part 2. Oxiranes: synthesis of γ'- and ε-hydroxy-β-enamino ketones and of α-tetrahydrofurylidene ketones, *J. Chem. Soc. Perkin Trans. 2,* 2095 (1992).
112. **Bartoli, G., Bosco, M., Cimarelli, C., Dalpozzo, R., De Munno, G., Guercio, G. and Palmieri, G.** Reaction of dianions of acyclic β-enamino ketones with electrophiles. 3. Nitriles: synthesis of pyridine and pyrimidine derivatives, *J. Org. Chem.* **57**, 6020 (1992).
113. **Chen, Y.L., Mariano, P.S., Little, G.M., O'Brien, D. and Huesmann, P.L.** Selectivity in β-enamino ketone anion generation and alkylation, *J. Org. Chem.* **46**, 4643 (1981).
114. **Bartoli, G., Cimarelli, C., Palmieri, G. and Rafaiani, G.** Stereoselective carbon-carbon bond formation: γ-Alkylation of lithium dianion of chiral cyclic β-enamino ketones, *Tetrahedron Asymmetry* 719 (1992).
115. **Inagaki, S., Nishizawa, Y., Sugiura, T. and Ishihara, H.** Functionalization of the methyl group of 2-methylindole by direct generation of a C,N-dianion, *J. Chem. Soc. Perkin Trans. I.* 179 (1990).
116. **Bryson, T.A., Roth, G.A. and Jing-Hau, L.** Alkylation of substituted pyrrole dianions: synthetic studies directed at the B/C ring system of CC-1065, *Tetrahedron Lett.* **27**, 3685 (1986).
117. **Kaiser, E.M. and Knutson, P.L.A.** New base-catalyzed preparation of certain quinolizinium salts, *Tetrahedron Lett.* 3583 (1975).
118. **Meyers, A.I. and Lowe, M.F.** The N-9, C-1 dianion of tetrahydro-β-carbolines. Regioselective alkylations leading to indole alkaloid systems, *Tetrahedron Lett.* **25**, 2641 (1984).
119. **Levy, J., Laronze, J-Y, Laronze, J. and Le Men, J.** A synthesis of tabersonine, *Tetrahedron Lett.* 1579 (1978).
120. **Hay, J.A., Portlock, D.E. and Wolfe, J.F.** Dimetalated heterocycles as synthetic intermediates. IV. Dilithio derivatives of 2-methylbenzimidazole, 2-benzylbenzimidazole, and related compounds, *J. Org. Chem.* **38**, 4379 (1973).
121. **Welch, W.M., Hanau, C.E. and Whalen, W.M.** A novel synthesis of 3-substituted indazole derivatives. *Synthesis* 937 (1992).
122. **Work, S.D., Bryant, D.R. and Hauser, C.R.** Condensations of the methyl groups of N-acetylbenzamide and diacetylimide by means of potassium amide in liquid ammonia, *J. Am. Chem. Soc.* **86**, 872 (1964).
123. **Wolfe, J.F. and Mao, C-L.** Formation and reaction of trisodio-N-acetylsalicylamide, *J. Org. Chem.* **32**, 3382 (1967).
124. **Wolfe, J.F., Trimitsis, G.B. and Hauser, C.R.** C-Aroylations of imides and N-aroylations of β-keto amides with aromatic esters by means of sodium hydride to form β-keto imides, *Can. J. Chem.* **46**, 2561, (1968).
125. **Taylor, J.D., Trimitsis, G.B., Hudlicky, T. and Wolfe, J.F.** Selective C-alkylation of phenylacetylureas through 1,3,5-trialkali salt intermediates, *J. Org. Chem.* **38**, 1236 (1973).
126. **Wolfe, J.F. and Mao, C-L.** Trialkali salts of N-acetyl-α-benzoylacetamide. Condensations with certain electrophilic reagents, *J. Org. Chem.* **32**, 1977 (1967).
127. **Wolfe, J.F., Mao, C-L., Bryant, D.R. and Hauser, C.R.** Formation and alkylation of di- and trialkali salts of 2,2'-diphenyldiacetamide, *J. Org. Chem.* **31**, 3725 (1966).
128. **Kashima, C., Yamamoto, M., Kobayashi, N. and Sugiyama, N.** Reaction of dicarboxamides with benzonitrile in liquid ammonia, *Bull. Chem. Soc. Japan* **47**, 1805 (1974).
129. **Taylor, J.D. and Wolfe, J.F.** New method for the preparation of 5-alkyl- and 5,5-dialkyl-1,3-thiazolidine-2,4-diones through dianion intermediates, *Synthesis* 310 (1971).
130. **Zask, A., Jirkovsky, I., Nowicki, J.W. and McCaleb, M.L.** Synthesis and antihyperglycemic activity of novel 5-(naphthalenylsulfonyl)-2,4-thiazolidinediones, *J. Med. Chem.* **33**, 1418 (1990).

131. **Wolfe, J.F. and Rogers, T.G.** Dianions derived from glutarimide, 3,5-morpholinedione, and 3,5-thiomorpholinedione as useful new synthetic intermediates, *J. Org. Chem.* **35**, 3600 (1970).
132. **Bryant, D.R. and Hauser, C.R.** Condensations at the α-carbon of succinimide through its dipotassio salt prepared by means of potassium amide, *J. Am. Chem. Soc.* **83**, 3468 (1961).
133. **Zask, A., Nowicki, J.W. and Jirkovsky, I.** Synthesis of 3-mercapto-2(5H)-furanones via reaction of dilithio-2,4-thiazolidinedione with α-haloketones, *Tetrahedron Lett.* **34**, 2719 (1993).
134. **Zask, A.** Preparation and reaction of dilithio-2,4-oxazolidinedione with α-halo ketones. A versatile synthesis of 3-hydroxy-2(5H)-furanones, *J. Org. Chem.* **57**, 4558 (1992).
135. **Belletire, J.L. and Mahmoodi, N.O.** Dianion-based methodology for the preparation of 2,3-disubstituted butyrolactones, *Synth. Commun.* **19**, 3371 (1989).
136. **Hermann, C.K.F., Campbell, J.A., Greenwood, T.D., Lewis, J.A. and Wolfe, J.F.** Aromatic heteroannulation via metalation-cyclization of N-acyl-2-chlorobenzenesulfonamides and N-acylbenzenesulfonamides, *J. Org. Chem.* **57**, 5328 (1992).
137. **Ludt, R.E. and Hauser, R.E.** The effect of tetramethylenediamine on the metalation of N-methyl- and N-phenylbenzylamine with n-butyllithium. Deuteration and electrophilic condensations of intermediate lithioamines. Cyclodehydrations to give N-substituted isoindolines, *J. Org. Chem.* **36**, 1607 (1971).
138. **Weber, R.W. and Cook, J.M.** Synthesis of 9-methoxy-1,6-diazaphenalene, *Heterocycles* **19**, 2089 (1982).
139. **Murray, T.P., Hay, J.V., Portlock, D.E. and Wolfe, J.F.** Dimetalated heterocycles as intermediates. V. Dianions derived from certain 2-hydroxy-4-methylpyrimidines, 2-amino-4-methylpyrimidines and related compounds, *J. Org. Chem.* **39**, 595 (1974).
140. **Smith, J.G. and Veach, C.D.** The reaction of the disodium adduct of benzophenone anil, *Can. J. Chem.* **44**, 2245 (1966).
141. **Smith, J.G. and Simpson, G.E.F.** Rearrangements during the acylation of vicinal dianions, *Tetrahedron Lett.* 3295 (1971).
142. **Smith, J.G., Ho, I. and Simpson, G.E.F.** Acylation of vicinal dianions. Formation of products by rearrangement and proton transfer, *J. Org. Chem.* **40**, 495 (1975).
143. **Smith, J.G. and Simpson, G.E.F.** Acylation of vicinal dianions. Further observations concerning proton transfer and rearrangements during reaction, *J. Org. Chem.* **41**, 2878 (1976).
144. **Smith, J.G. and Mitchell, D.J.** Single electron transfer as a mechanistic feature in the alkylation of vicinal dianions, *J. Am. Chem. Soc.* **99**, 5045 (1977).
145. **Mehrotra, K.N., Singh, I.S. and Roy, J.** Cyclization of dianions from 2-imino-1,2-diphenylethanone to oxazolines, oxazocines, and oxazonines, *Bull. Chem. Soc. Japan* **58**, 2399 (1985).

Chapter 5

DIANIONS FROM S,C-BIS DEPROTONATIONS

5.1 INTRODUCTION

Dianions that result from first removal of a sulfhydryl or RSH proton (S,C-deprotonation) can be considered as analogues of hydroxyl-based O, C-deprotonation (Chapter 3), although far fewer investigations of S,C-dianion intermediates have been conducted. In part, this is due to limited access to certain thiols, and the reactivity of the resultant anionic sulfur, which competes for alkylation. However, certain thiol-based dianions and the correct choice of reaction conditions offer great advantages toward molecule manipulation. Several articles and books have reviewed the utility of sulfur-containing compounds in organic synthesis,[1-3] one specifically covering the use of sulfur-containing dianions.[4] In this section, the various types of sulfur-carbon bis-deprotonations that lead to new carbon-carbon bond formation and novel reactions (including some that were previously reviewed), will be summarized.

There are two predominant classes of molecules that form dianions resulting from first deprotonating an acidic-SH moiety; thiols (RSH) and thioic acids (RC(S)SH). Thiol-based dianions can be further divided into three major categories: (a) "vicinal delocalized," which result from allyl mercaptan type bis-deprotonations (Eq. 5.1), (b) "vicinal fixed," which result

Thiol-based dianions

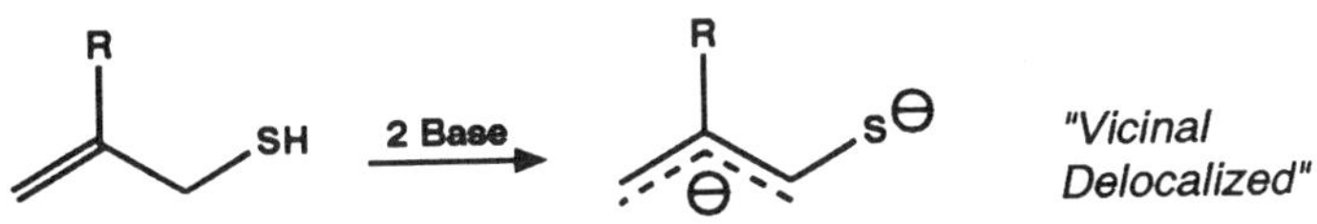

(5.1)

(5.2)

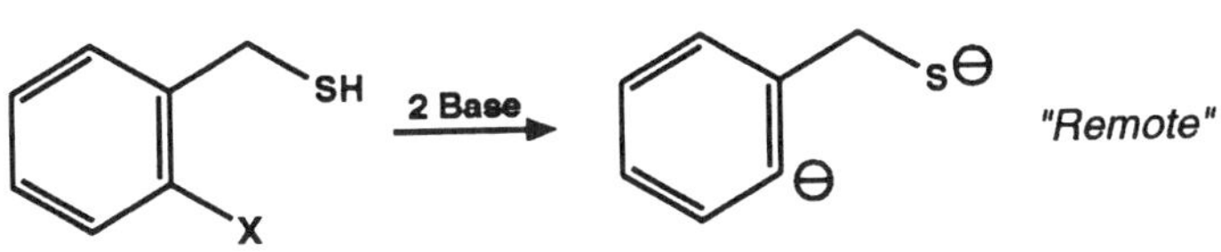

(5.3)

EWG = electron withdrawing group; X = halogen

Dithioic Acid-Based Dianions

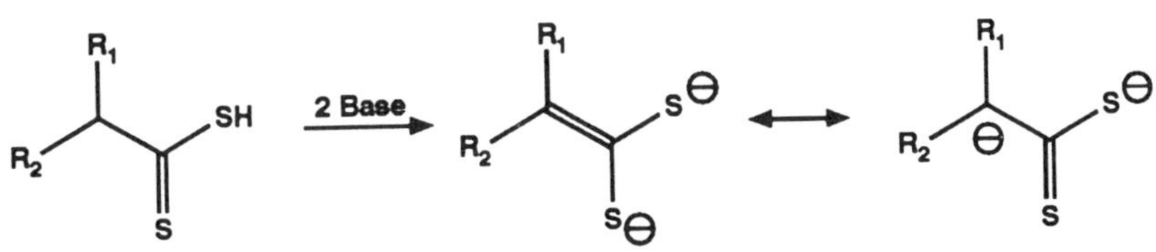

(5.4)

from stepwise thiol and α-stabilized C-H deprotonations (Eq. 5.2), and (c) "remote," which form from sequential thiol abstraction and metallation at a carbon position elsewhere in the molecule (shown for metal-halogen exchange, Eq. 5.3). "Vicinal delocalized" dianions afford resonance delocalized molecules that may undergo reaction at the vicinal (α) or terminus (γ) position, whereas "vicinal fixed" dianions, stabilized by a neighboring electron withdrawing group (e.g., carboalkoxy, aryl, etc.), undergo most reactions at the vicinal position. The dithioic acid-based dianions, existing predominantly in the dithiomethylene form (Eq. 5.4), may be considered analogous to α-carboxylate dianions although the resultant chemistry is quite different. Specifically, dithiomethylene dianions alkylate solely at the sulfur atoms.

5.2 THIOL-BASED DIANIONS

5.2.1 Generation of thiol-based S,C-dianions

The reactivity and regiochemical outcome (product ratio) of reactions with dianions of allyl mercaptan or methallyl mercaptan are affected by the countercation, temperature and additives (e.g., HMPA or TMEDA). Similar differences in experimental conditions for the proton abstractions of allylic alcohols were previously noted (Chapter 3). Owing to the absence of a enolate-type, electron-stabilizing group, the dianions of allylic mercaptans may be prepared with alkyllithium bases. For example, the yellow-red dilithio allyl mercaptan solutions were generated with 2 equivalents of n-BuLi that varied in HMPA/TMEDA content in THF at 0°C for 4 h. Alkylations were conducted at 0°C or -100°C. Ensuing studies found that the magnesium derivative of the dianion, formed by adding $MgBr_2$ to the preformed dilithio species, showed dramatically enhanced α-reactivity. Conditions for the preparation of the dilithio- and magnesium-allyl mercaptan dianions are provided in Section 5.4.

For vicinal fixed dianions, a second stabilizing moiety is present, and an increased variety in the choice of base is possible. When the stabilizing group is an ester or lactone, the vicinal dianion is best generated with nonnucleophilic bases, for example, 2.2 equivalents of both LDA and TMEDA in THF at -78°C. When the stabilizing group is aryl, the conditions outlined for allyl mercapto dianions were employed.

Remote S,C-dianions, thus far relegated to metal-halogen exchange, have been generated by reaction of the desired mercapto halo-precursor with 2 equivalents of n-BuLi. In almost all transformations involving S,C-dianions, the carbanion site is reacted with an electrophile followed by conversion of the sulfhydryl anion to the thiolether by reaction with methyl iodide or a related suitable blocking group. This strategy both suppresses oxidation to the disulfide, and permits further elaboration of the sulfur-containing functional group.

5.2.2 Reactions of "vicinal delocalized" S,C-dianions

Dianions formed from allyl mercaptan (R = H; thioacrolein) and methallyl mercaptan (R = Me) readily react with electrophiles (Eq. 5.5). When the dianion is generated with 2 BuLi, reaction occurs predominantly at the γ-position, although the amount of additives greatly affects the product ratio.[5] For example, the reaction of the allyl mercaptan dianion with n-butylchloride (2 equiv.) affords α:γ ratios from 4:6 to 15:85 depending upon the amount of HMPTA and potassium t-butoxide added.[5] Alkyl halides, aldehydes, ketones, and epoxides all were successfully condensed with these dianions. One exceptional experimental variation proved to expand the versatility of these dianions. When the dilithio cations were exchanged for a magnesium, the regiochemistry was completely reversed giving greater than 9:1, α:γ selectivity.[6] The range of electrophiles was expanded to α,β-unsaturated carbonyls, which undergo 1,2-addition. Overall, the regiochemistry can be controlled by the countercation (Table 5.1).

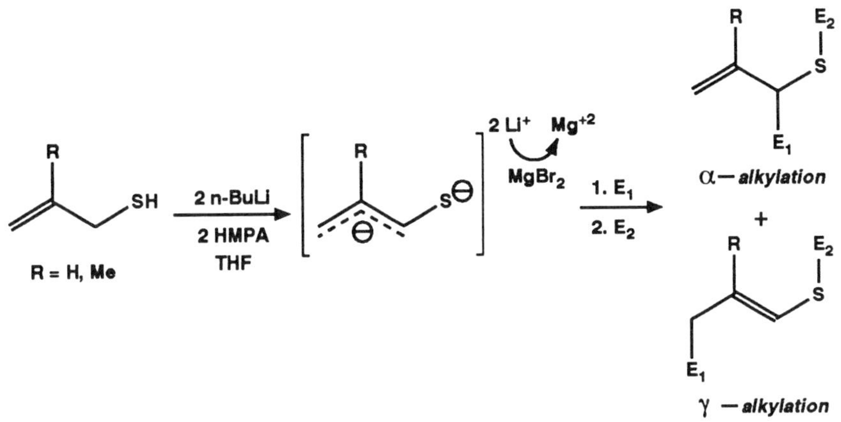

(5.5)

Table 5.1
Regiochemistry of Vicinal Delocalized S,C-Dianions Reactions (Eq. 5.5)

Cation	R	E_1	E_2	α:γ ratio[a]	Yield (%)	Ref.
$2Li^+$	H	n-BuCl	n-BuCl	45:55	88	5
$2Li^+$	H	n-$C_5H_{11}Cl$	MeI	35:65	80	5
$2Li^+$	H	PhCHO	MeI	18:82	83	5
$2Li^+$	Me	PhCHO	MeI	27:73	80	5
$2Li^+$	H	EtCHO	MeI	26:74	69	5
Mg^{+2}	H	EtCHO	MeI	91:9	75	6
$2Li^+$	Me	EtCHO	MeI	46:54	80	5
Mg^{+2}	Me	EtCHO	MeI	88:12	75	6
$2Li^+$	H	PhC(O)Me	MeI	30:70	95	5
$2Li^+$	H	MeSSMe	MeI	20:80	70	5
$2Li^+$	H	n-decyl-Br	n-decyl-Br	23:77	90	5
$2Li^+$	Me	n-octyl-Br	MeI	31:69	68	5
Mg^{+2}	H	i-PrCHO	MeI	98:2	87	6
Mg^{+2}	H	cyclopentanone	MeI	94:6	76	6
Mg^{+2}	H	methyl vinyl ketone	MeI	100:0	73	6

[a] Ratios vary depending upon amount of HMPTA and t-BuO^-K^+ added; see refs.

Successful control over the regiochemistry led to further utility of the allylic thiolethers as precursors in oxirane synthesis.[6,7] Sequential reaction of a carbonyl compound and MeI with the magnesium salt of the dianion leads to β-hydroxy methyl thiolethers (Eq. 5.6). The thioether moiety is then alkylated with another equivalent of MeI (or Meerwein's reagent) to give the sulfonium salt, which undergoes an intramolecular ring closure upon treatment with base to afford vinyl oxiranes in 48-82% yield (Eq. 5.6).

Ingenious recognition of the divinyl β-hydroxythiolethers (formed from addition of the dianion to α,β-unsaturated carbonyls) as precursors for a Cope rearrangement led to another useful extension of S,C-dianions. Under mild conditions, the deprotonated, divinyl derivatives (Eq. 5.7) undergo rearrangement in refluxing THF over 1-7 h in 30-40% yield. The rearrangement also occurs at room temperature when HMPA is added (60-70%).

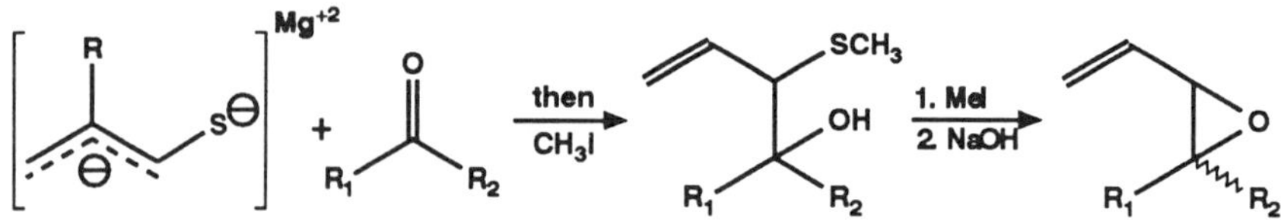

(5.6)

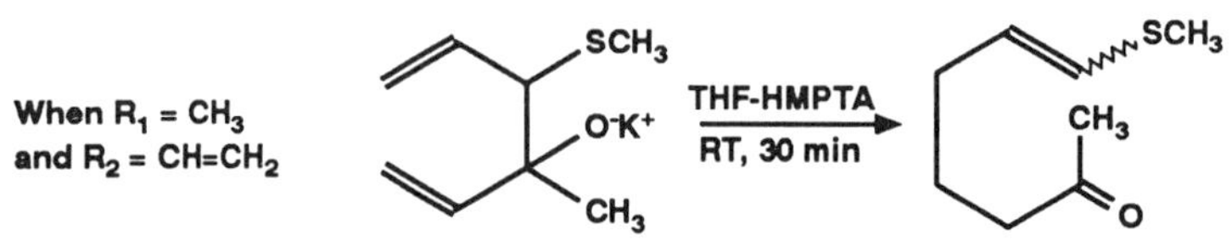

(5.7)

5.2.3 Reactions of vicinal "fixed" S,C-dianions

Benzyl mercaptan (thiobenzyl alcohol) falls in this S,C-dianion category by virtue of the fact that the phenyl ring serves to both stabilize and localize the reactivity of the carbanion site. However, this dianion also holds many features in common with the vicinal delocalized S,C-dianions (Section 5.2.2.), and should be viewed similarly.[5]

Reaction of benzyl mercaptan (Li+ salt) with n-BuLi at -5°C in the presence (2.5 h) or absence (4.0 h) of TMEDA forms the orange-colored dianion.[5,8] The benzyl mercaptan dianion can be regarded as a thiobenzaldehye dianion. This dianion undergoes reaction with carbonyls, trimethylsilyl chloride, alkyl halides, epoxides, 1,3-dibromopropane and epichlorohydrin, the latter two sequences yielding tetrahydrothiophenes (Eq. 5.8).[5,8]

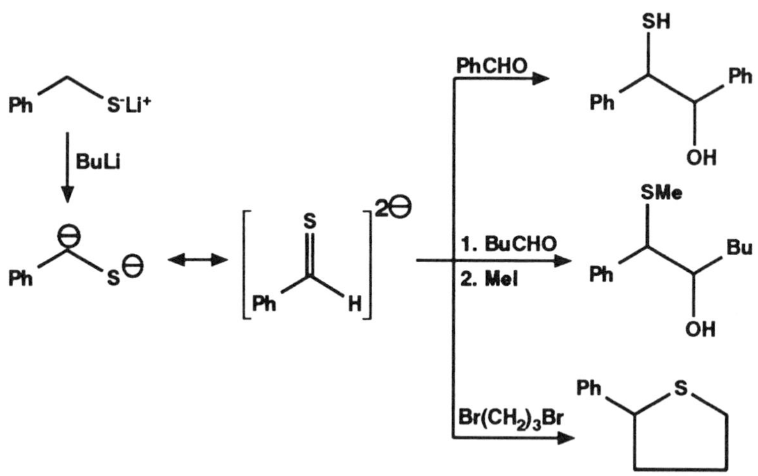

(5.8)

Carboxylester groups provide both α-carbanion stabilization and a versatile functional group for further manipulation following reaction of the fixed S,C-dianion. The most widely used examples of these dianions, ethyl α-mercaptoacetate,[9,10] ethyl α-mercaptoproprionate,[9,10] and α-mercapto-γ-butyrolactone,[11] all give C,S-bis-alkylation products. When the acyclic dianions are condensed with carbonyls, α-mercapto-β-hydroxyesters are formed, which following reaction with ethyl chloroformate, give good yields of α-alkylidene esters or lactones as shown in Table 5.2 (Eq. 5.9).[4,9-14]

The process proceeds through first formation of a thiirane, which undergoes spontaneous or thiophile-promoted (e.g., EtO_3P) sulfur extrusion to form the alkene. Although aldehydes react with the ethyl α-mercaptoacetate dianion to give low diastereoselectivity, the alkene products were formed in good yield and greater than 85% (E)-stereoselectivity. The alkene stereochemistry was found to be independent of the original diastereomer stereochemistry. The mechanism of the reaction has been studied in detail.[14]

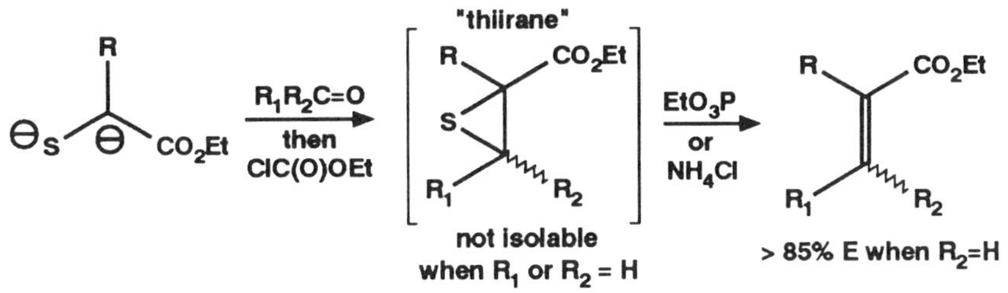

(5.9)

Table 5.2
Synthesis of α,β-Unsaturated Esters by Reaction of Ethyl α-Mercaptoester Dianions with Carbonyl Compounds (Eq. 5.9)

R	R_1	R_2	E:Z ratio	Yield (%)	Ref.
H	n-Pr-	H	100:0	37	12
H	n-Pentyl-	H	92:8	51	14
H	n-Hexyl-	H	95:5	19	12
Me-	n-Hexyl-	H	58:42	55	12
H	Cyclohexyl	H	100:0	35	14
H	Ph-	H	100:0	57	12
Me-	Ph-	H	53:47	45	12
H	Me-	Et	43:57	64	9,12
Me-	Me-	Et	23:77	59	12
H	$-(CH_2)_5-$		n/a	60	9,12
Me-	$-(CH_2)_5-$		n/a	43	9,12
H	Ph-	Me	100-97:0-3	58-61	12,14
H	$(Me)_2C{=}CHCH_2CH_2-$	Me	38:62	70	9

Better stereoselective outcomes were observed following condensation of the dianion of α-mercapto-γ-butyrolactone with aldehydes giving greater than 80:20 E/Z ratios (Eq. 5.10) of the α-alkylidene lactone (Table 5.3).[11-14]

The non-symmetric ketones, 2-butanone and acetophenone gave approximate 1:1 and 100:0 E/Z ratios, respectively. The monoanion of the thiocarbonate of α-mercapto-γ-butyrolactone was prepared and found to afford the same products in comparable yield and stereoselectivity (Eq. 5.11).[12]

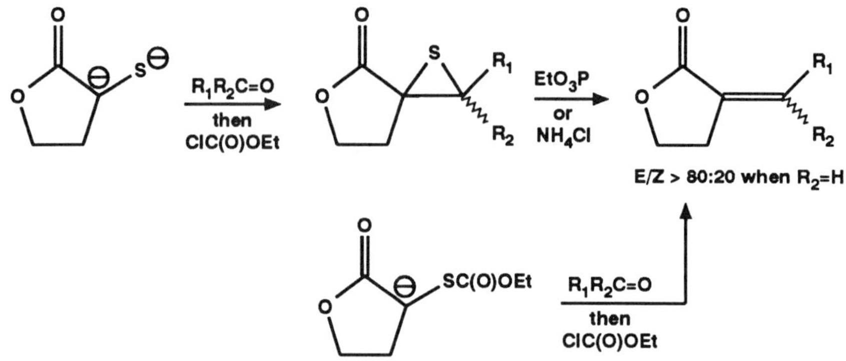

(5.10)

(5.11)

Table 5.3
α-Alkylidene-γ-butyrolactones from the reaction of α-Mercapto-γ-butyrolactone Dianion with Carbonyl Compounds (Eq. 5.10)

R_1	R_2	E:Z ratio	Yield (%)	Ref.
H	H	n/a	7	12
Me-	H	92:8	30	12
n-Bu-	H	95:5	69	11
Ph-	H	100:0	54	11
i-Pr-	H	95:5	61	11
Cyclohexyl-	H	95:5	59	11
Me-	Me-	n/a	50	11
Ph-	Me-	100:0	40	11
$-(CH_2)_4-$		n/a	43	11
(E)-n-Pr-	H	78:22	45	13

5.2.4 Reactions of remote S,C-dianions

Few reactions of remote S,C-dianions have appeared. However, Equation 5.12 shows that sulfur-based remote dianions are quite useful intermediates. In this instance, the authors took advantage of the sulfur nucleophilicity (following condensation with a carbonyl at the remote site) to form a spiro-isobenzotetrahydrothiophene via the intermediate hydroxy sulfide.[15]

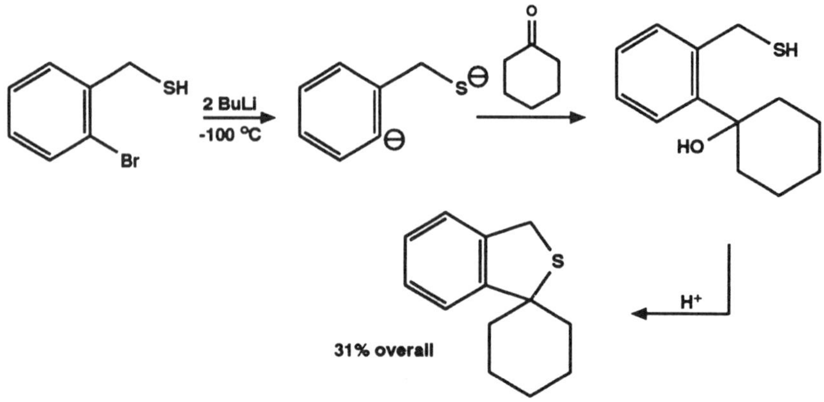

(5.12)

(Z, Z)-Tricyclohexaprenol and the (E,E)-isomer have been postulated as precursors to tricyclic geoterpanes, and as a cholesterol substitute in prokaryote membranes. The synthesis of the (Z,Z)-isomer was accomplished using the dianion of ethyl-2-mercaptoacetate (Scheme 5.1).[16] The dianion coupling sequence afforded a favorable 2:1 mixture of Z/E, α,β-unsaturated esters, whereas Wittig-type coupling reactions failed.

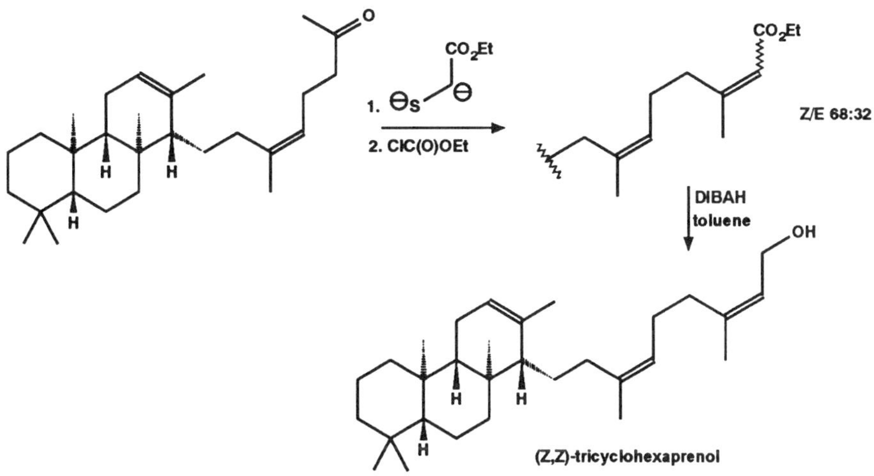

Scheme 5.1. Synthesis of (Z,Z)-tricyclohexaprenol

5.3 THIOIC-ACID BASED DIANIONS

5.3.1 Generation of thioic acid-based S,C-dianions

Dithioic acid dianions (dithiomethylene dianions) can be prepared by two distinct methods; (a) bis-deprotonation of an dithioic acid, or (b) reaction of a preformed carbanion with carbon disulfide followed by treatment with a second equivalent of base (Eq. 5.13). LDA (2.1 eq.) is used most often to generate dianions from thioic acids.[17]

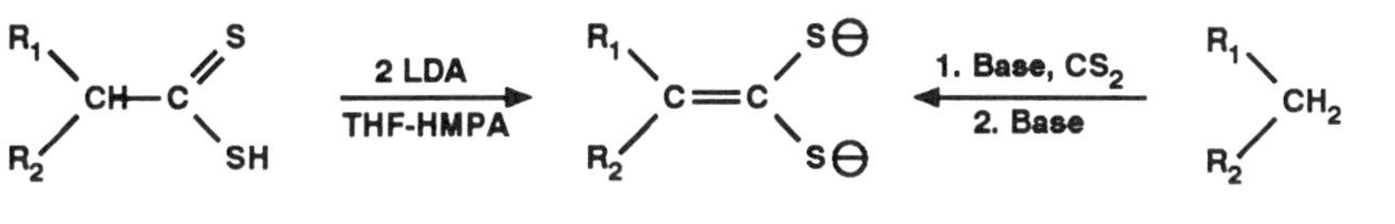

(5.13)

R_1 = alkyl, RC(O)-, ROC(O)-, etc.
R_2 = H, alkyl, etc.

Alternatively, enolates of ketones and esters react with carbon disulfide to form the intermediate thioate anion, which can be deprotonated by lithium amide bases (e.g., cyclohexyl(i-Pr)N-Li) in THF to form the dithiomethylene dianion (Eq. 5.13).[18-21] Little work has been conducted with carbonyl compounds because these dianions undergo high yielding alkylation reactions at the sulfur atoms. The dithioacetal products are extremely versatile materials (Sect. 5.3.2.).

5.3.2 Reactions of thioic acid-based S,C-dianions

Thioic-acid dianions, formed by the two methods indicated in Section 5.3.1., alkylate at both sulfur atoms to furnish dithioacetals in good to excellent yield. Dithioacetals can be converted to aldehydes, thiolesters, α-haloesters, carboxylic acids,[23] or can undergo stepwise organocuprate Michael additions in route to α-isopropyl- or α-tert-butyl carbonyl compounds (Scheme 5.2).[1-4,18-21]

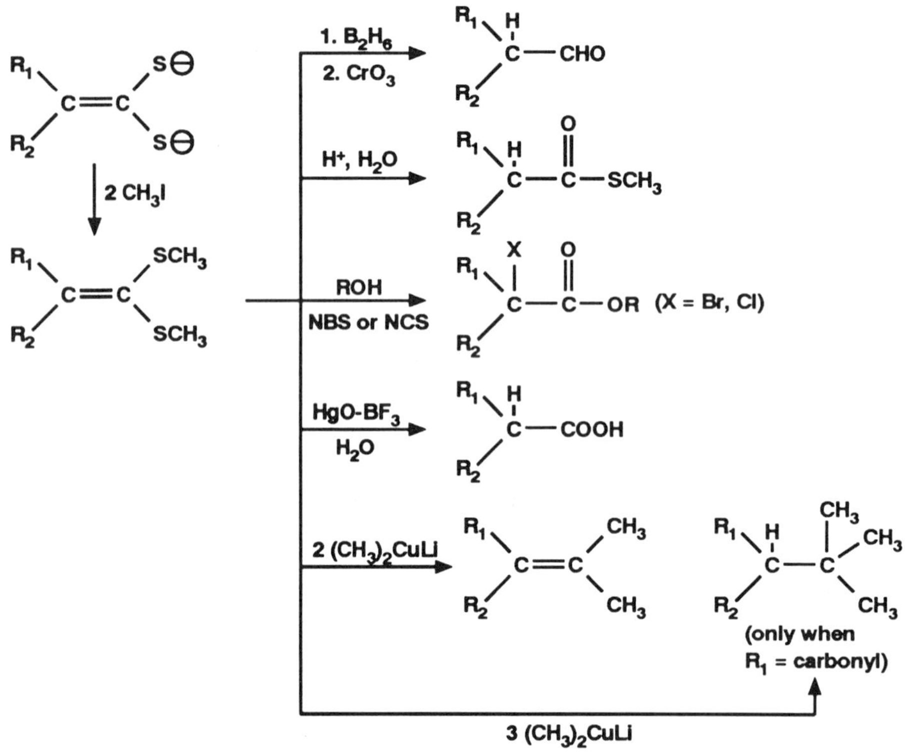

Scheme 5.2. Synthetic utility of dithioacetals

5.4 EXPERIMENTAL PROCEDURES

4-Methylthio-5-hexen-3-ol (Eq. 5.5) - reaction of propanal with the magnesium-based dianion of allyl mercaptan (from Pohmakotr, Geiss and Seebach).[6]

A mixture of allyl mercaptan (0.8 mL; 10 mmol) and TMEDA (3.2 mL; 20 mmol) in 28 mL THF at 0°C was stirred with n-BuLi (21.0 mmol; 1.6 N) for 4 h resulting in an orange-colored solution of the dilithio reagent. Freshly prepared magnesium bromide-etherate complex (3.8 mL; 10 mmol) was added, resulting in a brightening of the color. After stirring for 15 min, the temperature was reduced to -78°C and propanal (0.58 g; 10 mmol) was added. The solution was stirred for 1 h, warmed to 0°C, and treated with methyl iodide (0.7 mL; 10 mmol), which deposited a colored precipitate after 1-2 h. The solution was partitioned between ethyl ether (or pentane) and water, neutralized with 1.0 N HCl, and the product extracted into the organic layer. The organic layer was washed twice with water, once with brine, and dried over sodium sulfate to give 1.32 g of product (91%; α/γ = 9:1) that distills at 80°C @ 0.02 Torr.

Ethyl-3-methyl-2-pentenoate (Eq. 5.9) - reaction of 2-butanone with the dianion of ethyl mercaptoacetate (from Tanaka et al.).[12]

To a solution of lithium diisopropyl amide (44 mmol) and TMEDA (44 mmol) in 40 mL THF at -78°C was added ethyl mercaptoacetate (2.4 g; 20 mmol) in 7 mL THF. The solution was stirred for 1 h, followed by the addition of 2-butanone (22 mmol). The mixture was stirred an additional 2 h at which time ethyl chloroformate (20 mmol) in 5 mL THF was added portionwise while maintaining the temperature at -78°C. After 30 min, the temperature was raised to room temperature, stirred for 1 h, and quenched with aqueous NH_4Cl (10 mL). The reaction mixture was poured into dilute HCl, and extracted four times with ethyl ether. The organic layers were combined, washed twice with brine, dried over sodium sulfate, and concentrated to afford 64% of the product as a 43:57 E/Z mixture. Bp = 93-97°C @ 60 Torr.

(E)-α-Benzylidene-γ-butyrolactone (Eq. 5.10) - reaction of α-mercaptobutyrolactone dianion with benzaldehyde (from Tanaka et. al.).[13]

Under a nitrogen atmosphere, n-butyllithium (44 mmol) was added to a solution of diisopropylamine (4.5 g; 44 mmol) and TMEDA (5.1 g; 44 mmol) in THF (40 mL) at -78°C. After 20 min, α-mercapto-γ-butyrolactone (2.4 g; 20 mmol) in 7 mL THF was added via syringe. The solution was stirred for 1 h, and benzaldehyde (40 mmol) in 3 mL THF was added, stirred for 2 h, followed by ethyl chloroformate (2.2 g; 20 mmol) in 5 mL of THF. The reaction mixture was stirred for 30 min, the cooling bath removed and stirred an additional 1 h. The reaction was quenched with 10 mL saturated NH_4Cl and extracted with ethyl ether (4 x 50 mL). The organic layers were combined, washed with brine, saturated $NaHCO_3$, and brine again. The ether layer was dried over sodium sulfate, and purified by column chromatography (silica; chloroform) to afford 1.89 g (54%) as colorless crystals of the (E)-isomer. Mp = 117.5-118.5°C.

REFERENCES

1. **Block, E.** *Reactions of Organosulfur Compounds*, Academic Press, New York, 1978.
2. **Block, E.** Organic sulfur compounds in organic synthesis: some recent advances, *Aldrichimica Acta.* **11**, 51 (1978).
3. **Oae, S.** *Organic Sulfur Chemistry: Structure and Mechanism*, CRC Press, Boca Raton, Fl., 1991 and references therein.
4. **Tanaka, K. and Kaji, A.** The synthetic utility of sulfur-containing dianions, *Sulfur Rep.* **1**, 97 (1980) and references therein.
5. **Geiss, K-H., Seebach, D. and Seuring, B.** Doppelmetallierung von Benzyl- und Allylmercapten; das Thiobenzaldehyd- und Thioacrolein-dianion, *Chem. Ber.* **110**, 1833 (1977).
6. **Pohmakotr, M., Geiss, K-H. and Seebach, D.** Anwendungen der Magnesiumderivate von Thioacrolein-dianionen zur Synthese schwefelfreier Verbindungen (Vinyl- und Divinylepoxide, Acylcyclopentene, *Chem. Ber.* **112**, 1420 (1979).
7. **Seebach, D., Geiss, K-H. and Pohmakotr, M.** A facile entry to vinyl and divinyl oxiranes, *Angew. Chem. Int. Ed. Engl.* **15**, 437 (1976).
8. **Seebach, D. and Geiss, K-H.** Doubly metalated thiobenzyl alcohol-thiobenzaldehyde dianion, *Angew. Chem. Int. Ed. Engl.* **13**, 202 (1974).
9. **Tanaka, K., Yamagishi, N., Tanikaga, R. and Kaji, A.** Dianions of ethyl α-mercaptoacetate and ethyl α-mercaptoproprionate: A new method for an efficient synthesis of α,β-unsaturated esters from carbonyl compounds, *Chem. Lett.* 471 (1977).
10. **Vedejs, E., Arnost, M.J. and Hagan, J-P.** Conformational effects on the 2,3-sigmatropic shift of sulfur-stabilized ylides and enolates involving bicyclic transition states, *J. Org. Chem.* **44**, 3230 (1979).
11. **Tanaka, K., Uneme, H., Yamagishi, N., Ono, N. and Kaji, A.** Dianion of α-mercapto-γ-butyrolactone: An efficient reagent for the stereoselective synthesis of α-alkylidene-γ-butyrolactones from carbonyl compounds, *Chem. Lett.* 653 (1978).

12. **Tanaka, K., Yamagishi, N., Tanikaga, R. and Kaji, A.** New methods for synthesis of α,β-unsaturated carboxylic esters from carbonyl compounds using monoanions of dithiocarbonates, and dianions of ethyl α-mercaptoacetate and ethyl α-mercaptoproprionate, *Bull. Chem. Soc. Japan* **52**, 3619 (1979).
13. **Tanaka, K., Uneme, H., Yamagishi, N., Tanikaga, R. and Kaji, A.** New method for stereoselective synthesis of α-alkylidene-γ-butyrolactones using monoanion of O-ethyl S-(tetrahydro-2-oxo-3-furanyl)thiocarbonate and dianion of α-mercapto-γ-butyrolactone, *Bull. Chem. Soc. Japan* **53**, 2910 (1980).
14. **Matsui, S.** Mechanism of formation of α,β-unsaturated esters in the reaction of ethyl mercaptoacetate dianion with carbonyl compounds, *Bull. Chem. Soc. Japan* **57**, 426 (1984).
15. **Parham, W. E., Egberg, D. C., Sayed, Y. A., Thraikill, R. W., Keyser, G. E., Neu, M., Montgomery, W. C., and Jones, L. D.** Spiropiperidines. 1. Synthesis of spiro[isobenzofuran-1(3H), 4′-piperidin]-3-ones, spiro[isobenzofuran-1(3H), 4′-piperidines], and spiro[isobenzotetrahydrothiophene-1(3H), 4′-piperidines], *J. Org. Chem.* **41**, 2628 (1976).
16. **Ladenburger, C. and Heissler, D.** Synthesis of (Z, Z)-tricyclohexaprenol, *J. Chem. Res. (S)* 330 (1991).
17. **Ziegler, F. E. and Chan, C. M.** Preparation of ketene thioacetals from dithioic acid dianions, *J. Org. Chem.* **43**, 3065 (1978).
18. **Corey, E. J. and Chen, R. H. K.** α-Dithiomethyleneketones. Generation and application to synthesis, *Tetrahedron Lett.* 3817 (1973).
19. **Marino, J. P. and Kostusyk, J. L.** Deprotonation studies in α-oxoketene dithioacetals. I., *Tetrahedron Lett.* **27**, 2489 (1979).
20. **Marino, J. P. and Kostusyk, J. L.** A direct synthesis of substituted thiophenes via α-oxoketene dithioacetals. II., *Tetrahedron Lett.* **27**, 2493 (1979).
21. **Johansen, O. H. and Undheim, K.** Conjugate addition of lithium dimethylcuprate to an α,β,β-trialkylthio-unsaturated ketone, *Acta Chem. Scand., Ser. B.* **B33**, 460 (1979).
22. **Konen, D. A., Pfeffer, P. E., and Silbert, L. S.** α-Anions. VIII. Dithioesters and ketene mercaptals from aliphatic acids and esters. *Tetrahedron* **32**, 2507 (1976).
23. **Martin, S. F.** Synthesis of aldehydes, ketones, and carboxylic acids from lower carbonyl compounds by carbon-carbon coupling reactions, *Synthesis* 633 (1979).

INDEX

A

Acetoacetone 1,3-C,C-dianion, 31-44
Acetoacetate 1,3-C,C-dianion, 31-44
Acetone 1,3-C,C-dianion, 42
N-Acylsulfonamide N,C-dianions, 221
Alkanoic acid-based 1,3-O,C-dianions
  acylation, 99-100
  addition reaction by RLi, 90
  alkylation with,
    alkyl halides, 91-93
    aziridines, 94
    epoxides, 93
    Michael acceptors, 94
  condensation,
    aldehydes/ketones, 94-99
    influence of cation, 97
    influence of chiral auxiliary, 99
    regioselectivity with enals and enones, 98
    stereoselectivity of, 96-98
  cyclic alkanoic acid dianions, 102
  α-heteroatom functionalization, 100-101,103
  α-heteroatom substituted, 110
  generation of, 90-91
  oxidative coupling, 102-103
  α-oxo-substituted, 109
  α-thio-substituted, 106-109
    alkylation, 106-107
    condensation, 108-109
    lactone synthesis, 108-110
  use in preparation of,
    α-peroxocarboxylic acids, 103
    ketones, 103
    lactones, 95-96
Alkenoic acid-based 1,3-O,C-dianions
  1,3- vs. 1,5-reactivity, 111
  condensation, 113
  heterocyclic variation, 116
  influence of cation, 112
  preparation via Birch reduction, 117
    reactions, 117
  reactions, 111-114
Alkynoic acid-based 1,3-O,C-dianions
  alkylation, 118,119
  condensation, 118
  pyranone synthesis, 119
Amine-based N,C-dianions, 222
Aniline-based N,C-dianions, 222
Azines as 1,6-C,C-dianions, 64
Aziridines, reaction with
  α-carboxylate dianions, 94

B

Beckman fragmentation, 163
Benzoic acid-based O,C-dianions, 119-124
Benzofuran carboxylic acid dianions,
  reactions, 121-122
Benzyne, 194
α-Bromoketones, 13-15
Butadiene C,C-dianions, 66-68
Butene C,C-dianions, 66-68
Butynoic acid O,C-dianion, 117-118

C

Carbamate-based N,C-dianions, 208-211
  generation, 209
  heterocyclic containing, 211
  remote, 209-211
    ring metalation to prepare, 210
  stereochemistry of reactions, 210
  vicinal, 209
β-Carbonyl α-carboxylate dianions
  reactions, 104-105
    stereoselectivity of, 105
    use of imidazolide, 104
Carboxamide N,C-dianions, 186-197
  α,α'-type, 187-189
    cyclic, 189
    reaction with electrophiles, 188
    N-TMS variation, 188
    thioether substituted, 189
  preparation of, 186-187
  remote, 191-197
    chiral auxiliary and, 192
    condensation, 192,193,195,196
    heterocyclic, 196
    reaction with benzyne, 194
    resonance delocalized, 193
    ring metalation of,
      acetanilides, 195
      heteroaromatic, 196
  syntheses,
    furanone, 192,193
    dihydropyran, 193
    stabilized by sulfone, 191-193

tin-metal exchange, 197
vicinal, 190-191
epimerization via, 190
from N-acylamino esters, 190
reaction with aldehydes, 190
hippurate form, 190-191
Carboxylic acid-based O,C-dianions, 88-130
resonance contributors, 88
types, 88-90
α-Carboxylate dianions (see alkanoic acid-based 1,3-O,C-dianions)
C,C-dianions, 9-86
C-13 study, 11
1,1-C,C-dianions (see geminal)
1,2-C,C-dianions (see vicinal)
1,3-C,C-dianions (see also β-dicarbonyl, sulfinyl-, sulfonyl- and phosphorus-stabilized 1,3-dianions), 27-54
miscellaneous, 42-44
acetone, 42
α-cyanoacetone
imine-based, 43,44
stereoselective alkylation, 43
α-nitroketone, 44
phenylacetone, 42
1,4-C,C-dianions (see remote)
α,α-C,C-dianions (see geminal)
controversy of geminal, 11
deprotonations, 10
examples, 10
hybridization, 11
stabilizing groups, 10
types, 12
Claisen rearrangement, 58-59,139,146,198
Cresol-based O,C-dianions, 159-160
Crotonic acid dianion, 111-115
Cyanoacetone 1,3-C,C-dianion, 43
Cyclic dianions,
β-dicarbonyl, 34-36
vicinal, 23-27
Cyclopentadiene stabilization, 26,27
1,2-Cyclohexanedione, 1,4-dianion, 59
1,3-Cyclohexanedione, 1,3-dianions, 35

D

(π)-Delocalized C,C-dianions, 65-70
allylic, 65-68
butene-based, 66-68
alkylation, 66-67
butadiene-based, 66-68
pyridine synthesis, 67
benzylic, 68-70
biphenyls as substrates, 70
cyclophane synthesis, 69
lutidine as substrates, 69-70
xylene C,C-dianions, 69
resonance contributors, 65
Dianions,
charge, 2
designations, 3-6, 9-11
nomenclature, 3
types, 1,9
β-Dicarbonyl, 10, 27-44
acylations, 40
TMEDA effect, 40
addition reactions, 40-42
carbon dioxide, 40,41
imines, 40,41
Michael reaction, 41,42
nitriles, 40,41
alkylation, 31-37
epoxides, 36-37
Lewis acid promoted, 37
tetrahydrofuran synthesis, 37
intramolecular, 36
regiochemistry, 34-36
condensations, 37-40
acetoacetate, 37-40
acetoacetone, 37-40
2-pyranone synthesis, 39
stereoselectivity, 39-40
thioester variation, 39
cyclic, 34-37
cyclohexanedione dianions, 35
generation of, 29,31
heterocyclic variation, 35
resonance contributors, 28
total syntheses using, 29-31
types, 27,28
1,4-Diazines (pyrazines),
use as 1,5-dianions, 64-66
Dihydropyrans, 52
Dioxenium cation, 135
Distal enolate, 133-135
Dithiane,
stabilize O, C-dianion, 108,145
stabilize remote C,C-dianions, 60,64
Dithiolane,
to stabilize remote dianion, 64,129
Dyotopic rearrangement, 96

E

Enaminone-based N,C-dianions, 212-215
  cyclic, 215
  generation, 212-213
  kinetic vs. thermodynamic, 212
  reaction with,
    alkyl halides, 213
    carbonyls, 213
    epoxides, 214
  synthesis of,
    pyridines, 214
    pyrimidines, 214
Enones, synthesis, 50-54
Epoxides, reaction with
  1,1-dianion, 19,25
  α-carboxylate dianions, 93

F

Furanoic acid dianions, 120-121
  condensation reactions of metalated, 120
  reactions of lateral deprotonated, 121
Furans,
  via tetronic acid dianion, 39
  ring metalation, 63,129

G

Geminal C,C-dianions, 11-20
  aldol reactions of, 14
  carbonyl-stabilized
  α-halocarbonyl, 13-15
    Hoffmann reaction of, 14-15
    homologation using, 14-15
    β-ketophosphonate synthesis, 15
  nitrile stabilized, 16
  nitro stabilized, 16-18
    resonance stabilized, 17
    competition with vicinal form, 17
    intramolecular chelate, 17
    diastereoselective reaction, 18
  resonance contributors, 12
  sulfonyl-stabilized, 18-20
    bis-alkylation, 18
    α-heteroatom stabilized, 20
    intramolecular cyclization, 19
    prep. of vinyl sulfone, 18
    prep. of γ-hydroxysulfones, 19
    reaction with bis-electrophiles, 19
Glutarate 1,3-C,C-dianions, 57

H

Hoffmann reaction, 14-15
Hydrazone-based N-C-dianions, 203-208
  N-carboalkoxy type, 204-205
  reactions, 204,205
  synthesis of,
    alkanes, 203
    azoles, 204,205
    enones, 205
β-Hydroxycarbonyl O,C-dianions, 133-139
  addition to vinyl nitro compounds, 136
  Claisen rearrangement, 139
  lactam variation, 138
  lactone variation, 137-138
    stereoselective reactions, 137
  proximal and distal enolate, 133
    reactions of, 134-135
  stereoselective alkylation, 136
β-Hydroxydithioester 1,3-O,C-dianion, 146
Hydroxyl-based O,C-dianions, 130-156
  allylic-based, 146-151
    1,3 vs. 1,5-system, 149
    reaction with,
      aldehydes/ketones, 149
      alkyl halides, 150
      RLi (addition), 147-148
      TMS-Cl, 150
    transposition of, 151
    vicinal-classification, 148
  generation of, 131
  non-deprotonation methods, 154-156
    reductive lithiation, 156
    RLi addition to,
      epoxide, 155
      carbonyl, 155-156
  propargylic based, 152-153
    reaction with,
      enones, 153
      RLi, 152
      CO_2, 152
  types, 130
  vinylic-based, 151-152
    by metal-halogen exchange, 151
    reaction with carbonyls, 151
    synthesis of tetrahydrofurans, 152
Hydroxyl-based 1,2-O,C-dianions (see vicinal O,C-dianions)
Hydroxyl-based 1,3-O,C-dianions (misc.), 145-146

dithioester as stabilizing group, 146
alkyne as stabilizing group, 146
β-Hydroxysulfenyl 1,3-O,C-dianions, 145
β-Hydroxysulfinyl 1,3-O,C-dianions, 139-145
alkylation, 143
chiral sulfoxide, 144-145
alkylation, 144
β-Hydroxysulfonyl 1,3-O,C-dianions, 139-145
stereospecific,
alkylation, 140-142
cyclic, 143
condensation, 141
reactions, 140

I

Imide-based N,C-dianions, 218-221
generation of, 218
reactions of acyclic, 218
reactions of cyclic, 219-220
synthesis of furanones via, 220
Indole-based N,C-dianions, 216-218
lateral metalation to form, 217
Intramolecular cyclization, 19
Isomerization of 1,1-C,C-dianion, 19-20

K

β-Ketophosphonium dianions, 51
condensation, 51
enone synthesis, 51
β-Ketophosphonate dianions, 51-53
acylation, 52
alkylation, 51
dihydropyran synthesis, 52
enone synthesis, 51-52
β-phosphonylacetoacetate dianions, 53
β-Ketosulfone 1,3-dianions, 44-50
acylation, 45-48
alkylation, 45-48
epoxides, 47
condensation, 45-47
oxidative coupling, 45-46,48
SO_2 extrusion, 45-46
tetrahydropyran synthesis, 46
β-Ketosulfoxide 1,3-dianions, 44-50
alkene synthesis, 49
alkylation, 48-50
asymmetric at sulfur, 50
bis-reaction, 48-49
condensation, 48-50
cyclic, 49-50
tandem Michael-Claisen reaction, 49

L

β-Lactam, 60,110,126,189
Lateral deprotonation, 119-124
of cresol, 159-160
2,6-Lutidine dianion, 69-70

M

Metal chelate structure,
and stereoselectivity, 39,161
Metal-halogen exchange, 13-15,62,151

N

N,C-dianions, 185-233 (see also acylsulfonamide-, amine-, aniline-, benzazole-, carbamate-, carboxamide-, enaminone-, hydrazone-, imide-, pyrrole-, sulfonamide-, sulfonimide-, sulfonylhydrazone-, and thioamide-based N,C-dianions)
preparation of, 186-187
Nicotinic acid O,C-dianions
condensation reactions, 120

O

O,C-Dianions, 87-184
types, 87-88
1,3-O,C-dianions (see also α-carboxylate and α,α'-O,C-dianions), 91-111
Ortho-directing group, 157-158
Oxazole acid-based O,C-dianions, 124
Oxime-based O,C-dianions, 160-166
heterocycle preparation via, 164-165
reaction with,
alkyl halides, 162-163
carbonyl compounds, 162,165
disulfides, 163
esters, 165
imines, 164
nitriles, 165
remote, 166
ring metalation to form, 166
syn/anti isomerism, 161-162
Oxy-Cope rearrangement, 115

P

Phenol-based O,C-dianions, 156-160
  ring metalation to form, 157
Phenylacetylene,
  1,4-C,C-dianion, 62
Phosphonate, synthesis, 15,20
Phosphorus-stabilized 1,3-C,C-dianions, (see also β-ketophosphonium and β-ketophosphonate) 51-54
Piperidones, synthesis, 41
Propargylic dianions (see also remote 1,3-C,C-dianions, alkyne-based), 56-57
Propyne C,C-dianion, 56
Propynoic acid O,C-dianion, 117-119
Proximal enolate, 133-135
Pummerer reaction, 108
Pyranones
  1,3-C,C-dianions, 35
  synthesis of, 39,113,119,129
Pyridines,
  synthesis, 67
Pyrrole-based N,C-dianions, 216
Pyrrolidin-3-one 1,3-C,C-dianions, 35

R

Remote carboxamide dianions (see carboxamide N,C-dianions, remote)
Remote carboxylic acid O,C-dianions, 124-130
  1,4-(O,C)-type (homoenolate)
    chiral sulfoxide in, 125-126
    reactions, 125-126
    sulfone stabilized, 125
  1,5-(O,C)-type (bis-homoenolate)
    addition to imine, 128
    imide-based 127
    reactions, 127-129
    sulfone-stabilized, 127-128
  1,Z-(O,C)-type (remote)
    use of dithiolane, 129
    ring metalation of furan to form, 129-130
Remote 1,3-C,C-dianions
  alkene based, 56
    vinylsulfone type, 56
  alkyne-based, 56-57
    dialkylation, 56
    heteroatom substituted, 57
  exo-stabilized, 57-58
    glutarates, 57
    1,3-dinitro compounds, 57-58
  types, 55
Remote 1,4-C,C-dianions, 58-62
  endo-stabilized, 58-60
    cyclic, 59
    dithioester, 59
    intramolecular condensation, 59,60
      asymmetric, 60
    β-lactam synthesis, 59-60
  exo-stabilized, 60-61
    ketone-sulfone, 61
    disulfone, 60
    ketone-resonance-delocalized, 61
      regiochemistry of reaction, 61
    bis-thioketal, 60,61
  ring-metalated, 62
Remote 1,5-C,C-dianions,
  endo and exo, 62-64
  Fisher carbene complex stabilized, 63
Remote 1,Z-C,C-dianions, 64-65
Remote hydroxyl-based O,C-dianions, 153-154
  reaction with,
    alkyl halides, 154
    carbamates, 154
  sulfone stabilized, 153-154
  via tin-metal exchange, 154
Remote C,C-dianions, 54-65
  nomenclature and designations, 54-55
Ring metalation (see also individual molecule and dianion types), 62
  aromatic acid O,C-type, 119-124
  carbamate N,C-based, 210
  carboxamide N,C-based, 195

S

Shapiro reaction, 205-207
1,2-Succinate ester dianions, 22-23
Sulfenyl-stabilized 1,3-C,C-dianions, 50
Sulfinamide-based N,C-dianions, 199
Sulfinyl-stabilized 1,3-C,C-dianions, 44-50
  alkylation, 48
  nomenclature, 44-45
  regiochemistry of deprotonation, 45
Sulfonamide-based N,C-dianions, 199-203
  α,α′-, 199-201
    cyclic, 200-201
    reactions with electrophiles, 200-201
    synthesis of oxathiazines, 200

remote, 201-203
lateral deprotonation, 202
ortho-metalation and, 201-202
entry to substituted thiophenes, 203
Sulfonylhydrazone-based N,C-dianions, 205-208
1,2-carbonyl transposition, 207-208
metalation and, 208
Shapiro reaction, 205-207
synthesis of α-methylene lactones, 207
Sulfonyl-stabilized 1,3-C,C-dianions, 44-50
alkylation, 45
α,α′ and α,γ-types, 44-45
condensation, 45
nomenclature, 44-45
regiochemistry of deprotonation, 45
Syn/Anti isomerism (oxime), 161

T

Tetronic acid 1,3-dianion, 39
Thiazole carboxylic acid O,C-dianions, 124
Thioamide-based N,C-dianions, 197-198
alkylation of 197
thio-Claisen, 197-198
Thioic acid-based S,C-dianions, 241-242
generation, 241
reactions, 242
Thiol-based S,C-dianions, 235-241
generation, 236
preparation of epoxides using, 238
preparation of alkenes via sulfur extrusion, 239
reaction with,
alkyl halides, 237
carbonyl compounds, 237
remote, 240-241
thioaldehyde resonance contributor, 238
types, 235
vicinal delocalized, 236-237
Thiophene,
oxidative coupling, 62
Thiophene carboxylic acid O,C-dianions
reactions of metalated, 123
reactions of lateral deprotonated, 123-124
Tin-metal exchange, 53,152,197
Toluic acid O,C-dianions, 119-120

V

Vicinal C,C-dianions, 21-27
acyclic, 21-23
carbonyl stabilized, 21
cyclic, 23-27
cycloalkyl 1,2-dicarboxylates, 24
Michael reaction with, 24
ring formation, 24,25
cyclopentadienyl-based, 26,27
α-ketoamide stabilized, 23
lactones synthesis, 22,24,25
of theoretical interest, 21
oxazoline-based, 23
reaction with carbonyls, 22-26
ring annulation via, 22
succinate esters, 22-23
succinimide, 26
Vicinal N,C-dianions (see individual N,C-types)
Vicinal O,C-dianions (α-hydroxy dianions), 131-133
kinetic vs. thermodynamic enolate, 132
reaction with,
alkyl halides, 132
TMS-Cl, 132
Vinylic 1,3-C,C-dianions (see also remote 1,3-C,C-dianions, alkene-based), 56

X

Xylene C,C-dianions (benzylic), 69-70

Y

Ylids, phosphorus-based dianions, 51-53